Industrial Chemical Process Analysis and Design

Industrial Chemical Process Analysis and Design

Mariano Martín Martín

Assistant Professor
University of Salamanca
Department of Chemical Engineering

ELSEVIER

AMSTERDAM • BOSTON • HEIDELBERG • LONDON
NEW YORK • OXFORD • PARIS • SAN DIEGO
SAN FRANCISCO • SINGAPORE • SYDNEY • TOKYO

Elsevier
Radarweg 29, PO Box 211, 1000 AE Amsterdam, Netherlands
The Boulevard, Langford Lane, Kidlington, Oxford OX5 1GB, United Kingdom
50 Hampshire Street, 5th Floor, Cambridge, MA 02139, United States

British Library Cataloguing-in-Publication Data
A catalogue record for this book is available from the British Library.

Library of Congress Cataloging-in-Publication Data
A catalog record for this book is available from the Library of Congress.

ISBN: 978-0-08-101093-8

For Information on all Elsevier publications
visit our website at https://www.elsevier.com/

Publisher: Joe Hayton
Acquisition Editor: Joe Hayton
Senior Editorial Project Manager: Kattie Washington
Production Project Manager: Kiruthika Govindaraju
Cover Designer: Greg Harris

Typeset by MPS Limited, Chennai, India

A la memoria de mi madre.
María Manuela Martín Criado.
D.E.P.

Contents

Preface

Introducing students to the analysis and evaluation of chemical processes involves integrating knowledge from all the fields within Chemical Engineering, namely, fluid mechanics, heat and mass transfer, reactor design, unit operations, and process safety and control, and it also requires a basis in simulation. Lately biology and biotechnology have also stepped inside that box. While many Chemical Engineering curricula leave this task for the sole module of Process Design, at some universities there is a specific module dealing with evaluating how traditional processes in the chemical industry have been developed, the driving forces that have changed the way of producing major chemicals (eg, sulfuric acid), and the evaluation of the units involved. Despite the tight schedules of three- or four-year curriculums required for a bachelor's degree in Chemical Engineering, history tells us that learning from mistakes and successes via case studies allows critical thinking and provides a strong background.

This book reflects the work of many years of teaching "Introduction to Chemical Processes and Technologies" at the University of Salamanca and as a visiting professor at University of Birmingham, University of Leeds, University of Maribor, Universidad de Concepción, and Carnegie Mellon University, as well as industrial experience in my position at P&G and as a consultant for biofuels companies. I have tried to select a few processes that use all the possible raw materials—air, water, biomass, minerals, and fossil fuels—to produce a number of basic chemicals, from oxygen, nitrogen, and hydrogen, to ammonia, methanol, nitric acid, sulfuric acid, soda ash, sodium chloride, bioethanol, penicillin, FT fuels, and biodiesel. Unlike many other references dealing with similar topics, this book is not merely descriptive, but uses chemical engineering principles to analyze the processes that transform the raw materials into products. Therefore, throughout the text one can find distillation column analysis (ie, air separation, ethanol and methanol purification), reactor designs (eg, ammonia converter, SO_2 converter, sugar fermentation, biodiesel production, ammonia and methanol synthesis), absorption columns (eg, ammonia purification, CO_2 capture, sulfuric acid production), membrane analyses (eg, desalination), electrolysis (eg, water and NaCl), adsorption processes (eg, air desiccation, PSA), evaporator and crystallizer designs, power cycles, and complete process analyses (eg, air liquefaction; nitric acid, sulfuric acid, hydrogen, methanol, ammonia, and sodium carbonate production).

The organization of the book is as follows. Chapter 1, The chemical industry presents an introduction to the evolution of human needs and its effect on chemical products and processes. Chapter 2, Chemical processes presents the basics in process development and analysis. It covers the principles of process synthesis, focusing on the heuristic approach, and presents flowsheeting, the role of security and environmental issues in process design decision-making, as well as a brief summary of the chemical engineering principles that will be used throughout the rest of the book. Chapter 3, Air starts with the first raw material. I selected air because of its abundance and life-supporting function. Air is used to obtain its components, and that process is analyzed from liquefaction to final separation. Air humidity is also evaluated. Chapter 4, Water showcases water as a raw material. Apart from presenting current concerns regarding its consumption and usage, the chapter is devoted to evaluating the use of salt water to produce freshwater, and to the use of salt in the production of major chemicals. Chapter 5, Syngas presents the use of carbon-based material for the production of syngas, evaluating all the steps, including purification. The gas is later used in the production of ammonia, methanol, and synthetic fuels. Chapter 6, Nitric acid and Chapter 7, Sulfuric acid cover the use of minerals and/or gases to evaluate two important chemicals of the inorganic industry. Chapter 6, Nitric acid is devoted to nitric acid production. It is the first complete case study, analyzing all the units from ammonia to the concentration of nitric acid. Chapter 7, Sulfuric acid studies the production of sulfuric acid. The two main methods are evaluated: lead chambers and the contact process. Finally, Chapter 8, Biomass discusses the use of biomass. This chapter is linked to Chapter 5, Syngas since biomass is a particularly carbon-rich raw material. However, current attention and efforts towards a more sustainable industry are not forgotten, as well as polymerization kinetics for synthetic rubber production as an attempt to produce materials that match the natural ones. Typical biofuels, specialty chemicals such as penicillin, and synthetic rubber are evaluated. I acknowledge that other chapters could have been added, but as the reader can see, the topics selected cover a wide range of raw materials and units.

Nowadays there is an unwritten rule that the use of software must accompany any module of an engineering degree. Thus, some of the problems presented require the use of software; therefore I have used EXCEL, GAMS, MATLAB®, and gPROMS, or process simulators like CHEMCAD, to provide solutions. I thank all the software companies for their input. The text does not aim to teach the use of any one of these specific packages—there are other books better-suited to that aim—but rather how to use software in general to solve the particular problems presented throughout the book. Furthermore, other software exists that can be used to solve the same problems.

This work would not have been possible without previous professors on the subject who decided not just to focus on process description, but also to perform a systematic analysis of the processes based on the principles of chemical engineering. In addition, it was crucial to use the constructive criticism of my students, who at some point asked me, How can we study this module? (Edgar Martín, December 2014) Where can we find problems to practice? To all of them, my sincere gratitude. In particular, special thanks go to those who reviewed the manuscript: previous MSc students at the University of Salamanca such as Alberto Almena, Verónica de la Cruz, Borja Hernández, José Antonio Luceño, Santiago Malmierca and Antonio Sánchez; and prof. I.E. Grossmann from Carnegie Mellon University. I would like to thank the software companies who provided their support, and the chemical companies who granted permission to present real flowsheets and unit schemes, which are of high value to students.

A note to instructors: This book can be used in two ways. For those who teach a module on processing raw materials into products, it provides a description of a number of processes with examples for classroom seminars and end-of-chapter problems for students to evaluate their understanding of the material. Most of the problems have been exam questions over a number of years at the University of Salamanca. An appendix with the solutions is provided at the back of the book. For those professors whose programs do not have such a module, the book can be used as reference for subjects such as chemical engineering principles, reactor engineering, and unit operations, and can also provide information and case studies for putting together an open problem of process design for a classical process design class at the senior level. Furthermore, examples and problems for the use of process simulators, as well as the use of other software packages, are presented and provided.

A note to students: This book provides not only the description of the processes and the physico—chemical principles of operation, but also a considerable number of detailed examples throughout the text, including computer code for solving some of them. There is a larger collection of problems at the end of each chapter to challenge your understanding of the material in the chapters, and for you to practice and get better insight into the processes covered. You can find the results of the problems at the back of the book for guidance.

Mariano Martín Martín
Salamanca, Spain, February 2016

The chemical industry

1

1.1 EVOLUTION OF THE CHEMICAL INDUSTRY

Chemicals and materials have been used and developed by mankind over centuries. Human history has traditionally been divided into eras directly related to the evolution of the use and processing of materials, that is, the stone and iron eras. Fig. 1.1 shows this evolution, from natural polymers and ceramics in 10,000 BC, to the development of synthetic materials via metals processing back in the Middle Ages, sponsored by the wars in Europe. In this chapter we will see how mankind's evolution and needs have guided the development of the chemical industry.

The journey starts at the beginning of society with the use and manipulation of natural products. The more demanding society became, the more complex products and efficient processes were needed. This chapter will be the link between all the chapters included in this book. We focus on chemical process analysis when and why they originated, and how they evolved over decades to improve the quality of products and meet the needs of industry at every turn. The story begins with the use of natural products and the principles of biotechnology for food-related activities. The main turning point was the moment the first chemical process as we know it today was put together, 1746. It corresponds to the lead chambers Roebuck used for the production of sulfuric acid within the NaCl industry.

For simplicity, this timeline is divided into eight stages, from prehistory to today. Note that a more detailed analysis would lead to a more segmented timeline, but for the sake of argument and the purpose of this introductory chapter, this will suffice (García et al., 1998, Ordóñez et al. (2006a, 2006b)).

1.1.1 PREHISTORY

In the beginning mankind was nomadic, satisfying its needs directly from nature. Little or no transformation was needed, and if anything, it was craftwork.

1.1.2 FIRST SETTLEMENTS

However, as society changed from nomadic habits to established settlements, needs also underwent similar drastic changes. The discovery of fire and its control allowed cooking and curing of meat for preservation as smoked meat.

Industrial Chemical Process Analysis and Design. DOI: http://dx.doi.org/10.1016/B978-0-08-101093-8.00001-X

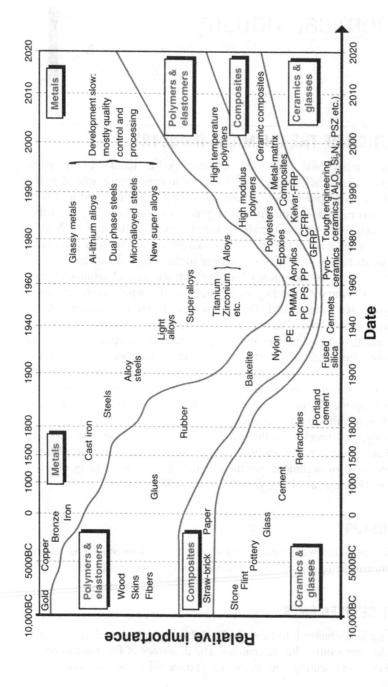

FIGURE 1.1

Evolution of engineering materials.

Reproduced with permission from: Ashby, M.F., 2004. Materials Selection in Mechanical Design, third ed. Butterworth-Heinemann, Oxford; Ashby (2004), Copyright Butterworth-Heinemann, third ed.

As expected, food was a primary concern for mankind. Other methods such as solar curing or the use of salts as preservatives, which already required chemicals or chemical transformation, were also used. Salt production was therefore a need, and the evaporation of seawater was the method of choice.

Cave paintings dating back to 10000–5000 BC were the first evidence of the establishment of mankind into groups, but they also represented the capability of producing pigments out of natural species. As a result, mankind no longer looked to nature to satisfy its needs, but started developing its own means. For instance, around 7000 BC the use of fire allowed the production of fired-clay pottery. Crop irrigation has been known since 5000 BC. Several hundred years later copper artifacts were also used. Stone tools were developed and used by 4000 BC, and the wheel can also be dated back to this time. Most of these tools were used to grow and produce food. There is evidence of the domestication of cattle from 4000 BC. Also by that time, early biotechnology appeared. For instance, bread was produced in Egypt from 3500 BC, the production of cheese dates back to 2000 BC, yogurt out of fermented milk was created by central Asian People in the Neolithic period, barley beer can be dated back to 2500 BC in ancient Egypt, and by the same time rice beer was being produced by the Chinese. Furthermore, butter and natural glue gave rise to certain chemical and biochemical craftwork.

Animal skins have been used for protection via the production of garments, clothing, shelter, carpets, or decoration. One of the main issues, as with any other natural resource, was its preservation. Leather tanning using tree bark is said to have originated among the Hebrews. The techniques developed were kept as professional secrets and passed down through generations from father to son under a halo of magical foundations.

The first well-established generations were born in the Mediterranean area. Grapevine and olive trees are widely available in these regions. Mankind used both resources in several ways. From olive trees, oil was obtained as a means to cook food, to preserve it, and to use it as fuel for lighting. From the grapevine mankind obtained juice, and via fermentation, wine, which later aged into vinegar. For ages, vinegar was the strongest acid known to man.

These established civilizations also developed gypsum (2500 BC), and pigments for house and personal decoration. In terms of personal care, perfumes, dyed textiles, and paints appeared, which indicates a certain degree of developed craftwork. As we can see, it was society that demanded such products to improve the quality of life.

1.1.3 ALCHEMISTS

Knowledge of how to obtain and process raw materials was scattered. The ancient Greeks developed certain basics to explain these transformations. Although the theories were not correct—that is, the theory of the four elements—they worked in terms of pursuing rational thinking about natural processes.

The rational thinking from the ancient Greeks, together with theocentric concepts from Arabs and Christians, resulted in what is known as alchemy, an Arab word. Although it was surrounded by a magic halo, alchemists developed and prepared a large number of new materials and chemicals such as acids, alkalis, salts, etc. That also allowed improvements in perfume and dye production, as well as metallurgic developments. The aim of this development was more mystical: the search for eternal youth and the philosopher's stone.

Around the 13th century, the center for knowledge moved from the East to the West. Universities, by putting together Greek knowledge, Arab and Jewish learnings, and Christian ideas, created the origins of the Renaissance, an intellectual revolution.

1.1.4 LOWER MIDDLE AGES

As civilization developed in Europe during the 12th and 13th centuries, textile production transferred to villages and towns, where workers in the same fields united in trade guilds becoming rather powerful. Thus, through the middle ages the use of iron spread not only as a material for agriculture and domestic tools, but as weapons too. Wars were the reason for the huge development of metallurgy. By this time mankind learned how to improve the taste of wine, bread, and cheese. Furthermore, the origin of a certain chemical industry was in place from basic natural species:

- weak acids (vinegar, lemon juice, acid milk)
- alkalis (carbonates from ashes, lime).

1.1.5 MIDDLE AGES

Around the 14th century, strong acids such as chlorhydric acid, nitric acid, the mixture of both, and sulfuric acid (oil of vitriol) appeared. The discovery of sulfuric acid is as important to the chemical industry as the discovery of fire, since it allowed the preparation of a number of salts and other acids.

Although the aim was still the search for eternal youth, along the way a number of chemicals with interesting properties were produced, such as sodium sulfate, a laxant, and chloride as a byproduct from the production of chlorhydric acid using sulfuric acid and sodium chloride:

$$H_2SO_4 + 2NaCl(salt) \rightarrow 2HCl + Na_2(SO_4)$$

Furthermore, commercial trade increased as a result of the ideas in the Renaissance and the geographic discoveries in the 13th and 14th centuries. Both contributed to the development of larger manufacturing centers. However, the demand was limited to a few privileged classes, nobles, and clergy, which slowed down technological development.

During this preindustrial period, a number of activities that can be considered to belong to the chemical industry were carried out in the fields of metallurgy, chemical medicine, and the production of glass, soap, powder, and inorganic acids (based on craftsmen's knowledge). For instance, Lazarus, Eeker, and Agrícola in the 16th century produced nitric acid from salt and ferrous sulfate.

1.1.6 INDUSTRIAL REVOLUTION

Transition from the craft-based production system to the industrial one required the identification and further understanding of the principles and foundations of nature. Lavoisier's work by the end of the 18th century can be considered as the beginning of chemistry as a modern science. Thus, the chemical composition of some of the most common products was becoming known by the beginning of the 19th century. The basis of the chemical processes has its foundations in John Dalton's atomic theory and Jöns Jacob Berzelius's work to develop the Periodic Table. Organic species were gaining attention thanks to the work of Friedrich August Kekule, founder of the Theory on Chemical Structure, and Stanislao Cannizzaro.

As a result, industry specialized. A particular industry will not cover the entire process from raw materials into final products as the craftmen used to do. The chemical industry is divided into three categories:

- Basic chemicals,
- Intermediate products,
- Consumer goods.

1.1.6.1 Siderurgy

This is the first industry that grew following the commercial trades. In the beginning, charcoal from wood was used as a fuel and reductor agent. However, its scarcity due to the diverse uses of wood, such as in the naval or construction industries, and its use as fuel, limited the progress. The use of mineral coal was unsuccessful for quite some time due to the problems related to the presence of sulfur, the amount of volatiles, and the fact that the increase in the mass during preheating was troublesome for the furnace. It wasn't until 1735 that Abraham Darvy first produced coke out of mineral coal. John Wilkinson improved the production process by building the coal heaps around a low central chimney built of loose bricks with openings for the combustion gases to enter. In 1802 the first battery of beehives was put together near Sheffield, and by 1870, 14,000 beehive ovens were already in operation in the coalfields of West Durham. During the 18th century, coke became the source for growth as it allowed the development of the stream engine, a basic component of the Industrial Revolution.

1.1.6.2 Textile industry

The steam engine was the driver for a number of craft-based industries since it allowed higher production rates, specialization, and division of work. This fact changed the way industry was conceived. Artisans lost their dominium over tasks and, in England, class differentiation vanished as a result of the Industrial Revolution. The first steam engine dates back to 1705 (Newcommen and Carley), but it was James Watt who from 1765 to 1868 increased the energy efficiency. Soon after, it was applied to power looms to weave sheds. As a result of the growth of the textile industry, a number of other industries also developed since larger amounts of dyes and bleaches were needed. Nature was not able to provide such quantities and the chemical industry grew to provide those chemicals: acids and alkalis.

1.1.6.3 Sulfuric acid industry

The production of acids focused its efforts on the oil of vitriol, since it allowed the production of other acids. The first method was lead chambers, initially proposed by John Roebuck in Birmingham in 1746. It consisted of large chambers internally covered by lead to protect the structure from corrosion, where sulfur oxides, steam, and oxides of nitrogen were put into contact. The presence of nitrogen oxides was required for the reaction to progress. In 1775 Lavoisier found the composition of the acid and a few years later, in 1778, Clement and Desormes showed that the nitrogen oxides were actually the catalysts for the production of sulfuric acid. It wasn't until 1828 that Gay-Lussac modified the original process by adding the tower (that nowadays is named after him) to recover the nitrogen oxides. The needs for more concentrated acid, in particular for its use in the Leblanc process, resulted in another step forward in sulfuric acid production. The Glover Tower was added to the process in 1859. The first facility using the enhanced process dates back to 1935.

Although the concentration of the acid produced using the modified lead chambers process was enough for the Leblanc process and for the production of orthophosphates for the fertilizer industry, it was not appropriate for the production of pigments and dyes, or for explosives, where nitration processes were the basis. The heterogeneous method, the contact process, was patented by Peregrine Phillips in 1831. It consisted of the oxidation of SO_2 gas to SO_3 using air with platinum as catalyst. However, there were still technical limitations to the use of this method. The purity of the gases required in the process and the initial low demand of the concentrated acid did not allow the industrial application of this method until 1870, when it became technically and economically feasible. By the end of the 19th century, the Badische Anilin und Soda Fabric (BASF) produced sulfuric acid at industrial scale using platinum first as catalyst. Platinum was substituted by V_2O_5 during the First World War.

Sulfuric acid production was for years an indicative index of the development of a country.

1.1.6.4 Sodium carbonate industry—soda processes

At the beginning of the 19th century, Leblanc suggested the production of sodium carbonate from salt and sulfuric acid. This process was the turning point for the world's chemical industry. In particular, we need to highlight the fact that for the first time the economic feasibility of the process was obtained only by reusing the byproducts hydrochloric acid (HCl) and sulfur (S).

HCl could be recovered by absorption. In 1836 William Gossage designed the Gossage Tower to recover HCl from the gas phase based on its solubility in water. He realized that contact area was key for the process. From that solution, chlorine (Cl_2) was obtained from oxidation. Cl_2 was (and still is) used as a disinfectant and bleach. In the case of S recovery, the Claus process patented in 1883 allowed sulfur recovery from the wastes of the Leblanc process.

The main drawback of the Leblanc process was the low purity of the sodium carbonate, which drove the development of the Solvay process by Ernest Solvay in the 1860s. The process breakthrough was the so-called Solvay Tower, which solved the technical problems of the first process patented by H.G. Dyan and J. Henning in 1834.

1.1.6.5 Coal gas industry

William Murdoch from 1792 to 1798 used coal gas for illumination of his own house. He was hired by the Boulton and Watt firm, and they began a gaslight program to scale up the technology. However, electricity soon replaced it. From the coal gas industry, as well as in the production of coke, there were a number of wastes and byproducts, discarded at first, that could be used. From the ammoniacal effluents, ammonia could be recovered and later sold as ammonia sulfate, a fertilizer. On the other hand, the growth of the organic chemical industry was due to the other residue, the coal tar, which was a raw material for a wide range of products.

The first organic compound artificially synthesized from inorganic materials was urea. It was obtained by treating silver cyanate with ammonium chloride by Friedrich Wöhler in 1828. From coal tar, Ruge in 1843 separated phenol, aniline, and other compounds. Aniline was the raw material for the first synthetic dye, aniline blue, patented in 1856 by Perkin.

Explosives were the other main chemicals, produced in 1885. Trinitrophenol was adopted by France and England, and trinitrotoluene was selected by Germany. Both substituted nitrocellulose and nitroglycerine which were highly unstable.

By that time (1892) calcium carbide (CaC_2) was accidentally obtained in an electric furnace by Thomas L. Willson while he was searching for an economical process to make aluminum. The CaC_2 was the raw material for the production of calcium cyanamide, a fertilizer, and for ammonia production via alkali hydrolysis. On the other hand, the decomposition of CaC_2 with water generated acetylene, which later was used as the origin of many other organic compounds (ACS, 1998).

The development of the chemical and pharmaceutical industries, and the production of dyes, placed Germany at the top of the world's chemical industry for decades. The discovery of the dynamo in 1870 was the origin of electrochemistry.

1.1.7 INDUSTRIAL SOCIETY

The increasing quality of life in developed countries created new needs in the transportation and communication sectors, as well as a significant increase in the demand of some others. For instance, society demanded services including water, gas, power, and better buildings and houses. All of them implied larger energy consumption. The chemical industry was energy intense and thus it guided, the development of the power industry, improving the use of hydraulic resources, coal, and later, crude oil.

1.1.7.1 Nitric acid industry

It was the need for pigments and explosives that increased the demand for nitric acid. Before First World War (WWI) it was produced from sodium nitrate (Chile saltpeter or Peru saltpeter) and sulfuric acid. However, Chile saltpeter was the only natural source for nitrogen-based fertilizers, and thus both industries competed for the same raw materials. Therefore, there was a need for a different source of nitric acid for the chemical industry.

By 1840 Frédéric Kuhlmann studied the production of nitric acid via ammonia oxidation over platinum. The process did not reach industrial scale until the beginning of the 20th century when Ostwald and Eberhard Brauner found the operating conditions for an acceptable yield. The use of ammonia was an alternative path towards nitric acid, however, ammonia was also basic to the fertilizer industry.

The problem was solved in 1913 when the Haber−Bosch process—named after Fritz Haber, a chemist, and Carl Bosch, an engineer with the Basiche firm—was discovered. The reaction, carried out at high pressures and temperatures over metallic oxides, was a challenge for the field of materials and reactor design, representing:

- Development of heterogeneous catalysis.
- Production of pure N_2, obtained from air involving cooling, condensation, and air rectification. In essence, air fractionation and the Linde process were developed behind this need.
- Production of pure and reasonably cheap hydrogen, by developing coal gasification technologies followed by the water was shift reaction to get rid of the carbon monoxide (CO) that was produced together with the hydrogen.
- Developed high-pressure and -temperature gas processing, in particular the problem of the high diffusivity of hydrogen and the material design to handle it.

1.1.7.2 Coal gasification

Coal gasification represented (in 1923) the establishment of carbon chemistry via the production of synthesis gas, syngas. Syngas mainly consisted of hydrogen and CO, the building block for a large number of compounds such as methanol, from which formaldehyde can be produced via oxidation and Bakelite by its polymerization (ACS, 1993).

Syngas was also a raw material for synthetic fuels. When Germany lost access to crude oil during the WWI, an alternative was presented that used syngas to obtain synthetic gasoline and diesel. The process was developed by Franz Fischer and Hans Tropsch, and patented in 1925. It was based on the hydrogenation of CO over catalysis. At the same time, the West lost its main supplier for pigments and intermediates, and thus the chemical industry grew by developing novel processes based on state-of-the-art technologies and improved organization and productive models. Thus the United States displaced Germany from its top position in the chemical industry.

1.1.7.3 Polymers

By the mid-1900s, the first modification of natural polymers such as cellulose and rubber was successful, making it possible to achieve new properties for novel applications. Furthermore, the unsuccessful experiments in processing several mixtures of small molecules resulted in viscous fluids or solids that were highly stable. The increased needs for modified macromolecular products provided an opportunity for the production of synthetic plastics from those small organic molecules, creating Bakelite (mentioned above), polyvinyl chloride, acrylic plastics, polystyrene, nylon, low-density polyethylene, etc., and making it possible to obtain tailor-made products with the proper properties. The chemical industry provided the raw materials and acetylene found a new use.

It was again as a result of a war, Second World War (WWII), that the supply of natural macromolecules was cut off. The only industry capable of substituting a natural species by a synthetic ones was, and still is, the chemical industry. That required research and development. For instance, when the United States could not access natural rubber, it took 2 years to develop synthetic rubber; by 1944 all vehicles used synthetic rubber for their tires. In 1954 the first catalysts capable of producing molecules with a certain 3D order were developed, the Ziegler−Natta. Novel and more specialized materials such as resins were thus developed.

1.1.7.4 Petrochemical industry

The source for most of the small molecules, the acetylene, was produced in an electric furnace from calcium carbide ($CaO + 3C \rightarrow CaC_2 + CO$). The high demand to satisfy the growing macromolecules market increased its price and the industry turned its focus to crude oil. Crude fractionation and refining was developed in the beginning of the 20th century in the United States.

The growth of the automobile industry and the demand of society for higher quality of life increased the demand for gasoline. Production via crude fractionation could not meet the demand, and therefore heavier fractions that could not be easily sold became the source for fuels (via cracking). We can trace cracking back to work by Burton and Humphries in 1912.

Apart from small molecules and gasolines, the production of aromatics became of great interest. Thus petrochemistry replaced carbochemistry due to the lower cost of raw materials and final products, and ethylene replaced acetylene as the building block of the chemical industry.

1.1.8 RENEWABLE AND NONCONVENTIONAL-BASED DEVELOPMENT

The easy access to cheap and abundant raw materials slowed down the research on other energy sources and raw materials. However, the growth in population and increasing energy needs over the years, together with the political stability of the main producers, has moved society to find alternative sources of energy. There has been an important development of renewable-based sources of chemicals and power from not only biomass such as energy crops (switchgrass, miscanthus) and nonedible sources of oil (including algae), but also solar and wind energy. Lately a number of processes have been developed to produce substitutes for crude-based gasoline and diesel, such as bioethanol and Fischer–Tropsch gasoline and diesel, dimethyl ether, dimethyl furan, furfural from biomass, biodiesel, from waste cooking oil, algae or nonedible oil, glycerol ethers from the byproduct of the biodiesel industry. Typically solar and wind have been used for the production of power using photovoltaics or concentrated solar power facilities, and wind turbines, respectively. The difficulties in storing solar and wind energy have also been a challenge that the chemical industry has accepted either by the development of batteries or the production of chemicals such as methane from CO_2 and electrolytic H_2.

Apart from renewable-based sources, two nonrenewable but abundant sources of methane have attracted attention: shale gas and methane hydrates. Shale gas is natural gas trapped in shale formations. The development of extraction techniques, horizontal drilling, and fracking are currently allowing the exploitation of the vast reserves found in the United States, Mexico, Argentina, and China. Methane hydrate is actually methane trapped in ice that has been generated from organic wastes in the continental limits or below the ice of frosted surfaces. Hydrates represent more than 50% of organic carbon, and currently Japan has made interesting progress in the industrial use of this methane (Martín and Grossmann (2015)).

1.2 CHEMICAL INDUSTRY IN FIGURES

World chemical turnover was valued at €3156 billion in 2013; that is divided across several regions, as presented in Fig. 1.2. We can distinguish five major

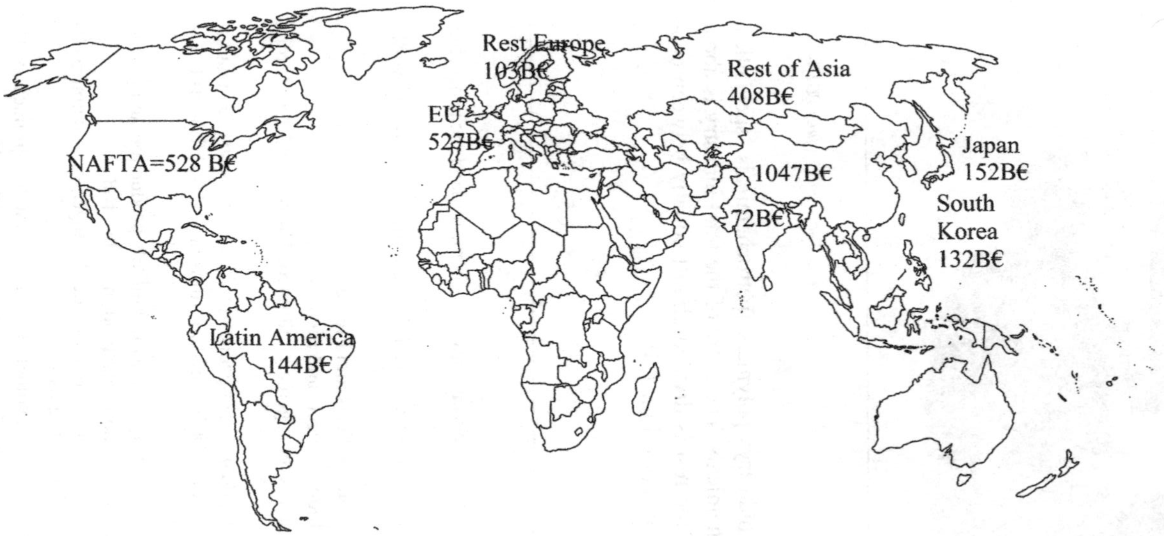

FIGURE 1.2

The chemical industry in sales.

Rest of Asia
408B€

Japan
152B€

South
Korea
132B€

1047B€

72B€

Rest Europe
103B€

EU
527B€

NAFTA=528 B€

Latin America
144B€

Cefic, Facts and Figures 2013–14.

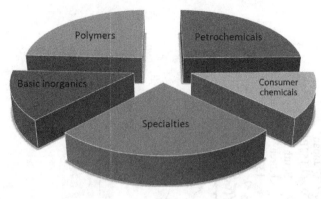

FIGURE 1.3

Businesses in the chemical industry.

Cefic, Facts and Figures 2013–14.

businesses within the chemical industry: polymers, petrochemicals, specialties, basic inorganics, and consumer chemicals. The share of the total industry is about 25% each for the first three, and the rest is divided almost equally between consumer chemicals and basic inorganics (see Fig. 1.3).

REFERENCES

ACS, 1993. <http://www.acs.org/content/acs/en/education/whatischemistry/landmarks/bakelite.html>.

Ashby, M.F., 2004. Materials Selection in Mechanical Design, third ed. Butterworth-Heinemann, Oxford.

Cefic, 2013. The European Chemical Industry Facts and Figures.

Cefic, 2014. The European Chemical Industry Facts and Figures.

García, D.J., 1998. La industria química y el ingeniero químico. Universidad de Murcia, Servicio de Publicaciones.

ACS, 1998. <http://www.acs.org/content/dam/acsorg/education/whatischemistry/landmarks/calciumcarbideacetylene/commericialization-of-calcium-carbide-and-acetylene-commemorative-booklet.pdf>.

Martín, M., Grossmann, I.E., 2015. Water–energy nexus in biofuels production and renewable based power. Sustainable Production and Consumption 2, 96–108.

Ordóñez, S., Luque, S., Álvarez, J.R., 2006a. Evolución de la ingeniería química (I) Historia de la tecnología química. January. 124–131.

Ordóñez, S., Luque, S., Álvarez, J.R., 2006b. Evolución de la ingeniería química (y II) Área de conocimiento Historia de la tecnología química. February. 198–207.

Chemical processes

2

2.1 INTRODUCTION TO PROCESS ENGINEERING

The word *design* refers to a creative process searching for the solution to a particular problem. When it comes to chemical processes, the aim is to come up with the stages and the units that transform a raw material or a number of raw materials into a desired set of products. At this point we must refer to the book by Rudd, Powers, and Siirola (1973), the first one to present this procedure as a systematic process. The field of chemical process engineering finds in that work its first textbook.

A *chemical plant* or facility comprises a number of operations and reactors that allow the transformation of raw materials into products of interest and their purification. The typical stages are as follows:

- Feedstock preparation,
- Reaction,
- Product Purification,
- Unconverted reactants recycle,
- Emissions control,
- Packing/Product transportation.

In Fig. 2.1 we present the scheme for the production of ethanol via biomass gasification. We see the different stages of any chemical process. Biomass is processed so that the size is appropriate for gasification. Next, the raw syngas is purified from hydrocarbons, hydrogen sulfide (H_2S), ammonia (NH_3), etc., and its composition adjusted in terms of the hydrogen (H_2) to carbon monoxide (CO) ratio. Subsequently, the gas reacts to produce not only the desired product, ethanol, but also methanol, propanol, and butanol. Furthermore, the conversion of the syngas is not complete, and the unreacted gases are separated and recycled, while the liquid products are separated via distillation. This process also has emission control units in the form of sour gas capture. Carbon dioxide (CO_2) and H_2S are removed from the gas stream because they are poisonous for the catalyst.

The *process engineer* is the person responsible for putting together the flowsheet that defines the path from the raw materials to the products. His main aim is to find the optimal way to obtain a product with the specifications that meet the demand at the lowest cost, under safe conditions, and with the lowest

Industrial Chemical Process Analysis and Design. DOI: http://dx.doi.org/10.1016/B978-0-08-101093-8.00002-1

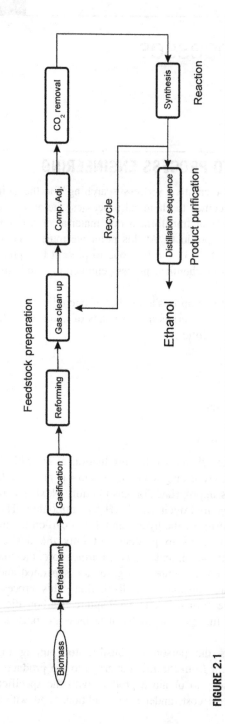

FIGURE 2.1

Bioethanol production process.

environmental impact. Therefore, process optimization has an important role. Before 1958, the good economic situation allowed a number of rules that resulted in units being overdesigned. In addition, environmental constraints were relaxed to help grow the economy. However, as times became tougher, the processes were designed and operated closer to their optimal specifications. Therefore, the process engineer had a more difficult task. Nowadays, overdesign is a burden for the profitability of the facility.

Making decisions on the best process is a tough problem in itself. First, there are multiple alternatives for almost every processing stage. Furthermore, there is no single design criterion, but a number of them. Economic evaluation is the one every manager will ask for first. However, society is concerned with the environment and the security of the people that work in the plant and close by. Therefore, environmental assessment, safe operation, and flexibility are other concerns that transform the original problem into a multiobjective one.

There are four main problems that the process engineer typically has to address:

- *New process or product*: The challenge is the lack of information, and therefore the main concern is the uncertainty in the outcome. As a result of the technical risk, industry is resistant to try novel strategies unless huge economic advantages are expected. Process scale-up is key to reducing risk (Zlokarnik, 2006).
  It consists of evaluating the performance of the process at the laboratory scale, where its technical feasibility can be shown, moving on to the pilot plant, and finally full industrial scale. Pilot plants of 1/30 scale are available at research facilities for novel oxycombustion-based processes (eg, CIUDEN in Spain).
- *Modifications of a process*: Because there is experience with a previously developed process, decisions are sometimes predicated on previous solutions, which may not provide the best options.
- *Building capacity*: Bottlenecks must be identified to determine the units to be modified and the parallel lines needed.
- *Intelligence*: We need to know why the competition can produce so cheaply and with such good quality.

The project consists basically of deciding:

1. Which product to obtain. A market study of needs determines this.
2. How much to produce. What is the market demand for the product? Size has always been an advantage in the chemical industry, which benefits from economies of scale. A larger plant is not proportionally more expensive. Sometimes the operation of the plant (ie, the inability to cool down a reactor, or raw material availability) limits the size of the plant.
3. Quality of the product. Society demands better products for more specialized operations.
4. Quality of the raw material. What is the tradeoff between the price for a purer material and the processing cost to purify it on-site?

5. Storage issues. How much must be stored (if any) and for how long? Do the raw materials age, and are there safety issues with storage?
6. Byproducts. Are they useful, and do they present environmental or safety concerns?
7. Waste treatment. Is this expensive?
8. Allocation of the plant. This is typically a supply chain problem. The agents involved are the market, the location of customers willing to buy the product, raw material allocation, determination of the price to supply the plant, and an inventory of raw materials and products for normal operation of the plant.

The process engineer must consider all of these aspects when dealing with a process. Process design can be divided into four stages.

2.1.1 PROBLEM DEFINITION: CONCEPT

At this point the needs, specifications, and process technologies must be identified to determine the feasibility of the process. The challenge is that it is an open problem, ie, there are a number of solutions. The tools that we have are the literature, encyclopedias, and brainstorming.

2.1.2 PROCESS SYNTHESIS: ALTERNATIVE TECHNOLOGIES

The second stage aims to put together a flowsheet for the process using the information gathered in the previous stage. The challenge is the large number of alternative technologies available and the trade-offs among them. The tools are the information regarding the performance of such technologies, including rules of thumb, as well as more advanced techniques such as Douglas's hierarchy and superstructure optimization.

2.1.3 MASS AND ENERGY BALANCES: ANALYSIS OF THE PROCESS

Typically the mass and energy balances of a process consist of a complex system of equations, including thermodynamics of the species and of the mixtures, reactor kinetics, and liquid–liquid and liquid–vapor equilibrium, etc. Over the last several decades a number of packages have been developed not only to solve systems of equations (eg, EXCEL, gProms, GAMS, Matlab, Mathcad), but to model processes rigorously (eg, CHEMCAD, ASPEN, ASPEN HYSYS, PRO II). Section 2.4 presents a brief summary.

2.1.4 DESIGN CRITERIA: EVALUATING THE ALTERNATIVES

This last stage is the most challenging one. After the analysis of the process we need to decide which alternative design to choose and present. There are a number of indexes or metrics to compare processes, such as Net Present Value;

a measure of process profitability; energy consumption; feasibility analysis; HAZOP safety analysis; and environmental assessment, including Life Cycle Assessment (LCA), controllability of the process, its flexibility, etc. Whether we comply with one (ie, the most profitable process) or we look for a robust solution (ie, a profitable process but one that is friendly to the environment and has no safety issues), the decision must be supportable before a Board of Directors.

Bear in mind that the decisions at the level of conceptual design represent 80% of the total investment cost of the plant. In other words, the flowsheet determines the units of the process and thus the cost of the process itself, but for adjustments and refinement.

2.2 PRINCIPLES OF PROCESS DESIGN

In this section we present to the reader methods to sketch the flow diagram for a process. Some of the methods will not be further pursued in this text, but it is important that the reader becomes acquainted with the words and terms in process design. All these methods aim to propose a systematic way of evaluating the large number of alternatives that the process engineer has to consider in order to come up with a process flowsheet.

A. Hierarchy Decomposition

In this method, proposed by Douglas in 1985 and later modified by R. Smith at the University of Manchester (UMIST by that time) (Smith, 2005), there are five decision levels.

- Level 1: Continuous process versus batch process.
- Level 2: Input—output structure.
- Level 3: Recycle.
- Level 4: Separation stage, and vapor and liquid recovery.
- Level 5: Heat recovery and integration.

B. Superstructure Optimization

This method consists of the formulation of a mathematical problem where all the alternative processes are modeled based on mass and energy balances, thermodynamic equilibrium, rules of thumb, experimental data, etc. The optimal flowsheet is extracted from the superstructure by solving the problem. For further details see Biegler et al. (1997), Kravanja and Grossmann (1990), and Yeomans and Grossmann (1999).

C. Evolutionary Methods

These are based on know-how and a previously available design. Using the original process as a starting point, it is modified to address its weaknesses and/or to increase capacity.

In order to the reader to have a better understanding of the processes that will be described in the following chapters, we are going to briefly comment on the first of the three methods. However, in some processes (eg, air liquefaction, lead chambers), evolutionary methods are behind the development of alternatives for improving the yield, as it will be shown in the following chapters.

2.2.1 DOUGLAS HIERARCHY (DOUGLAS, 1988)

2.2.1.1 Level 1: batch versus continuous process

We use a batch process under the following circumstances:

- Low production capacity (typically below 5×10^5 kg/y).
- Multiproduct plants (eg, cosmetics and detergents).
- Special market properties such as seasonality (eg, ice cream and buñuelos) or aging (eg, fresh food).
- When there are scale-up problems. For instance, in batch exothermic reactions, the larger volumes do not maintain the area-to-volume ratio, and heat transfer problems may appear.

2.2.1.2 Level 2: input–output structure

At this level we try to answer a number of questions related to the purity of the raw material: Should we purify them? The purity of the byproducts: Should we recycle or eliminate them? Do we need to purge part of the recycle to avoid build-up within the process? Furthermore, can the unreacted raw materials be recycled or not? We also need to evaluate the number of product streams of the process. Based on this black box analysis of the process, the preliminary economic potential of the alternatives can be determined in order to disregard the worst ones. To address all of these questions there are some basic rules:

- If the impurity is not inert, we need to remove it.
- If the impurity is in the feedstock and it is in the gas phase, we process it.
- If the impurity is in a liquid feedstock and it is a product or a byproduct, we process the feedstock though a separation stage.
- An impurity in large amount must be removed.
- If the impurity is an azeotrope, we process it.
- We must determine when it is easier/cheaper to eliminate the impurity from the feedstock or from the products.
- If the boiling point of the impurity is lower than $-48°C$, we need to purge it or use membrane separation.
- Oxygen, water, or air are used in excess as reactants.
- The number of product streams depends on the boiling point of the gases and liquids that we obtain. Solids are evaluated separately.
- Valuable byproducts are sold.
- Avoid buildup in the process.

2.2.1.3 Level 3: recycle

This decision level is used to evaluate the reaction step. The aim is to determine the reaction steps required, the recycle streams, and the need for an excess of one of the reactants. For that we need to know the kinetics and equilibrium governing the chemical reaction or reactions in order to figure out how it/they can be altered. Typically the reactions involve heat transfer, and sometimes the design of the reactor is heavily conditioned by how we can remove the heat generated. The effect of the reaction on the economic potential is addressed at this level.

Apart from know-how on the reaction step, a set of general rules can also help in decision-making:

- There exist optimal conditions for the operation of the reactor in terms of feedstock composition, pressure and temperature, and reactor configuration for heat removal as a result of the chemical reaction or reactions taking place.
- If there are several reactions that require different operating conditions, a number of reactors are needed.
- If the reaction is exothermic, but the increase in the temperature is lower than 10%, the reaction operates as adiabatic.
- For an endothermic reaction, heat can be provided directly for isothermal operation and low production capacity, or indirectly for higher production capacity.
- The aim is to reach 96—98% of the equilibrium conversion.

2.2.1.4 Level 4: separation structure

From this point on, the separation costs are included in the evaluation. The product stream from the reactor is evaluated as follows:

- If it is in the *liquid phase*, the separation is based on species volatility. If the relative volatility is larger than 1.1, distillation is selected, otherwise liquid—liquid separations can be considered.
- If we have a *gas—liquid mixture*, a flash separation is used to separate phases.
- If the reactor operates at high temperature, the products are cooled down and condensed. If the cooled liquid phase comprises only reactants, it is recycled. If it contains only products, it is sent to further purification. The gas phase is sent to vapor recovery. The reactants are recycled. We can avoid flash separation if the amount of vapor is small. In that case the multiphase stream is sent to liquid separation.
- If the product is in the *vapor phase*, the stream is cooled down to condense a fraction and separate the phases. The condensed liquid is processed as described above. Sometimes cooling is not enough and compression can be used to separate the phases. Otherwise, if the phases cannot be separated, partial condensation is considered.
- If the products in a stream have no value, the stream is not further processed.

Another method is the one proposed by Jaksland et al. (1995), which consists of two levels. In the first level, the mixture is characterized to identify the separation technique based on the vapor pressure ratio. In level two, we select the separation task, the sequence of separations, and determine the need for solvent and type.

In this stage we see that distillation is the most important operation to separate liquid mixtures. Distillation is an energy-intense operation. When a mixture of components is to be separated, the order of their separation determines the cost of the product. Therefore, much effort has been placed on systematically determining the cheapest distillation sequence. There are two main approaches: heuristic rules and mathematical optimization. For the sake of this text we focus on the first one and leave the reader with the reference for the second (Biegler et al., 1997).

Synthesis of distillation sequences (Rudd et al., 1973, Jiménez, 2003). Relative volatility, of the species involved determine the separation of the mixture. If it is 1, there is no way. If it is 1–1.1, it is difficult. If it is 1.1 or higher, it is easy. Thus, if the relative volatility of two species is lower than 1.1, distillation is not an option.

The problem we face here is the large number of options; for example, for three components, assuming sharp splits, there are two different flowsheets:

$$(ABC) - [(A) - (BC)] - [(B) - (C)]$$
$$(ABC) - [(AB) - (C)] - [(A) - (B)]$$

For four components we get five flowsheets. In general, let N be the number of components; the number of units, N_o, is given as

$$N_o = \frac{[2(N-1)]!}{(N-1)!N!} \tag{2.1}$$

For a one distillate—one bottom column, the general rules are as follows:

1. Remove corrosive or reacting component first.
2. Remove products as distillates.

Heuristics for Column Sequencing
H1. Forbidden splits:
   a. Don't use distillation if the relative volatility is lower than 1.05.
   b. If $(\alpha - 1)_{\text{extractive dist.}}/(\alpha - 1)_{\text{reg. distillation}} < 5$, use ordinary distillation.
   c. If $(\alpha - 1)_{\text{sep liq-liq}}/(\alpha - 1)_{\text{reg. distillation}} < 12$, use ordinary distillation.
   d. Consider absorption as a candidate if refrigeration is needed for distillation.

H2. The next separation is that with the largest relative volatility. Easy first, difficult last.

   This rule is based on the size of the column. The minimum number of stages in a distillation column can be computed as given by Eq. (2.2):

$$N_{min} = \frac{\log\left[\left(\frac{x_{i,D}}{1-x_{i,D}}\right)\left(\frac{1-x_{j,W}}{x_{j,W}}\right)\right]}{\log\alpha_{ij}} \tag{2.2}$$

where α is the relative volatility, and x_i is the molar fraction of component i in the feed, F, or the distillate, D. Therefore, the larger the relative volatility, the shorter the column.

H3. Remove the most abundant component first.

   The size of the column (its diameter) is a function of the vapor phase across it as given by Eq. (2.3), where V is the vapor flow and u the vapor velocity across the column, ρ the density, and K the design constant:

$$D = \sqrt{\frac{4V(m^3/s)}{\pi u(m/s)}} = \sqrt{\frac{4V(m^3/s)}{\pi K \sqrt{\frac{\rho_L - \rho_V}{\rho_L}}}} \tag{2.3}$$

   Furthermore, the flowrate of the vapor phase has to be reboiled. Thus, the energy at the reboiler increases for higher flowrate. If we reduce the flow processed in the column by separating the most abundant sooner, the following columns will be smaller and the energy consumption lower.

H4. If the α and concentration are not very different, use direct sequence (most volatile first).

   The justification behind this rule is again related to the energy consumption and the diameter of the column. Let's assume a mixture ABC equimolecular with similar relative volatilities between the components AB and BC, and that the feed is saturated liquid.

   Direct Separation: See Fig. 2.2 for the flow pattern across the column.

   Indirect Separation: See Fig. 2.3 for the flow pattern across the column.

   Therefore, the direct sequence is cheaper.

H5. Remove mass separation agent in the next column.

H6. Favor sequences that do not break desired products. We save a column or more.

   These rules must be applied in order of priority, and every time a decision on the separation of products is to be made.

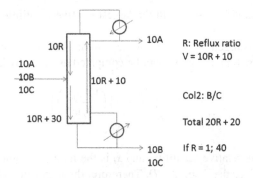

FIGURE 2.2

Direct separation sequence.

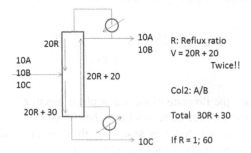

FIGURE 2.3

Indirect distillation sequence.

EXAMPLE 2.1

Let's look at an example for the separation of a mixture courtesy of Prof. I.E. Grossmann. Fig. 2E1.1 presents an example for the use of the heuristic rules.

The order for the volatilities of the species and the values are given below:

O(I) :ABCDEF
O(II) :ACBDEF
    α_I A/B 2.45, B/C 1.18, C/D 1.03, E/F 2.5
    α_{II} C/B 1.17, C/D 1.7

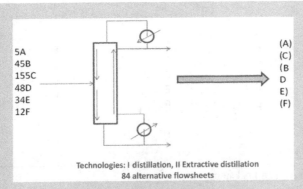

Technologies: I distillation, II Extractive distillation
84 alternative flowsheets

FIGURE 2E1.1

Distillation sequence example.

Solution
H1. Exclude $(C/D)_I$ since $\alpha_I = 1.03$

$(B/C)_I$ 1.18 vs $(C/B)_{II}$ 1.17 0.18/0.17 < 5 → exclude $(C/B)_{II}$

H2. *Separate largest* α

$$\alpha_{I(A/B)}\ 2.45.$$

Similarly,

$$\alpha_{I(E/F)}\ 2.5$$

H3. *Get rid of most abundant*

$$A/B = 5A \quad E/F = 12F$$

Small amounts.
H4. *Direct sequence favors* → $(A/B)_I$

The desired products are BDE & C (Fig. 2E1.2).

1. We can separate C/BDE using extractive distillation. However, H1 does not recommend the use of extractive distillation.
2. We can separate BC/DE. Again, H1 does not recommend the use of regular distillation since the relative volatility is lower than 1.05.
3. We separate B/CDE using regular distillation ($\alpha = 1.18$).

Next, we separate C/DE using extractive distillation ($\alpha = 1.7$):

$$\frac{(\alpha-1)_{II}}{(\alpha-1)_I} = \frac{1.7-1}{1.03-1} \geq 5$$

We mix B and (DE)

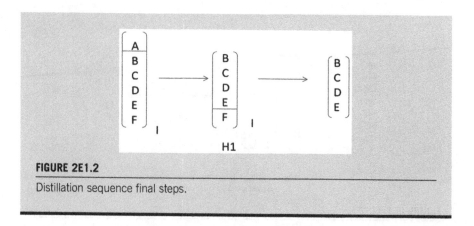

FIGURE 2E1.2

Distillation sequence final steps.

EXAMPLE 2.2

We have a mixture of propane (45.4 kmol/h), isobutane, (136.1 kmol/h), n-butane (226.8 kmol/h), i-pentane (181.4 kmol/h), and n-pentane (317.4 kmol/h) with relative volatilities of $C_3/iC_4 = 3.6$, $iC_4/nC_4 = 1.5$, $nC_4/iC_5 = 2.8$, $iC_5/nC_5 = 1.35$. Determine the sequence of distillation columns to separate the mixture.

Solution

Using H2 we separate $(C_3|iC_4 \ nC_4 \ iC_5 \ nC_5)$.
Using H2 we separate $(iC_4 \ nC_4|iC_5 \ nC_5)$.
Using H2 we separate $(iC_4|nC_4) \ (iC_5|nC_5)$.

2.2.1.5 Level 5: heat recovery and integration

Hierarchy decomposition considers heat integration as the last stage in process design. However, energy consumption represents one of the major economic burdens of any process, and therefore energy integration must be considered along with process synthesis. In fact, at Level 3 we already considered the operation of the reaction involving heat transfer. The aim of this stage is to identify the sources of energy within our process and reuse it in a systematic way. We consider two cases for heat integration due to their impact on the economy of the processes: the *heat exchanger networks* (HENs), to determine the minimum utility cost; and in the second step the design the structure of the network and *multieffect columns*, which have proved to be extremely useful to reduce energy consumption in bioethanol production processes.

Heat exchanger networks

Utilities Minimization Given:

- a set of hot streams that require cooling and a set of cold streams that need energy,
- flowrates and inlet/outlet temperatures are known,
- heat capacities are assumed constant,
- and the utility streams are known,

we aim to design the HEN that minimizes the operating and capital costs. Let us illustrate with an example.

EXAMPLE 2.3

Design a HEN with minimum consumption of utilities for the stream in Table 2E3.1.

Table 2E3.1 Data for Example 2.3

Streams	Fc_p (kW/°C)	T_{in} (°C)	T_{out} (°C)	Heat Flow (kW)
C1	15	360	600	
C2	13	300	450	5550
H1	10	600	320	
H2	20	540	320	7200
				−1650

From the first law of thermodynamics, we need 1650 kW of cooling. However, it is not possible to find a network with 1650 kW (see Fig. 2E3.1).

The procedure for computing the minimum utilities required is as follows:

1. Create temperature intervals. Assume a minimum ΔT_{min} (HRAT−heat recovery approach temperature), for instance, 10°C. For cold streams, there must be a stream that is at least $T + \Delta T_{min}$ above; ie, for the stream at 260°C, there should be another stream at 270°C. For hot streams, in order to cool them down, we need a stream 10°C below; ie, for the 93°C stream, we need another stream at 83°C.

2. Compute the heat available at each interval, k, and that required.
$Q_k = \sum_{i \in \text{cold or hot}} F_i \cdot c_{p,i} \cdot \Delta T_k$

3. Accumulate the heat available and that required; they are the cascades of hot and cold streams. $Q_{\text{available}} = \sum_k Q_k|_{\text{hot}}$; $Q_{\text{required}} = \sum_k Q_k|_{\text{cold}}$;

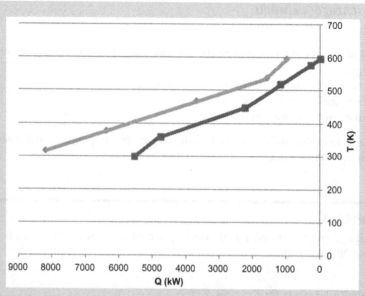

FIGURE 2E3.1

Hot and cold cascades.

4. Plot the cascades (see Fig. 2E3.1). The change in curvature is given when a second stream contributes to the composite hot and cold curves, respectively. In the figure it is possible to compute the minimum needs for utilities as the difference in the extremes of the curves when they touch. The required energy and the refrigeration needed are computed when both curves are at DT difference. To compute this point we continue the procedure.

5. Determine the Composite Curve (GCC) by accumulating the net energy at each interval k. The minimum value (most negative) is the energy required. By adding this value to the GCC on top we find the required energy, and at the bottom the cooling needs. Table 2E3.2 shows the results for the pinch.

The minimum utilities required are computed when the hot and cold composites are put close so that the distance between them is ΔT_{min}. Fig. 2E3.2 shows this point. We have horizontally displaced one of the curves so that the hot composite is above the cold one and the distance between them at the closest point in the y axis, the pinch, is ΔT_{min}. The horizontal distance at the top and at the bottom of the curves gives the minimum heat to be provided and the minimum energy to be removed from the system.

Table 2E3.2 Pinch Example 2.3 Results

Q available	Cascade	Hot	Temperatures Intervals	Cold		Q required	Cascade	Net	GCC	Adjusted	
	0		620		600		0		0	600	
600	600	H2	600	5	580	300	300	−300	−300	300	
2100	2700	H1	540	4	520	900	1200	−300	−600	0	
2700	5400		470	3	450	1050	2250	1050	450	1050	
1800	7200		380	2	360	2520	4770	180	630	1230	
			320	1	300	780	5550	1020	1650	2250	

(Hot stream bars labelled H1 and H2; cold stream bars labelled C1 and C2.)

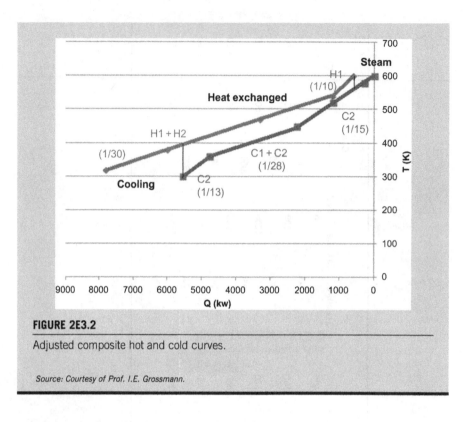

FIGURE 2E3.2

Adjusted composite hot and cold curves.

Source: Courtesy of Prof. I.E. Grossmann.

HEN Structure The idea is to minimize the area and the energy to be removed and that provided. Thus, a few heuristics that can be used are the following:

- High temperature streams are integrated first.
- Streams with high heat loads are integrated next.
- Search for larger gradients of temperatures.
- Look for larger global heat transfer coefficients.

Extreme alternatives:

a. Every hot stream with a cooling utility;
b. Every cold stream with a heating utility.
c. Recover as much heat as possible using the least amount of utilities;

Usually, there are many alternatives.
Near optimal HEN usually exhibits:

a. Minimum utility cost for a given DT_{min} (HRAT + 10K).
b. Fewest number of units.

Targeting (Pinch analysis) method:

1. Predict (a) and (b).
2. Develop a network that satisfies (a) and (b).

a. *As presented in the previous point*

For maximum heat exchange, place curves as close as possible within ΔT_{min} + HRAT. Note:

HRAT = 20K	600 kW heating	2250 kW cooling
HRAT = 10K	450 kW heating	2100 kW cooling
HRAT = 5K	375 kW heating	2025 kW cooling
HRAT = 0K	300 kW heating	1950 kW cooling (Absolute minimum)

1. Determine the excess or need of energy in each interval.
2. Cascade heat with 0 kW heating.
3. Cascade minimum heating:
    a. Minimum cooling.
    b. Temperature below zero entry-pinch,

Note that there will not be heat transfer across the pinch and we can divide the network into two above and below the pinch,

In the problem table we see that the maximum deficit of energy is 600 kW, which is the energy that needs to be provided.

b. *Minimum number of heat exchangers (N_{min})*

For each subnetwork assume at least one stream is exhausted in each match. According to Euler's theorem, for n nodes:

$$N_{min} = n(\text{hot}) + n(\text{cold}) + n(\text{utilities}) - 1$$

The following example illustrates this procedure:

EXAMPLE 2.4

Design a heat exchanger network (HEN) for the streams in Example 2.3.

Solution

In our case

Streams Above Pinch: H1, C1, steam	Nmin = 1 + 1 + 1 − 1 = 2
Stream Below Pinch: H1, H2, C1, C2, cooling water	Nmin = 2 + 2 + 1 − 1 = 4
	Total = 6

- We are looking for a network with 600 kW of heating and 2250 kW of cooling with the pinch at 540−520K.
- It should also have six units (two above the pinch and four below).

The subnetwork above the pinch is shown in Fig. 2E4.1.
Below the pinch, we have several alternatives. Fig. 2E4.2 shows the subnetworks where H2 and H1 are used to heat up C1 and C2, respectively.

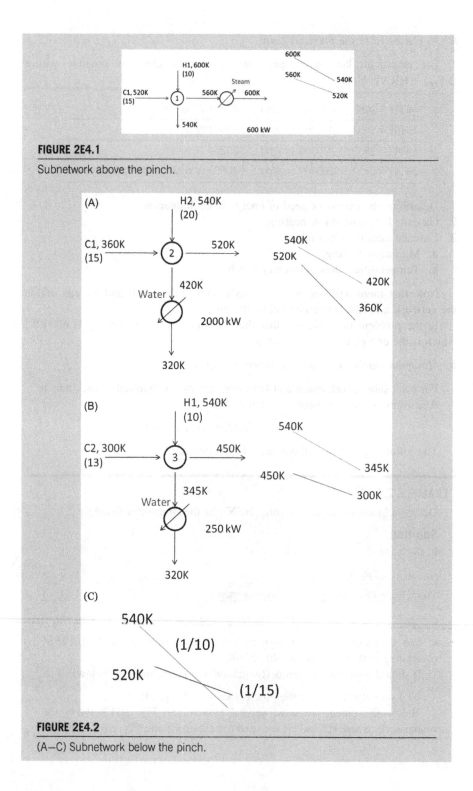

FIGURE 2E4.1

Subnetwork above the pinch.

FIGURE 2E4.2

(A—C) Subnetwork below the pinch.

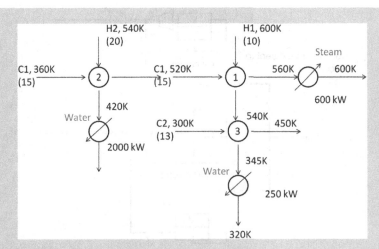

FIGURE 2E4.3

Optimal heat exchanger network.

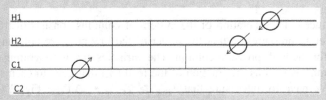

FIGURE 2E4.4

Matches diagram.

However, if we try to use H2 to heat up C1, we have temperature cross. Thus, the final network is of the form shown in Fig. 2E4.3.

The matches diagram is given by Fig. 2E4.4. We plot the hot and cold streams as lines assuming that the temperature decreases to the right. The matches are represented by lines connecting the streams that exchange energy. We place heat exchangers to represent the heat provided or removed using utilities.

Source: Courtesy of Prof. I.E. Grossmann.

Multieffect columns

In the case of the production of ethanol from corn grain, one of the most energy consuming pieces of equipment is the distillation column known as a "Beer column." It contributes to almost 50% of the energy of the process. Using multieffect columns the energy consumption is reduced by nearly half, and the cooling needs

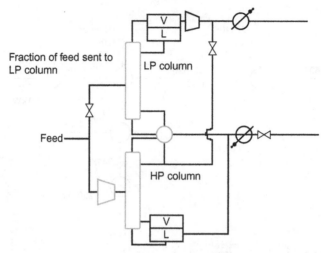

FIGURE 2.4

Scheme for a multieffect distillation column system.

are also reduced. The operation of the system of columns consists of treating the feed at different pressures so that the boiler of the low-pressure column acts as a condenser for the high-pressure column. The variables are the operating pressures of the columns as well as the fraction of the feed (α_i) that goes to each one of the columns. We have to make sure that $T_{\text{Reb}} > T_{\text{Cond}} + \Delta T_{\text{min}}$. Fig. 2.4 shows a scheme for a system of two columns operating as multieffect.

The formulation of the problem is given by the mass balance to the initial splitter, the temperature constrains so that heat can be transferred from the reboiler of the lower pressure columns to the condenser of the higher pressure columns, and the energy balances so that the heat exchanged between them matches and the mass and energy balances to each of the columns. A pressure increase from the lower pressure column to the higher pressure columns is enforced by a constraint (see Eq. 2.4).

Karuppiah et al. (2008) used this technology in the production of ethanol from corn, resulting in $3 million savings in steam just in the columns (see Fig. 2.5).

$$F = \sum_k F_k$$
$$Tb_k \leq Tc_{k+1} + dt, \quad k \in \text{COL}$$
$$Tb_k \leq Tc_k + dt, \quad k \in \text{COL}$$
$$Qb_k = Qw_{k+1}$$
$$P_{k+1} \geq P_k$$
$$M\&E_k$$

(2.4)

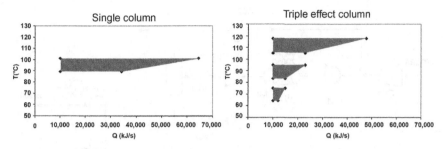

FIGURE 2.5

Energy and cooling savings when using multieffect columns.

From: Karuppiah, R., Peschel, A., Grossmann, I.E., Martín, M., Martinson, W., Zullo, L., 2008. Energy optimization of an ethanol plant, AICHE J. 54(6) 1499–1525.

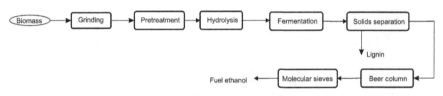

FIGURE 2.6

Block flow diagram.

2.3 FLOW DIAGRAMS

2.3.1 TYPES

Flow diagrams are a way to present information from the process Westerberg et al. (1979). Depending on the audience they are intended for, different formats and standards are applied.

1. Process diagrams.

    They are simple representations of the process with no technical details. Their aim is diverse, from publicity and advertising to providing a technical sketch.

    a. Block flow diagrams. They consist of a number of boxes representing each of the units or groups of units. The boxes are connected through arrows. See Fig. 2.6.

    b. Block flow process diagram. It is used for simple representation of the entire process. Typically all the units of the process are represented and linked by arrows, indicating the direction of the flow. Some major operating conditions are added to highlight them. It is the basis for the engineering process diagrams. See Fig. 2.7.

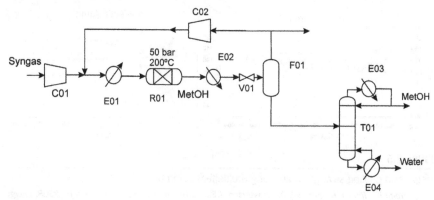

FIGURE 2.7

Block flow process diagram.

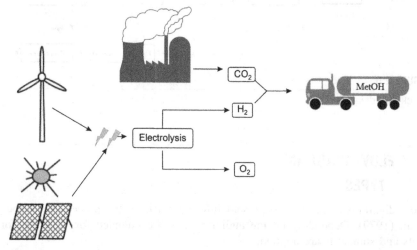

FIGURE 2.8

Graphic process diagram.

    c. Graphic process diagram. It is widely used in publicity or to sell the project to managers. It is free, and they are not standardized. See Fig. 2.8.

    d. Sankey diagram: It shows the flow of energy or mass involved in the process, typically as a whole. It allows identifying energy losses. See Fig. 2.9.

2. Process flow diagrams. They are the major source of information regarding a process, from the stream data to equipment sizing. The symbols that are used for the units and the way of reporting information on the operating conditions are standarized. Anyone in the field can interpret the diagram (see Fig. 2.10).

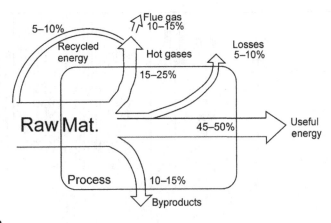

FIGURE 2.9

Sankey diagram.

The information on the streams must be provided here. See Section 2.4 for a detailed list of variables. A special type of engineering process diagram is the so-called Piping and Instrumentation Diagram. It includes everything from the size of the pipes and their material to the controllers and measuring devices needed for controlling the operation of the facility (see Fig. 2.11). Within the flag for a controller, the variable that is being measured and the action are reported using a legend. Typically there is a system of three letters XYY; Table 2.1 shows these letters.

In this special diagram the information related to the equipment is provided. For instance, we need to describe the units in parallel or the spare units that are needed in the process. With respect to the units, we provide the size, the thickness (Schedule) of the pipes and the equipment, the materials of construction and the isolation, type, and thickness. In Table 2.2 the typical operating information for the units reported in the diagrams is known. We also need to provide control information such as indicators, controllers, recorders, etc. Finally, the utilities used in the process must be provided.

2.3.2 SYMBOLS: PROCESS FLOWSHEETING

Each of the units involved in a process flowsheet has its symbol. Fig. 2.12 shows the symbols for a number of typical equipment types. The units are referred to using their initials, such as C for compressors and turbines, E for heat exchangers, H for fire heaters, P for pumps, R for reactors, T for towers, TK for storage tank, V for vessels, and also a number, since typically there are various units of the same category in the plant.

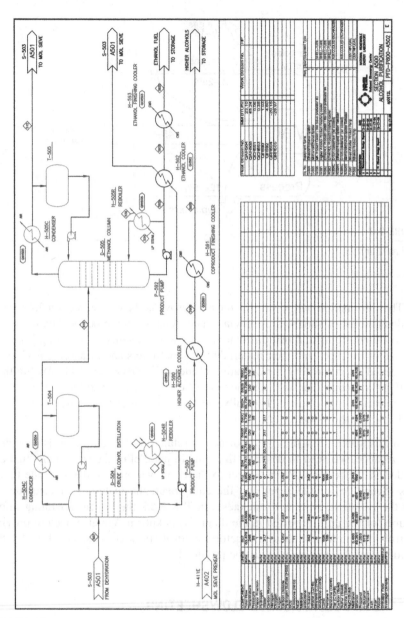

FIGURE 2.10

Process flow diagram.

*Reprinted with permission of the National Renewable Energy Laboratory, from page 99 of NREL publication:
http://www.nrel.gov/docs/fy07osti/41168.pdf, Accessed November 16, 2015; Phillips, S., Aden, A., Jechura,
J., Dayton, D., Eggeman, T., 2007. Thermochemical ethanol via indirect gasification and mixed alcohol
synthesis of lignocellulosic biomass. NREL/TP -510-41168 April 2007, (Phillips et al., 2007).*

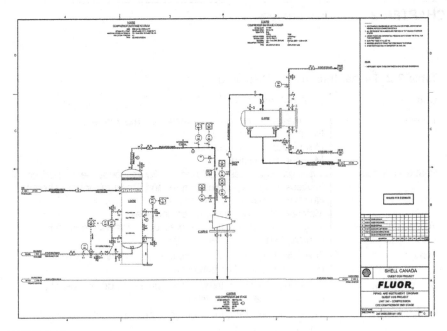

FIGURE 2.11

Piping and instrumentation diagram.

With permission from: Osaka Gas Chemicals., Ltd.

Table 2.1 Letters for Instrumentation

	First Letter (X)	Second or Third Letter (Y)
A	Analysis	Alarm
B	Burner flame	
C	Conductivity	Control
D	Density or specific gravity	
E	Voltage	Element
F	Flowrate	
H	Hand	High
I	Current	Indicate
J	Power	
K	Time or time schedule	Control station
L	Level	Light or low
M	Moisture or humidity	Middle or intermediate
O		Orifice
P	Pressure or vacuum	Point
Q	Quantity or event	
R	Radioactivity or ratio	Record or point
S	Speed or frequency	Switch
T	Temperature	Transit
V	Viscosity	Valve, damper, louver
W	Weight	Well
Y		Relay or compute
Z	Position	Drive

Table 2.2 Typical Information Displayed

Unit	Information
Towers	Size (height and diameter), pressure, temperature Number and type of trays Height and type of packing Materials of construction
Heat exchangers	Type: gas–gas, gas–liquid, liquid–liquid, condenser, vaporizer Process: duty, area, temperature, and pressure for both streams Number of shell and tube passes Materials of construction: tubes and shell
Vessels and tanks	Height, diameter, orientation, pressure, temperature, materials of construction
Pumps	Flow, discharge pressure, temperature, ΔP, driver type, shaft power, materials of construction
Compressors	Actual inlet flow rate, temperature, pressure, driver type, shaft power, materials of construction
Fired heaters	Type, tube pressure, tube temperature, duty, fuel, material of construction

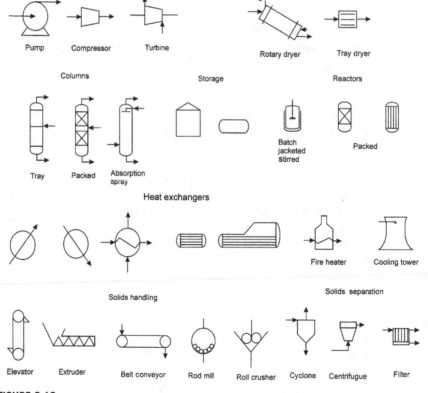

FIGURE 2.12

Symbols for typical equipment.

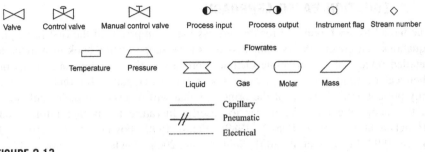

FIGURE 2.13

Control instruments and data legend.

Control instrumentation and the values provided in the flowsheet legend to report a flow rate using different units, mass or molar, pressure and temperature, etc. Fig. 2.13 shows the classic symbols for the control equipment, with lines representing the types of signals and flags for the information on the diagram.

2.4 MASS AND ENERGY BALANCES REVIEW

Mass and energy balances must provide the information for the streams displayed in the diagrams presented in Section 2.3. Typically the following information is to be included:

1. The mass flow of all units per unit of time. Volumetric flow rates should be avoided. Select the units that result in a reasonable range; for instance, from 0 to 1000.
2. Mass fraction of all streams (%).
3. Sometimes, molar flows (and molar fraction) can be interesting.
4. The stream temperature (°C or K).
5. The stream pressure. If it does not change much, show only once.
6. Enthalpy of flow (J, MJ, GJ, etc.).

Sometimes there is an error in the global mass balance, especially if there are experimental data or correlations used to perform the balances. There is a maximum: *even if the balances hold, it does not mean that they are correct.* Finally, the presentation of the results must accurately display the number of significant digits. There are two approaches to performing the mass and energy balances for a facility: equation-based and modular. The first one is the option of choice in most of this text, and a review of the principles is provided below. However, the use of computers in chemical engineering has spread, and modular simulators have been developed, such as CHEMCAD or ASPEN. For an introduction to the use of software for solving chemical engineering problems, we refer the reader to Martín (2014).

2.4.1 EQUATION-BASED APPROACH

The model for each unit and for the process itself comprises all the first principle equations that govern the system. It is out of the scope of this book to provide a detailed review of the principle chemical processes, those of unit operations or chemical reaction engineering; other books are better suited for this. Here we only present a brief review of the principles that will be used for the calculations in this book. We refer the reader to specific literature for further information (Houghen et al., 1959; Baasel, 1989; Walas, 1990; Fogler, 1997; Perry and Green, 1997; McCabe et al., 2001; Seider et al., 2004; Towler and Sinnot, 2012).

2.4.1.1 Mass and energy balances

They are based on the thermodynamic principles of conservation:

Mass balance with no chemical reaction, ie, mixers, separators, heat exchangers:

Mixer, heat exchanger,

$$\sum_{i \in \text{inlet}} m_{i,j} = m_{\text{out},j} \forall j \tag{2.5}$$

Splitter,

$$m_{\text{out}-k,j} = (\xi_k) m_{\text{in},j} \quad \forall j$$
$$\sum_k \xi_k = 1 \tag{2.6}$$

Mass balance with chemical reaction

a. Elementary balance:

$$\sum_{\text{in}} n_i = \sum_{\text{out}} n_i \tag{2.7}$$

b. Stoichiometry:

$$aA + bB \rightarrow cC + dD$$
$$n_A = n_{A,\text{ini}} - a/a \cdot n_{A,\text{ini}} \cdot X$$
$$n_C = n_{C,\text{ini}} + \frac{c}{a} n_{A,\text{ini}} \cdot X \tag{2.8}$$

where X is the conversion for the reaction. We define conversion as the fraction of the limiting reagent that has reacted. The reagent in excess is defined with respect to the stoichiometry of the reaction and the limiting reagent. Selectivity, S, refers to the fraction of the converted reagent that produces a certain product in multiple reaction systems.

c. Electrochemical reactions:
Redox potential:

$$E_{Reaction} = E_{semi\_reduction} - E_{semi\_oxidation}; \; E > 0 \text{ Spontaneous} \tag{2.9}$$

Nerst equation:

$$E = E^\circ - \frac{R \cdot T}{n \cdot F} \cdot \ln[Q] \tag{2.10}$$

Power:

$$e \cdot F \cdot V \equiv I \cdot t \cdot V \tag{2.11}$$

Charge flow:

$$e \cdot F \equiv I \cdot t \tag{2.12}$$

Equivalencies:

$$1 \text{ C} = 1 \text{ A} \cdot \text{s}; \; F = C/\text{mol}; \; 1 \text{ C} = V \cdot F, \; V = J/C$$

2.4.1.2 Energy balances

* *Enthalpy diagrams*: Mollier, Hausen, enthalpy composition. In these diagrams the Lever rule is useful for computing mixture composition and enthalpies.
* *Specific enthalpy (flow enthalpy–sensible heat*; we can also correct c_p with pressure):

$$H(T) = H_{T_{ref}}^T = \int_{T_{ref}}^T c_{p,i}(T)dT = \int_{T_{ref}}^T \left(A + BT + CT^2 + DT^3 \right)dT \approx \overline{c_p}(T - T_{ref}) \tag{2.13}$$

* *Enthalpy of reaction*:
  Heat of formation. We compute it as given by Eqs. (2.14) and (2.15), depending on the aggregation stage of reference of the chemical for gases and liquids, respectively.
  Gas:

$$\Delta H_{f,T} = \left(H_T^{T_{ref}} + \Delta H_{f,T_{ref}} \right) \tag{2.14}$$

Liquid:

$$\Delta H_{f,T} = \lambda(T) + H_{T_{ref}}^T + \Delta H_{f,T_{ref}} = c_{p,liq}(T - T_{ebul}) + \lambda(T_{ebul}) + H_{T_{ref}}^{T_{ebull}} + \Delta H_{f,T_{ref}} \tag{2.15}$$

Enthalpy of mixtures. For acids and alkalis in water we need to consider the energy involved in the dilution or concentration, heat of dilution.

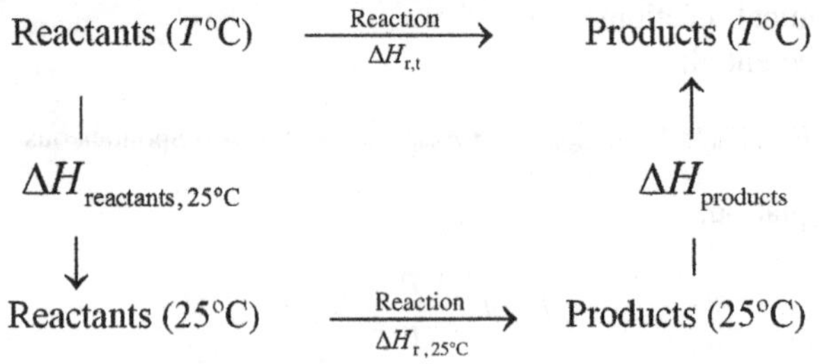

Reactants $(T^\circ C)$ $\xrightarrow[\Delta H_{r,t}]{\text{Reaction}}$ Products $(T^\circ C)$

$\Big|$ $\Delta H_{\text{reactants},25^\circ C}$ $\qquad\qquad$ $\uparrow$ $\Delta H_{\text{products}}$

$\downarrow$ $\qquad\qquad$ $\Big|$

Reactants $(25^\circ C)$ $\xrightarrow[\Delta H_{r,25^\circ C}]{\text{Reaction}}$ Products $(25^\circ C)$

FIGURE 2.14

H_r at temperature T.

However, in most cases heats of mixing are negligible and enthalpies can be considered as additive.

Heat of reaction: Assuming the heat of formation at reference temperature for the species in the final aggregation state, we have the heat of reaction:

$$\sum_{j=\text{in}}\sum_i n_i \left(H_T^{T_{\text{ref}}} + \Delta H_{f,T_{\text{ref}}}\right)_i + Q_{\text{reaction}} = \sum_{k=\text{out}}\sum_i n_i \left(H_T^{T_{\text{ref}}} + \Delta H_{f,T_{\text{ref}}}\right)_i \qquad (2.16)$$

To compute the enthalpy of the components, the Hess principle holds (see Fig. 2.14). Alternatively it can be computed using heat of combustion of the species.

- *Polytropic compression.* We compute the work involved by integrating Eq. (2.17). Typically, a compression ratio not larger than 4 or 5 is used based on rules of thumb.

$$W = -\int PdV; \quad PV^\gamma = cte \qquad (2.17)$$

$$W\,(\text{Comp}) = (F)\cdot \frac{8.314\cdot k\cdot(T_{\text{in}} + 273.15)}{((\text{MW})\cdot(k-1))}\frac{1}{\eta_s}\left(\left(\frac{P_{\text{out}}}{P_{\text{in}}}\right)^{\frac{k-1}{k}} - 1\right) \qquad (2.18)$$

- *Adiabatic energy balance*:

$$\sum_{j-\text{in}}\sum_i H(T) + Q_{\text{in}} + Q_{\text{gen}} + W_{\text{in}} = \sum_{k-\text{out}}\sum_i H(T) + Q_{\text{out}} + Q_{\text{losses}} + W_{\text{out}} \qquad (2.19)$$

2.4.1.3 Equilibrium relationships
2.4.1.3.1 Chemical equilibrium

To compute the composition of a stream under chemical equilibirum, the equilbrium constants can be formulated as a function of pressures. In this case we talk about apparent constants, or as a function of the fugacity, if the data are available. For simplicity, we present two examples where as a function of the partial pressures:

a. There is no variation in the number of moles $(CO + H_2O \leftrightarrow H_2 + CO_2)$,

$$kp_{WGS} = \frac{P_{H_2}P_{CO_2}}{P_{CO}P_{H_2O}} = \frac{y_{H_2}y_{CO_2}}{y_{CO}y_{H_2O}} = \frac{n_{H_2}n_{CO_2}}{n_{CO}n_{H_2O}} = \exp(-3.798 + 4160/T) \qquad (2.20)$$

b. There is variation in the number of moles $(CH_4 + H_2O \leftrightarrow 3H_2 + CO)$,

$$kp_{SR} = \frac{P_{H_2}^3 P_{CO}}{P_{CH_4}P_{H_2O}} = \frac{y_{H_2}^3 y_{CO}}{y_{CH_4}y_{H_2O}} P_t^2 = \frac{n_{H_2}^3 n_{CO}}{n_{CH_4}n_{H_2O}n_{TOT}^2} P_t^2 = \exp(30.42 - 27106/T) \qquad (2.21)$$

2.4.1.3.2 Phase equilibrium
Gas–liquid equilibrium
Humidification
Molar moisture (Dalton's Law):

$$y_m = \frac{P_v}{P_T - P_v} = \frac{\varphi P_v^{sat}}{P_T - \varphi P_v^{sat}} \qquad (2.22)$$

Specific moisture:

$$y = \frac{Mw_{cond}}{Mw_{dryair}} \frac{P_v}{P_T - P_v} = \frac{Mw_{cond}}{Mw_{dryair}} \frac{\varphi P_v^{sat}}{P_T - \varphi P_v^{sat}} \qquad (2.23)$$

Relative moisture:

$$\varphi = \frac{P_v}{P_v^{sat}} \qquad (2.24)$$

Vapor pressure for water: Antoine correlations. We can extend it to other species such as ammonia, methanol, etc.

$$\ln(P_v \text{ (mmHg)}) = 18.3036 - \frac{3816.44}{227.02 + T \, (^\circ C)} \qquad (2.25)$$

Specific volume:

$$v = \frac{RT}{P}\left(\frac{1}{Mw_g} + \frac{y}{Mw_v}\right) \qquad (2.26)$$

$$\text{Heat capacity(per kg of dry air)} = c_{p,g} + c_{p,v}y \qquad (2.27)$$

$$\text{Enthalpy(reference temperature } 0^\circ C = (c_{p,g} + c_{p,v}y)T + \lambda_{T=0^\circ C} \cdot y \qquad (2.28)$$

Flash Calculations We formulate the mass balances and the equilibrium relationships as follows:

$$F = V + L \qquad (2.29)$$

$$Fz_i = Vy_i + Lx_i \tag{2.30}$$

$$y_i = K_i\,x_i \tag{2.31}$$

where F, V, and L are the flows; and z_i, y_i, and x_i are the molar fractions of component i in the feed, the vapor phase, and the liquid phase, respectively. K_i is the equilibrium constant computed as P_v/P_{total}, Raoult's Law. Combining the equations (i corresponds to noncondensables, j are the condensables):

$$y_i = \frac{z_i}{\frac{V}{F} + \left(1 - \frac{V}{F}\right)\cdot\left(\frac{1}{K_i}\right)} \tag{2.32}$$

$$x_i = \frac{z_j}{(K_j - 1)\cdot\left(\frac{V}{F}\right) + 1} \tag{2.33}$$

$$\begin{aligned} K_i \gg 1 &\Longrightarrow V\cdot y_i \approx F\cdot z_i \\ K_j \ll 1 &\Longrightarrow L\cdot x_i \approx F\cdot z_j \end{aligned} \Longrightarrow y_i = \frac{f_i}{V} = \frac{f_i}{\sum f_i}, \quad x_i = \frac{y_i}{K_i} = \frac{f_i}{K_i \cdot \sum f_i} \tag{2.34}$$

This simplification holds if there are no K values from 0.1 to 10. The gas composition of the condensables and uncondensables is given by:

$$l_i = L\cdot x_i = \frac{f_i \cdot \sum f_j}{K_i \cdot \sum f_i} \tag{2.35}$$

$$v_i = f_i - l_i = f_i\left(1 - \frac{\sum f_j}{K_i \sum f_i}\right) \tag{2.36}$$

The liquid composition of the condensables and uncondensables is given by:

$$v_j = \frac{K_j \cdot f_j \cdot \sum f_i}{\sum f_j} \tag{2.37}$$

$$l_j = f_j\cdot\left(1 - \frac{K_j \cdot \sum f_i}{\sum f_j}\right) \tag{2.38}$$

Distillation Columns The operation is based on the differences in the relative volatility of the species (Fig. 2.15). The computation of the dew and bubble points is carried out as per Eqs. (2.39) and (2.40):

Dew points,

$$\frac{1}{P_T} = \sum_{i=1}^{n} \frac{y_i}{P_{v,i}} \tag{2.39}$$

Bubble points,

$$P_T = \sum_{i=1}^{n} x_i P_{v,i} \tag{2.40}$$

where P is the pressure, x is the molar fraction in the liquid phase, and y is the molar fraction in the gas phase.

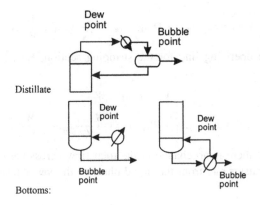

FIGURE 2.15

Distillation column temperatures.

The operation of the column is characterized by the number of stages (see Eq. 2.41) and the minimum reflux ratio that allows the operation (Eq. 2.42). Typically, the operating reflux ratio is around $1.1-1.5$ times the minimum:

Minimum number of trays:

$$N_{min} = \frac{\log\left[\left(\dfrac{x_{i,D}}{1-x_{i,D}}\right)\left(\dfrac{1-x_{j,W}}{x_{j,W}}\right)\right]}{\log \alpha_{ij}} \tag{2.41}$$

Minimum reflux ratio (Fenske Equation):

$$\left(\frac{L}{D}\right)_{min} = R_{min} = \frac{1}{\alpha_{ij}-1}\left[\frac{x_{i,D}}{x_{i,F}} - \alpha_{ij}\frac{(1-x_{i,D})}{(1-x_{i,F})}\right] \tag{2.42}$$

where α is the relative volatility, x is the molar fraction of distillate, D, W, of residue, feed, F, and L is the liquid flow returning to the column as reflux. The actual number of trays can be easily computed based on the McCabe iterative procedure. We perform mass balances to the column as given by Eqs. (2.43) and (2.44):

Global balance to the column:

$$F = D + W \tag{2.43}$$

Balance to the most volatile component:

$$F \cdot x_f = D \cdot x_D + W \cdot x_w \tag{2.44}$$

The operating lines in both sections of the column are computed by performing mass balances to the condenser and the reboiler section, including them. For the rectifying section:

$$V_n = L_{n-1} + D \tag{2.45}$$

$$V_n y_n = L_{n-1} x_{n-1} + D \cdot x_D \rightarrow y_n = \frac{L_{n-1}}{V_n} x_{n.-1} + \frac{D}{V_n} x_D \tag{2.46}$$

To obtain the operating line for the stripping section, Eqs. (2.47) and (2.48) are used:

$$V_m = L_{m-1} - W \tag{2.47}$$

$$V_m y_m = L_{m-1} x_{m-1} + W x_W \rightarrow y_m = \frac{L_{m-1}}{V_m} x_{m-1} + \frac{W}{V_m} x_W \tag{2.48}$$

Assuming that there is no change in the total flow across the column based on the small mass transfer rate from the liquid phase to the vapor phase, the operating lines become:

$$y_n = \frac{L}{V} x_{n.-1} + \frac{D}{V} x_D \tag{2.49}$$

$$y_m = \frac{L'}{V'} x_{m-1} + \frac{W}{V'} x_W \tag{2.50}$$

The feed can be in different aggregation states as compressed liquid, saturated liquid, a vapor liquid mixture, saturated vapor, and superheated vapor. We define the liquid and vapor fractions as Φ_L and Φ_V, respectively:

$$\Phi_L = \frac{L' - L}{F} \tag{2.51}$$

$$\Phi_V = \frac{V - V'}{F} \tag{2.52}$$

Thus the feed line, q line, is given by:

$$q = \frac{\Phi_V - 1}{\Phi_V} x_{n.} + \frac{1}{\Phi_V} x_f \tag{2.53}$$

Alternatively, Gilliland's equation, Eq. (2.54), can be used, where R is the reflux ratio (L/D) and N the number of trays:

$$\frac{N - N_{min}}{N+1} = 0.75 \left[1 - \left(\frac{R - R_{min}}{R+1} \right)^{0.5688} \right] \tag{2.54}$$

Absorption Columns The operation is based on the gas—liquid equilibrium. To evaluate it, we consider two alternatives, whether ideal (Raoult's Law) or nonideal (Henry's Law) hold:

Raoult's Law (ideal),

$$y_A P_T = x_A P_{v,A} \tag{2.55}$$

Henry's Law (nonideal),

$$y_A P_T = H C \tag{2.56}$$

where y and x are the molar fractions of A in the vapor and in the liquid phases, P_T is the total pressure and $P_{v,A}$ the vapor pressure, C represents concentration, and H is Henry's constant. From a mass balance $L \cdot x_o + V y_2 = L x_1 + V \cdot y_1$ and Henry's Law $y_1 = H x_1$,

$$\left(\frac{y_{n+1} - y_1}{y_{n+1} - H x_o}\right) = \frac{A^{n+1} - A}{A^{n+1} - 1} \quad \text{with } A = \frac{L}{V} \cdot \frac{1}{H} \tag{2.57}$$

Thus the theoretical number of trays is computed as follows (Kremser−Brown−Sounders):

$$N_{\min} = \frac{\log\left[\left(\dfrac{y_{n+1} - H x_o}{y_1 - H x_o}\right)\left(\dfrac{A - 1}{A}\right) + \dfrac{1}{A}\right]}{\log A} \tag{2.58}$$

Gas Law Throughout the text we will consider the behaviors of the gases as ideal (Eq. 2.59). However, operating at high pressures different thermodynamic models may be needed, including compressibility factors or more complex ones such as Peng Robinson, etc.

$$PV = nRT \tag{2.59}$$

Liquid−liquid equilibrium

Partition coefficient is given by:

$$k_i = \frac{C_{E,i}}{C_{R,i}} \tag{2.60}$$

where C is the concentration of component i in the extracted phase, E, and in the refinate, R. Triangular diagrams will also be used in the text.

2.4.1.4 Mass, heat, and momentum transfer

A number of units such as membranes, evaporators, and the flow in reactor pipes will be analyzed using principles of mass, heat, and momentum balances.

Mass transfer:

Molecular: Ficks Law	Convective interphase
$N_A = -D_A A \dfrac{dC_A}{dz}$ (2.61)	$N_A = k_c A(C^* - C)$ (2.62)

Heat transfer:

Molecular: Fourier's Law	Convective interphase
$q = -kA\dfrac{dT}{dz}$ (2.63)	$Q = UA\Delta T$ (2.64)

Momentum transfer:

Newton's Law: Flow in pipes (Newtonian)   Darcy's Law: Flow through porous media

$$\tau = -\mu \frac{dv}{dz} \qquad (2.65) \qquad\qquad v = -\frac{K}{\mu}\frac{dp}{dz} \qquad (2.66)$$

2.4.1.5 Kinetics and reactor design

Reactors are modeled using differential mass and energy balances. Furthermore, for plug flow or packed bed reactors, pressure drop is another variable. Here we briefly present the equations used to model a batch reactor, such as bioethanol production or packed bed reactors, like the ones in ammonia, methanol, or sulfuric acid production (see chapters: Nitric acid, Sulfuric acid, Biomass). The balances are based on differential mass and energy balances:

$$\text{In flow} + \text{Generation} - \text{Out flow} = \text{Accumulation} \qquad (2.67)$$

2.4.1.5.1 Batch reactors
Mass balance:

$$(-r_A) = -\frac{1}{V}\frac{dN_A}{dt} \qquad (2.68)$$

Energy balance:

$$\left(N_{A0}\left(\sum_{i=1}^{n} \Theta_i c_{p,i} + X\Delta c_p\right)\right)\frac{dT}{dt} = UA(T_a - T) + (-r_A)\left[\Delta H_R(T_R) + \int_{T_R}^{T} \Delta c_p dT\right] \qquad (2.69)$$

2.4.1.5.2 Plug flow/packed bed reactors
Mass balance:

$$-\frac{dF_A}{dW} = (-r_A) \qquad (2.70)$$

Energy balance:

$$F_{A0}\left(\sum_{i=1}^{n} \Theta_i c_{p,i} + X\Delta c_p\right)\frac{dT}{dW} = UA(T_a - T) + (-r_A)\left[\Delta H_R(T_R) + \int_{T_R}^{T} \Delta c_p dT\right] \qquad (2.71)$$

Ergun Equation:

$$\frac{dP}{dW} = -\frac{G(1-\phi)(1+\varepsilon X)}{\rho_b A_c \rho_0 g_c D_p \phi^3}\left[\frac{150(1-\phi)\mu}{D_p} + 1.752G\right]\frac{P_0}{P}\frac{T}{T_0} \qquad (2.72)$$

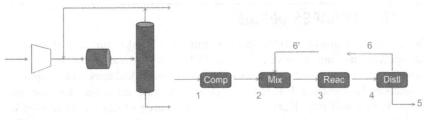

FIGURE 2.16

Modular simulation: Tearing process.

2.4.1.6 Design equations for the units

There are a number of books within the chemical engineering field that gather design rules of thumb for a large number of equipment types that help estimate energy consumption, operation of gas—liquid contact equipment, etc, eg Walas (1990).

Putting together the equations results in a large and nonlinear problem that can be solved using state-of-the-art software and solvers such as GAMS, gPROMS, and MATLAB.

2.4.2 MODULAR SIMULATION

This is the case for commercial software such as CHEMCAD or ASPEN plus, which have the units modeled as modules. Any time a property (ie, density, enthalpy, etc.) must be computed, a thermodynamic package is called that returns the value. The best performance for this software is the sequential solution of the flowsheet, since the models for each unit are very efficient and connecting them is easy. Dealing with recycling involves tearing of the streams. This technique can be automatic within the software, or can be done iteratively. For instance, for a flowsheet like the one given by Fig. 2.16, we compute stream 2, guess stream $6'$, compute 3, 4, 5, and 6, and iterate until $6' = 6$.

2.5 OPTIMIZATION AND PROCESS CONTROL

Industrial processes must be profitable, and that requires optimization of the use of raw materials and energy. It is not the aim of this book to present the use of optimization for the design of chemical processes since there are a number of works better suited (Biegler et al., 1997). However, it is necessary to familiarize the reader with the existence and current use of mathematical optimization techniques in decision-making as a way to unveil nonobvious trade-offs. Furthermore, process control also deals with feasible operation of processes. Therefore, a process that is unstable or difficult to control is of no use in the chemical industry. We refer the reader to those courses and textbooks that address these issues in detail.

2.6 SAFE PROCESS DESIGN

The chemical industry suffers from a number of risks due to the practices and species that are involved (Trenz, 1984). In particular, the main safety problems are related to leaks, either energy or mass leakages, fire, explosion, or uncontrolled chemical reactions. There are two concepts that we must define: hazard and risk. Hazard represents the source of danger. It is what can cause the accident. Risk measures the probability and the consequences of the accident. In other words, the possibility that an accident occurs from a particular danger, and that it causes damages and injuries, as well as impact. Thus we can compute:

$$\text{Risk} = \text{Frequency} \cdot \text{consequences} \tag{2.73}$$

Typically, the main reasons for accidents in the chemical industry are related to mechanical problems in the units; up to 44% of accidents are due to this fact. Twenty-two percent of accidents are because of operator error, while 12% are of unknown cause. The fourth major reason, which accounts for 11% of ocurrences, is due to problems in the process. On the other hand, only 1% is due to sabotage or arson. Fig. 2.17 shows these numbers. The hardware involved in the accidents is, in most of the cases, the piping system (up to 28% of the times). Seventeen percent of accidents are due to storage tanks, 10% due to the piping system of the reactor, and 5% due to holding tanks. In general, the storage of chemicals represents a major danger. Fig. 2.18 shows the main hardware involved in accidents.

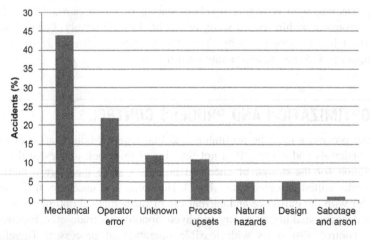

FIGURE 2.17

Reasons behind accidents in the chemical industry.

Data courtesy of Dr. Santoso.

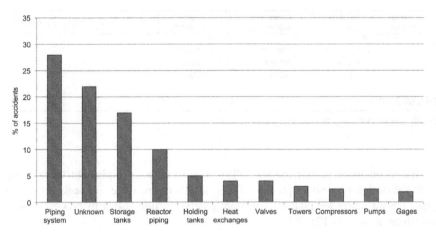

FIGURE 2.18

Equipment involved in accidents in chemical plants.

Data courtesy of Dr. Santoso.

With this information at hand, industry has worked hard from different points of view to address safety issues. They can be classified into two approaches (Trenz, 1984).

Traditional approach: When a process is already out there, there is not much that can be done to change it. Mitigation and prevention measures are the most effective techniques.

Current approach: It is called inherently safe design. Safety is another criteria along the decision-making process that leads to the design of a process. The basic idea is to remove the sources of danger. It is based on the talk that Trevor Klentz presented to the Society of Chemical Industry in 1977. In consists of four main pillars:

1. *Minimization (Intensification)*: This involves reducing the use of dangerous species and storage as well as using smaller equipment. Following these ideas, we reduce the consequences of potential explosion fires and leakages of toxic materials, which improves the effectiveness of protective systems.
2. *Substitution*: The idea is to use safer materials (less or nonflammable) and less or nontoxic species. In this substitution we have to evaluate whether the new material or species may cause or be responsible for new hazards.
3. *Moderate*: Most of the time it is not possible to substitute the dangerous materials. Thus, we should try to reduce the quantity of energy available within the process, and use smaller storage tanks and less severe process conditions.
4. *Simplify*: The simpler the process, the smaller the possibility for anything to fail. The smaller the amount of equipment, the fewer units that can fail, and the fewer opportunities for human error.

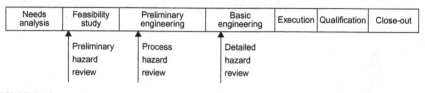

Needs analysis	Feasibility study	Preliminary engineering	Basic engineering	Execution	Qualification	Close-out
	Preliminary hazard review	Process hazard review	Detailed hazard review			

FIGURE 2.19

Safety integration in process design.

Based on these principles, we follow a procedure to implement them along with the design of a novel process. Typically, there are three moments along the plant design when we take into account safety considerations in process design decisions, as seen in Fig. 2.19.

2.6.1 PRELIMINARY HAZARD REVIEW

The first moment when safety must be included in the decision-making process for process design is during the *feasibility study*. At this point we should do the following:

List the process alternatives as a function of their safety.
Review the consequences on employees, the public, and the environment.
Review the impact of loss of utilities.
Compute the DOW index of fire and explosion. The *DOW index* is a quantitative method that provides an objective measure of the actual fire, explosion, and reactivity potential of the equipment and its content. It can help define the areas of potential losses.

$$DOW = \text{Material factor} \cdot \text{Process Unit hazard factor}$$

Material Factor corresponds to the intrinsic rate of potential energy release caused by fire or explosion produced by combustion or chemical reaction. *Process Unit Hazard Factor* incorporates all factors that are likely to contribute to the occurrence of fire or explosion incidents. The ranking is given below:

1–60	Light
61–96	Moderade
97–127	Intermediate
128–158	Heavy
≥ 159	Severe

2.6.2 PROCESS HAZARD REVIEW (PRELIMINARY ENGINEERING)

When the flowsheet is being put together, the next safety evaluation must be carried out considering the following factors:

Identification of safe operating limits.
Identification of causes of unacceptable hazards.
Mitigation of any unacceptable risks in the design.

2.6.3 DETAILED DESIGN REVIEW (BASIC ENGINEERING)

At this point, the aim is to identify potential problems associated with safety, reliability, and operability (HAZOP). Critical hazards, if identified, shall be quantified to ensure acceptable low risk. Design must be checked against all internal company standards to ensure compliance.

The *HAZOP* is a structured technique to identify hazards resulting from potential malfunctions in the process. However, it is essentially a qualitative process (see Fig. 2.20). The advantage of using it is that if performed early

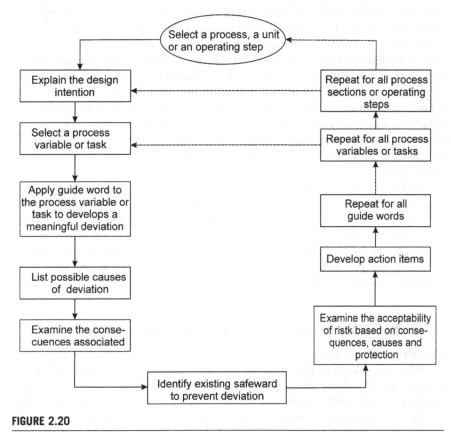

FIGURE 2.20

Hazop analysis flowchart.

enough in the design stage, costly modifications after the plant is built can be avoided.

The typical composition of the HAZOP team to carry out the analysis consists of process engineers, chemists, design engineers, instrumentation engineers, production supervisors, and safety and environmental engineers.

Process design has to handle all the possible operating conditions, not only those ideal or typical, including startup, shutdown, process upsets, seasonal changes in the incoming raw materials, and atmospheric conditions. Therefore, it must be tested to validate it at different scales, from plant trials to pilot plant or laboratory scale, so that the design specifications are met. A workplace is a safe place, but it is not only for safety reasons that we must consider safety issues during process design; safe designs save money.

2.7 PROCESS SUSTAINABILITY

Sustainable development is a production model by which satisfying the needs of current society still maintains the rights of future generations to continue satisfying their needs. Reaching this level of thinking and compromise has been possible due to the current concern for the environment and our legacy to future generations. It represents an equilibrium between the profitable operation of the process, its social acceptance, and its considerations for environmental protection. In this way, three pillars support the entire production scheme. On the one hand, we aim to *preserve the nature and environment*. Furthermore, *economic sustainability* aims for an efficient use of raw materials and support for the use of renewable ones. Finally *social sustainability* aims for dignity and respect for human rights in the market economy context. Therefore, we deal with a multiobjective problem that was to be applied to the entire supply chain, not only to a particular process. Novel metrics have been developed of late to account for these three pillars in the context of process design and the chemical supply chain. While economic evaluations are a widespread practice in the chemical industry, LCA has recently been developed to include environmental effects into the analysis, and models like Jobs and Economic Development Impact (JEDI) account for the effect of the process on the economic growth of the region.

As in the previous case for the safety of a plant, there are two main approaches to deal with the wastes generated in a chemical facility:

2.7.1 INTEGRATED PRODUCTION–PROTECTION STRATEGY

This approach is to be applied during process design. At this point it is possible to remove the source of pollution, and recycle byproducts and wastes within the process, in spite of the small amounts produced. Furthermore, we aim for high conversions and yield, improving the efficiency in the use of raw materials. Another possibility is to evaluate the process for changes in the species that improve the yield and reduce the wastes. In this section energy integration plays an important role in order to reuse the excess of energy while reducing the need for external energy input. Together with this integration, process intensification is gaining attention as a means to reduce equipment size. Furthermore, waste treatment should be considered in the economic evaluation as a way to stress the need to reduce the emissions and waste produced.

Novel design approaches including process modeling & optimization and research & development to improve each one of the stages are the methods of choice to perform this integrated approach. Dealing with different objectives and evaluating nonobvious trade-offs is a complex task that traditional schemes were not able to deal with.

2.7.2 MITIGATION STRATEGY

Whenever it is not possible to implement the previous approach, one or more of the following methods is typically attempted: waste treatment, wastewater treatment, gas treatment, sour gases capture, or solid waste management. All of these measures are expensive, they do not remove the source, and therefore they are not capable of avoiding the problems that persist. However, this approach is usually less capital intensive.

2.8 PROBLEMS

P2.1. Determine the minimum consumption of hot and cold utilities for the steams in Table P2.1.

Table P2.1 Stream Data

Streams	Q (kW)	T_{in} (°C)	T_{out} (°C)
A	800	60	160
B	500	116	260
C	300	160	93
D	800	255	138

P2.2. Determine the minimum consumption of hot and cold utilities for the steams in Table P2.2.

Table P2.2 Stream Data

Streams	Fc_p (kW/°C)	T_{in} (°C)	T_{out} (°C)
C1	15	360	600
C2	12	300	450
H1	11	600	320
H2	15	540	320

P2.3. A mixture of methanol ($Tb = 64°C$), ethanol ($Tb = 78°C$), propanol ($Tb = 97°C$), and butanol ($Tb = 117°$) with $\alpha_{Met/Et} = 2$, $\alpha_{Et/Prop} = 1.9$, and $\alpha_{Prop/butanol} = 3.5$ is to be separated into pure components. The molar rates (mol/s) are 6, 20, 5, and 4, respectively. Determine the optimal distillation sequence for ethanol as the main product.

P2.4. A five-component mixture, ABCDE, in order of decreasing volatility, is to be separated. The total flow rate is 1 kmol/s, and the composition is as follows (A B C D E) = (0.2 0.4 0.1 0.1 0.2):

Regular distillation,

$$\alpha_{AB} = 2, \ \alpha_{BC} = 1.7, \ \alpha_{CD} = 1.5, \alpha_{DE} = 1.04 \ (ABCDE);$$

Extractive distillation,

$$\alpha_{AC} = 2.5, \alpha_{AB} = 2.4, \alpha_{CB} = 2, \alpha_{BD} = 2, \alpha_{DE} = 1.7 \ (ACBDE).$$

P2.5. A mixture of four species needs to be fractionated. Determine the optimal sequence of columns. Two alternative technologies are available: regular distillation and extractive distillation. The initial flowrate is 1000 mol/h and the composition in molar fraction is as follows: A = 0.1; B = 0.2; C = 0.3; and D = 0.4. Additional data for the example:

Regular distillation,

$$\alpha_{AB} = 2, \alpha_{BC} = 1.5, \alpha_{CD} = 1.02, \ (ABCD);$$

Extractive distillation,

$$\alpha_{AB} = 2.1, \alpha_{BC} = 1.5, \alpha_{CD} = 1.4 \ (ABCD).$$

P2.6. Determine the minimum approximation temperature among the hot and cold streams so that the cold utility removes 1370 kW and the hot utility provides 390 kW for the streams in Table P2.6.

Table P2.6 Stream Data

Streams	Fc_p (kW/°C)	T_{in} (°C)	T_{out} (°C)
C1	15	360	600
C2	12	300	450
H1	11	600	320
H2	15	540	320

P2.7. Determine the minimum utilities consumption considering $\Delta T = 10°C$ for the streams in Table P2.7.

Table P2.7 Stream Data

Streams	Fc_p (kW/°C)	T_{in} (°C)	T_{out} (°C)
C1	2	20	135
C2	4	80	140
H1	3	170	60
H2	1	150	30

P2.8. Determine the utility requirements for the system represented in Fig. P2.8. The hot composite and the cold composite curves are given as (-) and (--), respectively. Assume $\Delta T = 25$ K.

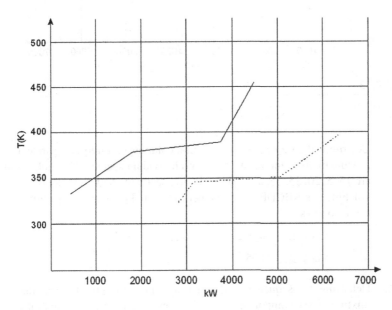

FIGURE P2.8

T–Q diagram.

P2.9. Determine ΔT for the system in Fig. P2.9 if the hot utility provides 1000 kW. In the figure (-) represents the hot composite curve, and (..), the cold composite curve.

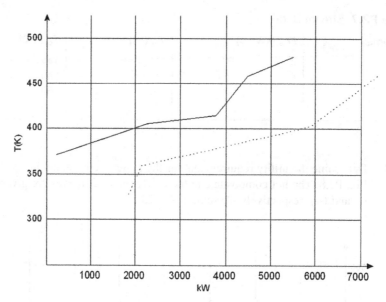

FIGURE P2.9

T–Q diagram.

P2.10. Determine the sequence of distillation columns required to separate a mixture of six components A–F into individual species. The molar flows are 15, 5, 20, 25, 55, and 10 kmol/h, respectively, and the order of volatilities is ABCDEF from higher to lower. The relative volatilities are as follows:

$$\alpha_{AB} = 1.05, \alpha_{BC} = 2.5, \alpha_{CD} = 1.85, \alpha_{DE} = 1.9, \alpha_{EF} = 2.5$$

$$\alpha_{AB,extractive} = 1.05$$

P2.11. Determine the sequence of distillation columns required to separate a mixture of six components A–F into individual species. The molar flows are 40, 5, 10, 25, 30, and 45 kmol/h, respectively. The order of volatilities is ABCDEF from higher to lower. The relative volatilities are as follows:

$$\alpha_{AB} = 1.05, \alpha_{BC} = 2.5, \alpha_{CD} = 1.05, \alpha_{DE} = 1.8, \alpha_{EF} = 2.5$$

$$\alpha_{AB,extractive} = 1.7, \alpha_{CD,extractive} = 1.8$$

P2.12. Compute the minimum requirement of utilities and design the optimal network among a set of cold (C) and hot (H) streams in Table P2.12.

Table P2.12 Stream Data

Streams	Fc_p (kW/K)	T_{in} (K)	T_{out} (K)
C1	15	320	560
C2	10	280	480
H1	10	600	300
H2	20	500	300

REFERENCES

Baasel, W.D., 1989. Preliminary Chemical Engineering Plant Design. van Nostrand Reinhold, Amsterdam.

Biegler, L., Grossmann, I.E., Westerberg, A.W., 1997. Systematic Methods of Chemical Process Design. Prentice Hall, Upper Saddle River, NJ.

Douglas, J.M., 1988. Conceptual Design of Chemical Processes. McGraw-Hill, New York, NY.

Fogler, S., 1997. Elements of Chemical Reaction Engineering, third ed. Prentice Hall, New York, NY.

Houghen, O.A., Watson, K.M., Ragatz, R.A., 1959. Chemical Process Principles. Vol 1. Material and Energy Balances. Wiley, New York, NY.

Jaksland, C.A., Gani, R., Lien, K.M., 1995. Separation process design and synthesis based on thermodynamic insights. Chem. Eng. Sci. 50 (3), 511–530.

Jiménez, A., 2003. Diseño de procesos en Ingeniería Química, Reverté. Barcelona.

Karuppiah, R., Peschel, A., Grossmann, I.E., Martín, M., Martinson, W., Zullo, L., 2008. Energy optimization of an Ethanol Plant. AICHE J. 54 (6), 1499–1525.

Kravanja, Z., Grossmann, I.E., 1990. PROSYN-an MINLP process synthesizer. Comp. Chem. Eng. 14 (12), 1363–1378.

Martín, M., 2014. Introduction to Software for Chemical Engineers. CRC Press, Boca Raton.

McCabe, W.L., Smith, J.C., Harriot, P., 2001. Unit Operations. McGraw Hill, New York.

Perry, R.H., Green, D.W., 1997. Perry's Chemical Engineer's Handbook. McGraw-Hill, New York.

Phillips, S., Aden, A., Jechura, J., Dayton, D., Eggeman, T., 2007. Thermochemical ethanol via indirect gasification and mixed alcohol synthesis of lignocellulosic biomass. NREL/TP -510-41168 April 2007.

Rudd, D., Powers, G., Siirola, J., 1973. Process Synthesis. Prentice Hall, Englewood Cliffs, NJ.

Seider, W.D., Seader, J.D., Lewin, D.R., 2004. Product and Process Design Principles. Wiley, New York, NY.

Smith, R., 2005. Chemical Process: Design and Integration. Wiley, New York, NY.

Towler, G., Sinnot, R.K., 2012. Chemical Engineering Design, Principles, Practice and Economics of Plant and Process Design, second ed. Elsevier, Singapore.

Trenz, T., 1984. Cheaper, safer plants, or wealth and safety at work: notes on inherently safer and simpler plants (1984) IchemE.

Walas, 1990. Chemical Process Equipment: Selection and Design Butterworth. Heinemann, Boston.

Westerberg, A.W., Hutchinson, H.P., Motard, R.L., Winter, P., 1979. Process Flowsheeting. Cambridge University Press.

Yeomans, H., Grossmann, I.E., 1999. A systematic modeling framework for superstructure optimization in process systhesis. Comp. Chem. Eng. 23, 555–565.

Zlokarnik, M., 2006. Scale-up in Chemical Engineering, second ed. Wiley-VCH, Germany.

Air

3.1 INTRODUCTION

3.1.1 COMPOSITION

Air is a plentiful raw material. Its composition includes nitrogen, oxygen, and noble gases (such as argon). Technically, air is assumed to have a composition of 79% nitrogen and 21% oxygen, as well as the moisture corresponding to the weather conditions. A more detailed composition can be found in Table 3.1. This composition can be altered by electric discharges so that certain molecules (such as H_2O, N_2, O_2, and CO_2) are dissociated and others formed (such as C_2H_2, H_2O_2, O_3, HNO_3, NH_3, NH_4NO_3, and NH_4OH). These species can be used as nitrogen fertilizer when it rains, but it is not enough to cover all needs.

3.1.2 USES

Air can be used as a comburant, as itself, or as a source of oxygen. The second main use is as a raw material for the production of N_2, O_2, and noble gases. Wastewater treatment is another alternative use. Air is bubbled in the pools of water treatment plants to decompose organics. Finally, it is possible, under the proper operating conditions, to produce NO as an intermediate in the production of HNO_3. For the production of NO we need high temperatures and fast cooling of the products. From NO, NO_2 is obtained via oxidation, and is later absorbed into water to produce HNO_3.

Nitrogen is a colorless, odorless, tasteless, nontoxic, and relatively inert gas. It does not sustain combustion or respiration, but it can react with metals such as lithium and magnesium to obtain nitrides. Furthermore, at high temperatures it combines with hydrogen, oxygen, and other elements. At atmospheric pressure, it condenses at 77.3K as a colorless liquid, and solidifies at 63.1K in the form of white crystals. As a cryogenic liquid, it is nonmagnetic and stable against mechanical shock.

It represents 78% of the atmosphere, but its weight contribution is small, 0.03% with respect to the mass of water and Earth. It can also be found as Chile saltpeter or Peru saltpeter. In order to use nitrogen in synthesis, purity over 99.8% is required. The main uses are the production of NH_3, $CaCN_2$, NaCN, or as an inert gas to pressurize aircraft tires, and purge pipelines, reactors, and storage

Industrial Chemical Process Analysis and Design. DOI: http://dx.doi.org/10.1016/B978-0-08-101093-8.00003-3

Table 3.1 Air Composition

Chemical	Volume Fraction	Mass Fraction
Nitrogen	78.14	75.6
Oxygen	20.92	23.1
Argon	0.94	0.3
Neon	1.5×10^{-3}	1×10^{-3}
Helium	5×10^{-4}	0.7×10^{-4}
Krypton	1×10^{-4}	3×10^{-4}
Hydrogen	5×10^{-5}	0.35×10^{-5}
Xenon	1×10^{-5}	4×10^{-5}

tanks. As a liquid, it can be used to capture CO_2 and in medical applications (Hardenburger and Ennis, 1993; Häussinger, 1998).

Oxygen is a colorless, odorless, tasteless, nontoxic gas. At atmospheric pressure it condenses at 90.1K as a blue liquid, and solidifies at 54.1K to form blue crystals. It is less volatile than nitrogen and thus it is obtained from the bottom of air distillation columns.

It is the more abundant species, representing 50% of the atmosphere, water, and cortex weight. High purity oxygen (above 99.5%) is used in metal cutting and welding. Lower purity oxygen (85–99%) and enriched air (below 85%) can be used as oxidants, to produce C_2Ca, for the partial oxidation of methane to syngas, for the catalytic production of HNO_3 from ammonia, for the oxidation of HCl to Cl_2, or as a bleach in the production of paper. Oxygen supports life and is used in water treatment (Hansel, 1993; Kirschner, 1998).

Noble gases are typically used to fill light bulbs, fluorescent pipes, and electronic devices. For instance, argon is used in metal welding as an inert atmosphere. Helium can be used in stratospheric balloons and diving apparatuses.

3.2 **AIR SEPARATION**

3.2.1 **HISTORY**

For centuries air was considered a gas element. In 1754 Joseph Black identified what he called "fixed air" (CO_2), so-called because it could be returned, or fixed, into the sorts of solids from which it was produced. Later, in 1766 Henry Cavendish produced a highly flammable substance that was eventually named hydrogen by Lavoisier. The word has its roots in its Greek meaning, "water maker." Finally, in 1772 Daniel Rutherford found that when he burned material in a bell jar, and then absorbed all the "fixed" air by soaking it up with a substance called potash, a gas remained. Rutherford referred to it as "noxious air" because it asphyxiated mice placed in the jar. This gas was what we today call nitrogen. The series of experiments culminated in 1774 when Joseph Priestley, inventor of carbonated water and the rubber eraser, found that "air is not an elementary substance, but a composition," or mixture, of gases. Using a 12-inch-wide glass "burning lens," he focused sunlight on a lump of reddish mercuric oxide in an inverted glass container placed in a pool of mercury. The gas emitted was "five or six times as good as common air." In subsequent tests, it caused a flame to burn intensely and kept a mouse alive about four times as long as a similar quantity of air. Priestley called his discovery "dephlogisticated air" based on the theory that it supported combustion so well because it had no phlogiston in it, and hence could absorb the maximum amount during burning. It was the French chemist Antoine Lavoisier who coined the name "oxygen"(American Chemical Society, 1994).

From the discovery of the components of air, research on gases continued. A number of important dates highlight key developments in the processes to liquefy and separate the components of air. For instance, the Joule-Thompson (Kelvin) effect was discovered in 1853 when it was realized that a compressed gas cools down when it expands, typically 0.25°C per atmosphere. There are some exceptions, such as hydrogen. In 1863 the concept of critical temperature was defined as the temperature above which it is not possible to liquefy CO_2 by compression. In 1877 James Dewar, in Faraday's laboratory, liquefied hydrogen. At the same time in Switzerland and France, oxygen was also liquefied. In our simplified review of history we reach 1895, when Karl von Linde proposed a procedure to liquefy air by isothermal compression followed by isobaric cooling and expansion. A number of modifications of this basic idea allowed higher yields.

3.2.2 CLASSIFICATION OF AIR SEPARATION METHODS

A. Physical separation

- Cryogenic:
  - Air distillation
- Noncryogenic:
  - Adsorption on molecular sieves
  - Diffusion across membranes

B. Chemical separation

This consists of reacting one of the components so that it is fixed. In this case the most reactive component is the O_2, and thus we obtain oxygen combined as a product and N_2 as a byproduct.

3.2.2.1 Physical separation

3.2.2.1.1 Cryogenic methods: air liquefaction and distillation

In order to separate air via rectification, a part of it should be liquid. We can only do that below the critical point, $T_{crit} = 132.5K$ and $p_{crit} = 37.7$ atm. The thermodynamic data show that air condenses at 81.5K (-192°C) at 1 atm, while if the pressure is increased up to 6 atm, it only has to cool down to 101K (−172°C). Remember that cooling 1 kg of air from 15°C to −196°C requires 415 kJ (Vian Ortuño, 1999). The evaluation of air liquefaction can be performed using several phase diagrams for air. In this chapter, process simulations will also be used for one particular case study. Bear in mind that liquid air must be handled with care. Liquid air burns the skin if it is in contact for a period of time. If the contact time is short, the quick evaporation of the liquid due to the skin's temperature protects the skin—the Leidenfrost principle (Agrawal et al., 1993).

Hausen diagram: It represents temperature on the y-axis versus entropy on the x-axis (Hausen, 1926). It has parametric lines for pressure and enthalpies. Below the bell line the gas–liquid mixtures can be seen. This is the region for air liquefaction. See Fig. 3.1.

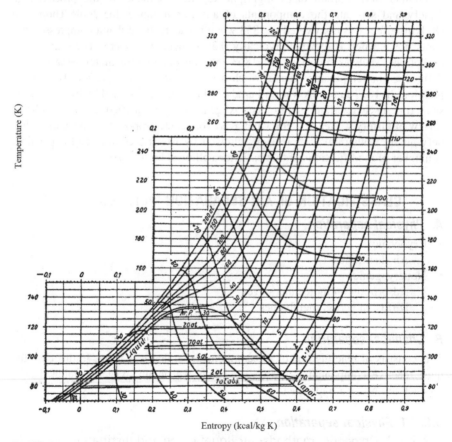

FIGURE 3.1

Hausen diagram.

Adapted with permission from: Hausen, 1926.

Hausen, H., 1926. Der Thomson-Joule-Effekt und die zustandsgrossen der Luft. C. Forsch.

Arb. Ing.-Wes. Heft 274.

Mollier diagram. This diagram dates back to 1904 when Richard Mollier plotted the total heat versus entropy. There are parametric lines for pressure and temperature, and discontinuous lines for constant specific volume. See Fig. 3.2.

The liquefaction process. This process, devised by von Linde, consists of four basic steps. First, the gas is compressed isothermally. Next, it is cooled down at constant pressure. The gas is then expanded, and finally, the gas and liquid phases are separated.

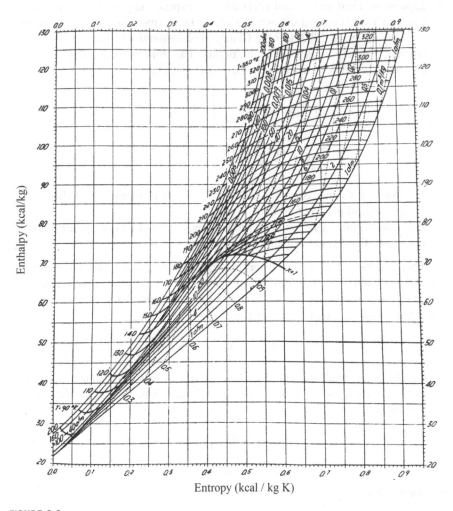

FIGURE 3.2

Mollier's diagram.

Adapted with permission from: Vancini, C.A., 1961. La sintesi dell'ammoniaca, Hoepli, Milano.

1. *Compression.* This represents the only work input to the system. From basic thermodynamics it can be proved that the minimum theoretical isothermal work assuming ideal gases is given by:

$$W = \Delta H - T_o \Delta S = - \int_1^2 P dV = RT_o \ln\left(\frac{P_2}{P_1}\right) \tag{3.1}$$

2. *Isobaric cooling.* The noncondensed fraction of the stream is used to cool down the compressed gas.
3. *Expansion.* There are several alternatives to expand the gas phase. Karl von Linde in 1895 proposed the use of a valve. This expansion occurs at constant enthalpy. Later, in 1902, George Claude used an isentropic expansion, wherein a large volume contraction occurs. The work generated in this expansion can be estimated using Eqs. (3.2) and (3.3):

$$W = \Delta H - T_o \Delta S = - \int_1^2 p dV = PV^k = K \tag{3.2}$$

$$W = \frac{nR \cdot k \cdot (T)}{((k-1))} \frac{1}{\eta_c} \left(\left(\frac{P_{out}}{P_{in}}\right)^{\frac{k-1}{k}} - 1 \right); \tag{3.3}$$

However, k is not constant, and thus the use of diagrams makes the computation of the work easier.
4. *Phase separation.* The nonliquid phase is recycled and its energy integrated to cool down the compressed air.

There are two variables that determine the efficiency of the liquefaction cycle. On the one hand, the air has to be compressed to ensure that we enter the two-phase region upon expansion (see Fig. 3.1). The work involved is the energy that the cycle consumes, and is given by Eq. (3.4):

$$W = \Delta H - T_o \Delta S = RT_o \ln\left(\frac{P_2}{P_1}\right) \tag{3.4}$$

On the other hand, the cooling obtained in the expansion along a constant enthalpy line could be estimated using Linde's equation, Eq. (3.5), where k has an average value of $2 \cdot 10^4\ \text{bar}^{-1} \cdot \text{K}$:

$$\Delta T = k \frac{P_2 - P_1}{T^2} \tag{3.5}$$

Therefore, we are interested in a high-pressure difference, $(P_2 - P_1)\uparrow$, and low-pressure ratio, $\left(\frac{P_2}{P_1}\right)\downarrow$

A simple computation for two cycles with the same pressure difference and different pressure ratio results in large energy savings (see Table 3.2). The first option consumes 3.6 times more energy.

Table 3.2 Energy Consumption in Linde's Cycle

	Cycle 1	Cycle 2
P_1 (atm)	1	50
P_2 (atm)	150	200
P_2/P_1	150	4
$P_2 - P_1$ (atm)	150	150
Energy (J)	360	100

EXAMPLE 3.1

Check the use of Linde's equation $\left(\Delta T = k\dfrac{P_2 - P_1}{T_1^2}\right)$ for an isenthalpic expansion with $k = 2 \cdot 10^4$ atmK, $P\,[=]$ atm and $T\,[=]$ K.

Solution

Take the enthalpy line of 90 kcal/kg. For different expansions, compute the final temperature (T_2) using Linde's equation. This value is compared to the one we read in Hausen diagram. Using the actual value for T_2, we compute the actual value for k. Table 3E1.1 shows the results. We see that the theoretical value for $k = 20{,}000$ atmK is only valid in specific regions along the isenthalpic line.

Table 3E1.1 Results

	T_1 (K)	P_1 (atm)	P_2 (atm)	T_2 (eq)	T_2 (diagram)	k real
$h = 90$ kcal/kg	233	200	150	214.5801	226	7600.46
	226	150	100	206.4213	214.5	11,747.48
	214.5	100	80	205.8063	210	10,352.31
	210	80	60	200.9297	200	22,050
	200	60	40	190	190	20,000
	190	40	30	184.4598	185.5	16,245
	185.5	30	20	179.6878	180	18,925.64
	180	20	10	173.8272	173.6	20,736
	173.6	10	5	170.2818	170	21,698.61
	170	5	2	167.9239	168.2	17,340
	168.2	2	1	167.4931	168.2	0

Linde–Hampson air liquefaction cycle

Description The simplest alternative is a cycle comprising the four stages described earlier. Fig. 3.3 shows the flowsheet for the system. First, an isothermal compression from 1 to 2. Second, the compressed air is cooled down at constant

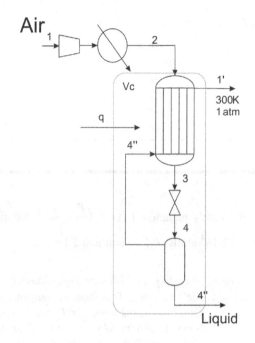

FIGURE 3.3

Linde's block diagram.

pressure in a heat exchanger using the nonliquid fraction as cooling agent. Once the compressed air is cold enough, it is expanded through a valve, typically down to atmospheric pressure. The air is then laminated into a liquid and a gas phase. If we now describe this cycle on the Hausen diagram, we start with atmospheric air at 300K and 1 atm at point 1.

The atmospheric air is compressed from 1 atm to 200 atm at constant temperature, point 2 in Fig. 3.4 for the Hausen diagram or Fig. 3.5 for the Mollier diagram. After each compression stage, the energy generated must be removed from the system. Next, a heat exchanger is used to cool down the air from point 2 to point 3. Subsequently, the air is expanded in the valve until atmospheric pressure, point 4. We now have a gas–liquid mixture that we separate, obtaining liquid air (4′) and gas air (4″) that is the cooling agent for the heat exchanger. Therefore, we see that energy is integrated within our process. This process is also called the "simple regenerative cycle," devised by Karl von Linde in 1895 in Munich. The actual yield to liquid air is low (Kerry, 2007).

Analysis An energy balance is performed on the volume within the curved cornered rectangle in Fig. 3.3. Using a basis of 1 kg we have:

$$1\text{kg } H_2 = f\, H_{4'} + (1-f)H_{1'} + q \tag{3.6}$$

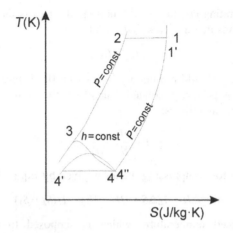

FIGURE 3.4

Linde's cycle in Hausen's diagram.

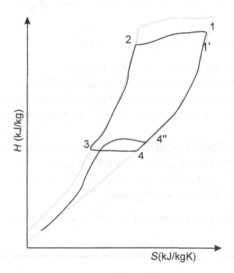

FIGURE 3.5

Linde's cycle in Mollier's diagram.

where f is the fraction of liquid air, q is the heat loss, and H_i is the enthalpy of point i. *Note that in cryogenic systems the energy losses correspond to the energy that enters the system.* Solving for f, we have Eq. (3.7):

$$f = \frac{H_{1'} - H_2 - q}{H_{1'} - H_{4'}} \tag{3.7}$$

Under ideal operating conditions, the nonliquid air exits the process under the same initial conditions that as it was fed into it.

$$H'_1 = H_1 \tag{3.8}$$

Furthermore, there should not be any losses ($q = 0$). Typically this means that the heat exchanger is perfectly isolated. Therefore, the optimal fraction of liquid air obtained would be as follows:

$$f = \frac{H_1 - H_2}{H_1 - H'_4} \tag{3.9}$$

Thus, the energy for compressing the air is given by Eq. (3.10):

$$W = \Delta H - T_o \Delta S = (H_2 - H_1) - T_o(S_2 - S_1) \tag{3.10}$$

where T_o is the feed temperature, which is supposed to be kept constant. Alternatively, we can compute it as ($PV = $ constant):

$$W = \Delta H - T_o \Delta S = -\int_1^2 P dV = RT_o \ln\left(\frac{P_2}{P_1}\right) \tag{3.11}$$

Note that there are a number of simplications:

- No pressure drops are considered in the analysis.
- When the air enters the two-phase region, an equilibrium is obtained and the liquid phase gets more concentrated in the less volatile species while the gas phase is enriched with the nitrogen; thus the exit streams do not have the same composition.
- The compression should be carried out in a multicompression system where at each stage the air is heated up and later cooled down. Thus, the path from 1 to 2 is not given by a straight line, but has peaks. Typically, the compression ratio is 4−5 per stage, which avoids surpassing 204°C.

Let us illustrate the performance of the cycle with an example.

EXAMPLE 3.2

Compute the liquid fraction obtained in a Linde cycle considering:

- point 1 in Fig. 3.4 is at 27°C and 1 atm, and
- point 2 is at 27°C and 200 atm.

The air is expanded to atmospheric pressure.

Solution

Molliere's diagram and Hausen's diagram are used in the solution of this example.

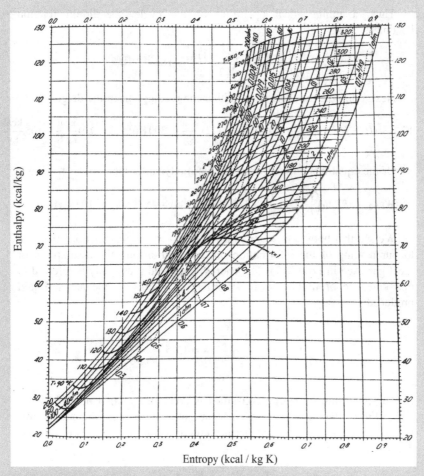

FIGURE 3E2.1

Reading of the air properties for example 3.2.

Using Molliere's diagram:
Look for the enthalpies and entropies of the different points of the cycle in Fig. 3E2.1, corners in the cycle.

$H_1 = 122$ kcal/kg	$S_1 = 0.9$ kcal/kgK
$H_2 = 114$ kcal/kg	$S_2 = 0.525$ kcal/kgK
$H_{4'} = 22$ kcal/kg	$S_{4'} = 0.0$ kcal/kgK

Thus, assuming ideal operation of the cycle, the liquid fraction becomes:

$$f = \frac{H_1 - H_2}{H_1 - H_{4'}} = \frac{122 - 114}{122 - 22} = 0.08 \frac{kg_{liquidair}}{kg_{air}}$$

Next, compute the work using either the readings from the diagram:

$$W_{1 \to 2} = (H_2 - H_1) - T_o(S_2 - S_1) = (114 - 122)\frac{kcal}{kg} - 300\,K(0.525 - 0.9)\frac{kcal}{kg\,K} = 105\frac{kcal}{kg}$$

Or the equations derived from the thermodynamic model:

$$W_{1 \to 2} = nRT\ln\left(\frac{P_2}{P_1}\right) = \frac{1}{29} \cdot 8.314 \cdot 300 \cdot \ln(200) = 456\,kJ/kg \equiv 109\,kcal/kg$$

We see that both values are similar, within the approximation error due to our reading of the information on the diagram and the assumption of ideal gas used in Eq. (3.11). The energy computed previously is referred to one kilogram of initial air. If we refer the energy to the actual liquid produced we have:

$$W_{1 \to 2} = \frac{105\,\dfrac{kcal}{kg}}{0.08\,\dfrac{kg_{liquid}}{kg_{air}}} = 1325\,\frac{kcal}{kg_{liquid}}$$

The minimum in energy consumed is computed as that needed to liquefy all the initial air, assuming constant temperature, since the energy provided to the system is an isothermal compression. The standard value is $T_o = 300K$. Thus:

$$W_{min} = W_{1 \to 4'} = (H_{4'} - H_1) - T_o(S_{4'} - S_1)$$

$$= (22 - 122)\frac{kcal}{kg} - 300\,K(0.0 - 0.9)\frac{kcal}{kg\,K} = 170\frac{kcal}{kg}$$

Therefore, the efficiency of the cycle is computed as the ratio between the minimum work, and the work used per kg of liquid air obtained:

$$\eta = \frac{W_{min}}{W_{1 \to 2}} = 0.13 \approx 13\%$$

Using Hausen's diagram:
In this case, the level rule is used to determine the temperatures before and after the expansion, T_3 and T_4. Thus, calling F the feed, L the liquefied air, and V the nonliquefied fraction, LF is the distance from point 4' to 4 and FV, that from 4 to 4".

$$f \cdot LF = (1-f) \cdot FV$$

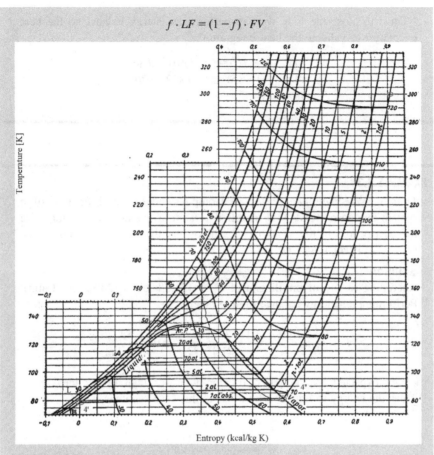

FIGURE 3E2.2

Lever rule in Hausen's diagram.

We measure on the diagram (Fig. 3E2.2) the distance between both edges of the two-phase region, where $LV = 6.5$ cm. Thus, we solve for LF:

$$LF = \frac{1-f}{f}FV = \frac{1-f}{f}(6.5 - LF) \Rightarrow LF = \frac{\dfrac{1-f}{f}6.5}{1 + \dfrac{1-f}{f}} = 5.98$$

Using the distance computed, we allocate point 4 in Fig. 3E2.2. Since the expansion is isenthalpic, $H_3 = H_4$, equal to 66 kcal/kg. Alternatively, we can compute the enthalpy by performing an energy balance to the phase separator once the liquefied fraction is known:

$$1\,\text{kg} \cdot H_4 = f \cdot H_{4'} + (1-f)H_{4''} = 0.08 \cdot 22 + (1 - 0.08)H_{4''}$$

where $H_{4''}$ and $H_{4'}$ are equal to 70 kcal/kg and 22 kcal/kg, respectively. Thus:

$$H_4 = 66\,\text{kcal/kg}$$

Just to close the loop we can perform an energy balance to the heat exchanger to validate the heat integration.

$$1 \text{ kg} \cdot (H_2 - H_3) = (1 - f)(H_{1'} - H_{4'})$$
$$1 \text{ kg} \cdot (114 - 66) = (1 - f)(122 - 70)$$
$$48 \approx 47.84$$

EXAMPLE 3.3

Linde's cycle with energy losses. Compute the liquefied fraction of a system that corresponds to a modified version of example 3.2. In this case the energy loss in the system is 2 kcal/kg and the temperature difference $(T_1 - T_{1'})$ is equal to 3°C.

Solution

According to any of the diagrams, $H_{1'} = 121.3$ kcal/kg (24°C and 1 atm). Performing an energy balance to the system:

$$f = \frac{H_{1'} - H_2 - q}{H_{1'} - H_{4'}} = \frac{121.3 - 114 - 2}{121.3 - 22} = 0.053 \frac{\text{kg}_{\text{liquid air}}}{\text{kg}_{\text{air}}}$$

As expected, the liquefied fraction is lower than in the ideal case. Thus, the energy to liquefy a kg of air in this case becomes:

$$W_{1 \to 2} = \frac{105 \dfrac{\text{kcal}}{\text{kg}}}{0.053 \dfrac{\text{kg}_{\text{liquid}}}{\text{kg}_{\text{air}}}} = 1981 \frac{\text{kcal}}{\text{kg}_{\text{liquid}}}$$

And the efficiency is given by:

$$\eta = \frac{W_{\text{min}}}{W_{1 \to 2}} = \frac{170}{1981} = 0.0858$$

The evolution of the liquefaction processes aims to increase the liquefied fraction, improving the energy efficiency of the processes and reducing the cost of investment. From the original Linde's cycle, there are several modifications. The first one considers the use of external cooling using ammonia. The cycle was renamed as the *Linde–Hampson cycle with precooling*. It is based on the fact that around $-33°C$ the isenthalpic lines have a steeper slope. In Fig. 3.6 the

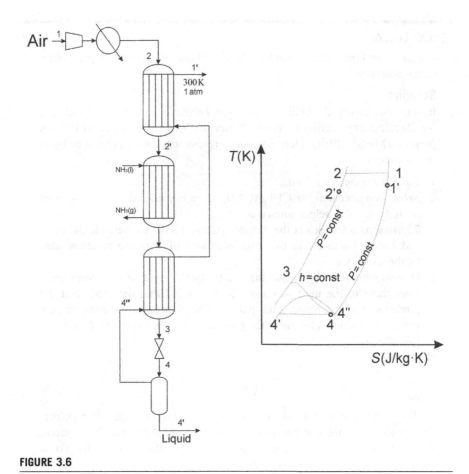

FIGURE 3.6

Linde–Hampson cycle with precooling.

flowsheet for the process and the scheme of the thermodynamic cycle are presented. Ammonia is used to cool down the compressed air. Therefore, there are three cooling steps. The first one cools the compressed air from room temperature just a few degrees. Next, ammonia is used to save cooling capacity from the nonliquefied fraction so that it is possible to cool the compressed air a bit further. The third cooling step uses the nonliquefied fraction for the final cooling of the compressed air to reach point 3 in the diagram. By doing this, the energy extracted from the system in the evaporation of the ammonia is somehow saved so that cold nonliquified air is still available to further reduce the temperature of the compressed air.

The example here does not use either of the diagrams presented above, but uses a process simulator, CHEMCAD, to compute the liquefied fraction.

EXAMPLE 3.4

Compute the liquefied fraction of a Linde–Hampson cycle with precooling using ammonia.

Solution

It is not the aim of the book to introduce the reader to the use of software for chemical engineering purposes; better textbooks can be found for that purpose (Martín, 2014). Here we only present the stages needed to build the flowsheet:

1. *Format Engineering Units.*
2. *Select components*: For CHEMCAD, air is considered as a component in itself; we also select ammonia.
3. *Thermodynamics*: Select the thermodynamic model to compute the mass and energy balances. In this case equations of state are recommended by the simulator.
4. *Drawing the flowsheet.* Use the palette to the right to drag your units from there to the main window. Bear in mind that the feeds and the products are the arrows in the palette. The compression cannot be performed in a single step since the pressure ration is large. We follow the rule of thumb:

$$\sqrt[n]{\frac{P_2}{P_1}} \leq 4$$

Thus, we use a four-stage compression system with intercooling. The efficiency of the compressors is assumed to be 90%, and they operate as polytropic ones. After each compression stage, the air is cooled down again to the initial temperature, 300K.

The most challenging part for the simulation is the recycling of the nonliquefied air. Since the energy balance must hold, we need to have cold available in the nonliquified fraction to cool down the compressed air. We use three heat exchangers that will allow two feeds and two products each. The compressed air is the main stream, and as cooling agents we have ammonia and the nonliquefied fraction. The intermediate heat exchanger is the one using ammonia. Finally, we use a valve to reduce the pressure up to 1 atm, and a flash to separate the liquid air from the nonliquefied fraction. Fig. 3E4.1 shows the flowsheet for the example.

Just for reference, Fig. 3E4.2A–C shows some of the characteristics of the units involved: a compressor, a heat exchanger, and the flash unit.

The following table shows the results of the problem. We see that we can liquefy up to 16% of the air, which is twice as much as the best value presented so far in the text (Table 3E4.1).

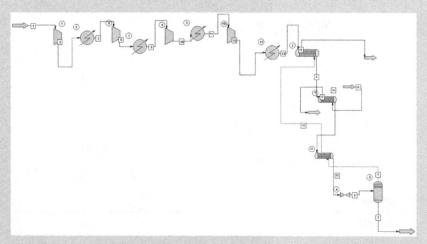

FIGURE 3E4.1

Linde–Hampson with precooling cycle simulation.

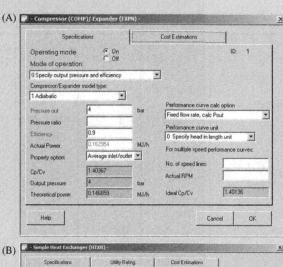

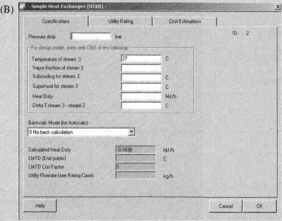

FIGURE 3E4.2

(A) Compressor unit. (B) Initial heat exchanger. (C) Flash.

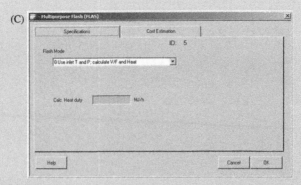

(C)

FIGURE 3E4.2

(Continued)

Table 3E4.1 Results From the Linde's Cycle With Precooling Simulation

Stream No.	7	16	17	6
Name				
- - Overall - -				
Molar flow kmol/h	0.0057	0.0023	0.0023	0.0288
Mass flow kg/h	0.1651	0.0400	0.0400	0.8349
Temp C	−194.6961	−33.4446	−33.4435	−194.6961
Pres atm	200.0000	1.0000	1.0000	1.0000
Vapor mole fraction	1.000	0.0000	0.7763	1.000
Enth MJ/h	0.0000	−0.16831	−0.12519	−0.18522
Tc C	−140.7000	132.5000	132.5000	−140.7000
Pc atm	37.7400	112.7848	112.7848	37.7400
Std. sp gr. wtr=1	0.862	0.619	0.619	0.862
Std. sp gr. air=1	1.000	0.588	0.588	1.000
Degree API	32.6531	97.1314	97.1314	32.6531
Average mol wt	28.9510	17.0310	17.0310	28.9510
Actual dens kg/m3	880.5709	681.4407	1.1173	4.6153
Actual vol m3/h	0.0002	0.0001	0.0358	0.1809
Std liq m3/h	0.0002	0.0001	0.0001	0.0010
Std vap 0 C m3/h	0.1278	0.0526	0.0526	0.6464

EQUIPMENT SUMMARIES

Heat Exchanger Summary

Equip. No.	3	12	13
Name			
1st Stream T Out C	1.0000	−30.0000	−111.0000
Calc Ht Duty MJ/h	0.0338	0.0431	0.1379
LMTD (End points) C	24.3340	13.4630	19.5290
LMTD Corr Factor	1.0000	1.0000	1.0000
1st Stream Pout atm	200.0000	200.0000	200.0000
2nd Stream Pout atm	1.0000	1.0000	1.0000

EXAMPLE 3.5

Compute the liquefied fraction obtained in the process given in Fig. 3E5.1 considering that the isenthalpic expansion occurs at 20 atm (60 kcal/kg). Determine the energy removed from the system using ammonia as the cooling agent.

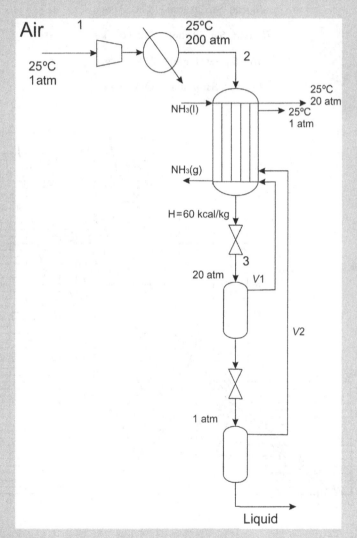

FIGURE 3E5.1

Multiple expansion cycle.

Solution

We perform a global energy balance from point 2 in Fig. 3E5.1 considering that we have two exists of nonliquid air, one at 20 atm and the other at 1 atm. Thus, following the notation on Fig. 3E5.1 and reading in Fig. 3E5.2 we have:

$$H_3(20 \text{ atm, } 160 \text{ K}) = 60 \text{ kcal/kg (problem data)},$$

$$H_{L1}(\text{liq. Sat. 20 atm}) = 43 \text{ kcal/kg},$$

$$H_{V1}(\text{vap. Sat. 20 atm}) = 71.5 \text{ kcal/kg},$$

$$H_{L2}(\text{liq. Sat. 1 atm}) = 22 \text{ kcal/kg},$$

$$H_{V2}(\text{vap. Sat. 1 atm}) = 68 \text{ kcal/kg}.$$

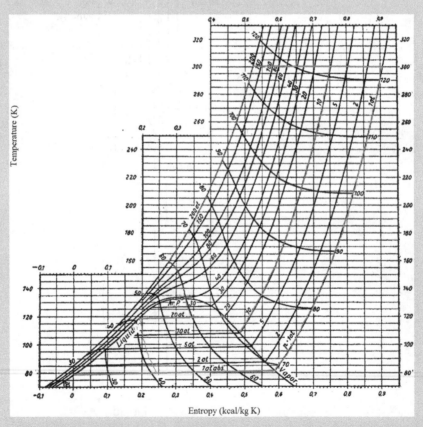

FIGURE 3E5.2

Thermodynamic scheme of the liquefaction process in example 3.5.

We have two expansions and thus we applied the level rule twice. For the first one,

$$F = L1 + V1$$
$$1\text{kg } H_3 = L1\, H_{L1} + V1\, H_{V1}$$

where $L1 = 0.4035$ kg, $V1 = 0.5964$ kg.

For the second expansion we have the following mass and energy balances:

$$L1 = V2 + L2$$
$$H_{L1}L1 = V2\, H_{V2} + L2\, H_{L2}$$

Solving the system of equations: $L2 = 0.2192$ kg, $V2 = 0.1842$ kg.

We now perform a global energy balance to compute the energy removed from the system, q:

$$1\text{kg} \cdot H_2 = V2\, H_{V2,\,out} + L2\, H_{L2,\,out} + q + V1\, H_{V1,\,out}$$
$$113.5 = 0.1892 \times 121 + 0.2192 \times 22 + q + 0.5964 \times 119$$
$$q = 15.42 \text{ kcal}$$

Alternatively, we can perform the energy balance to the heat exchanger alone:

$$H_2 + V2\, H_{V2} + V1\, H_{V1} = H_3 + q + V1\, H_{V1,\,out} + V2\, H_{V2,\,out}$$
$$113.5 + 0.1842 \cdot 68 + 0.5964 \cdot 71.5 = 60 + q + 0.5964 \cdot 119 + 0.1842 \cdot 121.$$
$$q = 15.41 \text{ kcal}$$

The problem can be stated differently. In case there is no precooling, compute the temperature at which we expand the stream.

Mass Balance to flash 1	$F = V1 + L1$
Energy Balance to flash 1	$F \cdot H_3 = V1\, H_{V1} + L1\, H_{L1}$
Mass Balance to flash 2	$L1 = V2 + L2$
Energy Balance to flash 2	$L1 \cdot H_{L1} = V2\, H_{V2} + L2\, H_{L2}$
Global energy balance	$F\, H_2 = V2\, H_{V2,\,out} + L2\, H_{L2,out} + V1\, H_{V1,\,out}$

Assuming $F = 1$, the system consists of five equations and five variables, where

$$H_2 = 113.5 \text{ kcal/kg,}$$

$$H_{L1}(\text{liq. Sat. 20 atm}) = 43 \text{ kcal/kg,}$$

$$H_{V1}(\text{vap. Sat. 20 atm}) = 71.5 \text{ kcal/kg,}$$

$$H_{V1,\,out}(\text{out, 20 atm}) = 123 \text{ kcal/kg,}$$

$$H_{L2}(\text{liq. Sat. 1 atm}) = 22\,\text{kcal/kg},$$

$$H_{V2}(\text{vap. Sat. 1 atm}) = 68\,\text{kcal/kg},$$

$$H_{V2,\,out}(\text{out, 1 atm}) = 121\,\text{kcal/kg}.$$

We solve the system using EXCEL. The results can be seen in Table 3E5.1. For further insight on the use of solver for chemical engineering problems we refer the reader to Martín (2014).

Table 3E5.1 Results

		H_2	H_3	H_{sat}	H_{out}
F	1	113.5	67.458693		
L1	0.14180025			43	
V1	0.85819975			71.5	121
L2	0.07706535			22	
V2	0.0647349			68	123
M balance flash 1		1	1	0	
E balance flash 1		67.458693	67.458693	0	
M balance flash 2		0.14180025	0.14180025	0	
E balance flash 2		6.0974106	6.0974106	0	
Global balance		113.5	113.5	0	

Claude's cycle

Description There are other alternatives to increase the liquefied fraction. In 1902 George Claude from the University of Paris used an expansion engine that was reversible and adiabatic so that the expansion was isentropic. Thus in the Claude cycle the air (or another gas) is compressed, typically to 40 atm. After cooling using nonliquefied fractions, a fraction is diverted and expanded. The expanded gas is mixed with the cold nonliquefied gas so that the mixture is used to cool down the compressed air. The diverted fraction is around 75% of the feed. See Fig. 3.7 for the flowsheet and the thermodynamic cycle. This technique was welcomed in the industry at the time since it required lower operating pressures and increased energy efficiency. The isentropic efficiency is defined as the ratio between the actual enthalpy difference in the expansion and the enthalpy difference for the ideal isentropic case (Fig. 3.8).

Analysis 1 kg of fed air undergoes the following path along the flowsheet. Performing an energy balance to the discontinuous line rectangle we have:

$$1\,\text{kg}\,H_2 + q = f\,g\,H_{4'} + [(1-f)\,g + (1-g)]\,H_{1'} + W_{\Delta S=0} \tag{3.12}$$

$$W_{\Delta S=0} = (1-g)\,(H_{3'} - H_{4'''}) \tag{3.13}$$

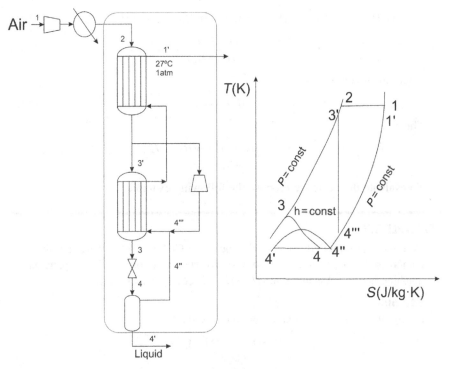

FIGURE 3.7

The Claude cycle for air liquefaction.

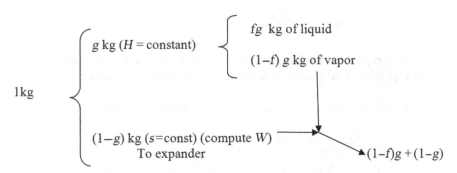

FIGURE 3.8

The Claude cycle for air liquefaction: analysis.

$$1 \text{ kg } H_2 + q = f g H_{4'} + [(1-f) g + (1-g)] H_{1'} + (1-g) (H_{3'} - H_{4''}) \tag{3.14}$$

$$f = \frac{(H_2 - H_{1'}) - (1-g)(H_{3'} - H_{4''}) + q}{g(H_{4'} - H_{1'})} \tag{3.15}$$

Under ideal conditions:

$$q = 0, \ H_{1'} \to H_1 \tag{3.16}$$

Thus:

$$f = \frac{(H_1 - H_2) + (1-g)(H_{3'} - H_{4''})}{g(H_1 - H_{4'})} \tag{3.17}$$

To evaluate the cycle we propose the following example.

EXAMPLE 3.6

Compute the liquefied fraction obtained in a Claude cycle where the air is compressed up to 40 atm, and the diverted fraction to the expansion machine is 75%. Assume that $(H_{3'} - H_{4''})$ is equal to 17 kcal/kg.

Solution

In any of the above presented diagrams we read:

$$H_1 = H_{1'} = 122 \text{ kcal/kg}$$
$$H_{4'} = 22 \text{ kcal/kg}$$
$$H_2 = 119 \text{ kcal kg}$$

Substituting in Eq. (3.17):

$$f = \frac{(H_1 - H_2) + (1-g)(H_{3'} - H_{4''})}{g(H_1 - H_{4'})} = \frac{122 - 119 + 0.75 \cdot 17}{0.25 \cdot (122 - 22)} = 0.63$$

$$fg = 0.63 \cdot 0.25 = 0.1575 \frac{\text{kg}_{\text{liquid air}}}{\text{kg}_{\text{air}}}$$

To compute the work required it is assumed that the power obtained in the expansion is accounted for as produced for the global energy balance:

$$W_{1 \to 2} = H_2 - H_1 - T_0(S_2 - S_1) = 119 - 122 - 300(0.655 - 0.905) = 72 \frac{\text{kcal}}{\text{kg}}$$

$$W_{\text{aportar}} = \frac{W_{1 \to 2} - W_{3 \to 4''}}{fg} = \frac{72 - 0.75 \cdot 17}{0.1575} = 376 \frac{\text{kcal}}{\text{kg}_{\text{liquid air}}}$$

$$\eta = \frac{W_{\text{min}}}{W_{\text{provided}}} = \frac{170}{376} = 0.45$$

Comparing these results with the previous ones, we see that the yield to liquid air is similar to the one obtained for the precooled Linde cycle, but the energy required is far lower.

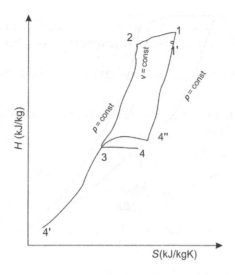

FIGURE 3.9

Thermodynamic representation of Phillips's cycle.

Phillips's cycle. The cycle is similar to the Linde's one, but in this case the cooling occurs at constant specific volume, which increases the efficiency. Fig. 3.9 represents the thermodynamic cycle using the Mollier diagram. For the evaluation of this cycle, it is the most appropriate diagram.

To illustrate the performance of this cycle, consider example 3.7.

EXAMPLE 3.7

Air at 27°C and 15 atm is compressed up to 200 atm, removing the heat generated. The air is next cooled at constant volume until it reaches the saturation point. A valve is used to expand the air to 1 atm. Compute the efficiency of the Phillips's cycle described.

Using Mollier's diagram we read the enthalpies and entropies of the different points of the process.

$H_1 = 121.6$ kcal/kg	$S_1 = 0.7236$ kcal/kgK
$H_2 = 114$ kcal/kg	$S_2 = 0.525$ kcal/kgK
$H_{4'} = 22$ kcal/kg	$S_{4'} = 0.0$ kcal/kgK

An enthalpy balance is formulated assuming ideal behavior of the cycle.

$$1 \text{ kg} \cdot H_2 = f \cdot H_{4'} + (1-f) \cdot H_1 \Rightarrow f = \frac{H_1 - H_2}{H_1 - H_{4'}}$$

$$f = 0.076 \text{ kg}_{\text{liquid}}/\text{kg}_{\text{air}}$$

This value is slightly larger than the one obtained with the Linde's simple cycle method. Next, the energy requirement is computed as follows:

$$W_{1\to 2} = (H_2 - H_1) - T_o(S_2 - S_1) = (114 - 121.6) - 300 \cdot (0.525 - 0.736) = 55.4\frac{kcal}{kg}$$

It is the smallest value seen so far when comparing the different cycles. As per kg of liquid air:

$$W_{1\to 2} = \frac{55.4\dfrac{kcal}{kg}}{0.076\dfrac{kg_{liquid\ air}}{kg_{air}}} = 729\frac{kcal}{kg_{liquid\ air}}$$

The minimum theoretical work is given as that required to obtain a kilogram of liquid air:

$$W_{min} = W_{1\to 4'} = (H_{4'} - H_1) - T_o(S_{4'} - S_1)$$

$$= (22 - 121.6)\frac{kcal}{kg} - 300\ K(0.0 - 0.736)\frac{kcal}{kg\ K} = 120.9\frac{kcal}{kg_{liquid\ air}}$$

And the yield results in:

$$\eta = \frac{W_{min}}{W_{1-2}} \cdot 100 = 16.58\% > example\ 3.2$$

Cascade cycle. This method can be dated back to 1890 when Kamerlingh-Onnes in Leyden used a series of refrigerants for his experimental work on cryogenics. Further development by Keesom in 1930 provided the highest efficiency in liquefaction so far. In this scheme, each refrigerant has a boiling point in decreasing order. Typically we use ammonia, ethylene, methane, and nitrogen. Their boiling points are 240K, 189K, 112K, and 77K, respectively. See Fig. 3.10 for the flowsheet. The advantage of the cycle is that it piece-wise approximates an ideal reversible cycle, resulting in an efficient thermodynamic process. The expansion takes place through a valve at low pressure, with a low increase in entropy. In spite of its efficiency, the refrigerant losses (leaks), the complexity of the operation, and the high investment have prevented its use in air separation (Kerry, 2007). Table 3.3 shows the energy involved per kg of liquid air obtained. We need 1.7 kg of initial feed of air, and the energy is 464 kcal/kg_{liquid}. The liquid fraction increases up to 0.59 kg_{liquid air}/kg air (Vancini, 1961).

Air distillation. Air liquefaction is the first stage in separating air into its components. When the vapor−liquid equilibrium is established, the vapor phase gets concentrated in the volatile component and vice versa. The boiling points of

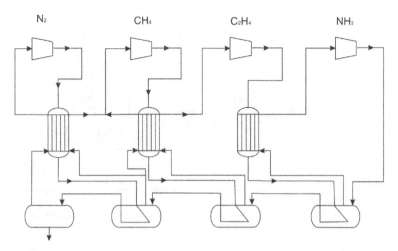

FIGURE 3.10

Cascade cycle for air liquefaction.

Table 3.3 Energy Consumption in the Cascade Cycle (Vancini, 1961)

Species	Energy Cost
1.70 kg Air	200 kcal
0.33 kg NH_3	44 kcal
0.96 kg C_2H_4	90 kcal
0.66 kg CH_4	130 kcal
	464 kcal

nitrogen ($T_b = 77.3$K) and oxygen ($T_b = 90.1$K) result in them being obtained from the top and bottom of the column, respectively (Latimer, 1967). Before feeding the air into the columns it is compressed, filtered, cleaned (using molecular sieves to remove water, CO_2, and other impurities), and then cooled down to almost liquefaction (Timmerhaus and Flynn, 1989).

Linde's Simple Column This is the simplest process to separate air into its components. It was industrially used at the beginning of the XX century. The system consists of a single column that operates slightly above atmospheric pressure. The feed to the column is cold air from a liquefaction process. The compressed air is cooled using the separated nitrogen and oxygen (see Fig. 3.11), and it is used as hot utility. It boils the oxygen, and is then expanded before it is fed into the column. The column operates as a stripping one. As the air rises through the column it gets concentrated in nitrogen, the most volatile of the two. A stream rich in nitrogen exits from the top while another

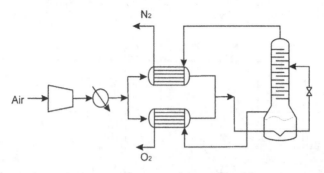

FIGURE 3.11

Linde's simple column.

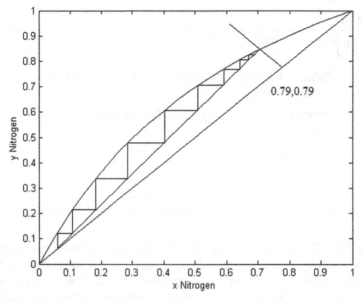

FIGURE 3.12

X–Y diagram for the simple column.

stream rich in oxygen is obtained from the bottom. The purity limit in the nitrogen is limited since no rectification section of the column is present (see Fig. 3.12). The slope of the q line depends on the fraction of liquid that we can obtain after cooling and expanding the air. The oxygen is recovered over the liquid, and has a composition of around 95% oxygen because it carries the argon, while for the top the nitrogen is recovered at 90% purity, as seen in the XY diagram.

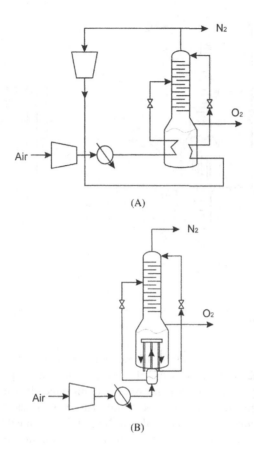

FIGURE 3.13

Simple column with recycle. (A) Recompresion, (B) Claude's device.

Simple Column With Recycle Recompression: The simple column design evolved to improve the purity of the nitrogen. To achieve that, part of the nitrogen from the top of the column is recompressed, cooled by using it as hot utility in the boiler of the column, expanded, and fed as if it were the reflux of the column. The compressed air is also used as hot utility in the reboiler. See Fig. 3.13A for the scheme of the system.

Claude's Column: Claude also presented his design for a distillation column. It had a dephlegmator kind of device so that the compressed air at 25 atm was fed to the column. In a tubular device submerged in liquid air, the compressed gas rises. Part of the vapor liquefies and returns to the feed chamber. The gas phase, rich in nitrogen, goes down via two lateral pipes, and also liquefies. The nitrogen is gathered at the annular space and is used as recycle for the column. The liquid oxygen is fed to the column after expansion. See Fig. 3.13B for a scheme of the device.

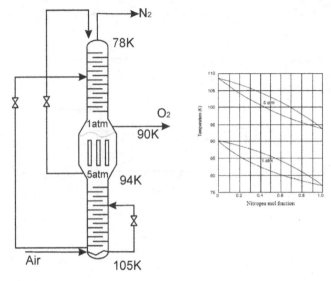

FIGURE 3.14

Linde's double column.

Linde's Double Column Linde's double column represented a technological breakthrough. The system consists of two columns that operate at different pressures (see Fig. 3.14). The low-pressure column typically works at 1 atm, and the higher-pressure column works at 5 atm. Air is used as hot utility in the high-pressure column so that it cools down. Next, it is expanded to obtain a liquid fraction and fed to the column. The bottoms of the column, recovered over the liquid of the high-pressure column reboiler, are expanded to 1 atm and fed to the lower-pressure column. The distillate of the high-pressure column is also expanded to 1 atm and used as reflux. This stream, rich in nitrogen, is fed at the top of the lower-pressure column. Therefore, only two streams are obtained from the system, both from the low-pressure column. From the bottoms, at about 90K, a stream rich in oxygen exits the system. The concentration is no more that 95%, due to the presence of argon in this stream. From the top, a stream rich in nitrogen at about 78K is produced.

Both columns are thermally coupled (see Fig. 3.14). The condenser of the high-pressure column acts as a reboiler for the higher-pressure column. For heat to be exchanged, a temperature gradient must be maintained. In this configuration, a temperature difference of about 4–5K is typically used. This temperature difference can be proven using the equilibrium diagrams. Typically there is no stripping region in the higher-pressure column since the operation of the upper column increases the oxygen concentration of the bottoms efficiently. The bottoms of the high-pressure column contain around 40% oxygen.

EXAMPLE 3.8

Evaluate the operation of Linde's double column. The low- and high-pressure columns work at 1 atm and 5 atm, respectively. The products are N_2, 98% molar, and O_2, with a purity of 96%. The air fed to the system is at 120K and 200 atm.

Solution

Antoine vapor pressure equation for N_2:

$$\ln(P_v) = \left(14.9542 - \frac{588.72}{(T(^\circ C) + 273.15 - 6.6)}\right)$$

Antoine vapor pressure equation for O_2:

$$\ln(P_v) = \left(15.4075 - \frac{734.55}{(T(^\circ C) + 273.15 - 6.45)}\right)$$

We assume ideal behavior for both cases and compute the XY diagram for the system as follows:

$$y_{N_2} = \frac{P_{Vap,N_2} x}{P_{Total}}$$

$$P_T = P_{O_2} + P_{N_2} = y_{O_2} \cdot P_{Total} + y_{N_2} \cdot P_{Total} = P_{Vap,O_2}(1-x) + P_{Vap,N_2} x \to x = \frac{P_{Total} - P_{Vap,O_2}}{P_{Vap,N_2} - P_{Vap,O_2}}$$

The operating line for the rectifying section is obtained by performing a mass balance to the higher section of the column:

$$V_n = L_{n-1} + D \quad \text{(Low pressure)}$$

$$V_n y_n = L_{n-1} x_{n-1} + D \cdot x_D \to y_n = \frac{L_{n-1}}{V_n} x_{n.-1} + \frac{D}{V_n} x_D$$

$$V_n + D_2 = L_{n-1} + D_1 \quad \text{(High pressure)}$$

$$V_n y_n + D_2 x_{D,2} = L_{n-1} x_{n-1} + D \cdot x_{D,1} \to y_n = \frac{L_{n-1}}{V_n} x_{n.-1} + \frac{D_1 x_{D,1} - D_2 x_{D,2}}{V_n}$$

The corresponding operating line for the stripping section is as follows:

$$V_m = L_{m-1} - W$$

$$V_m y_m = L_{m-1} x_{m-1} + W x_W \to y_m = \frac{L_{m-1}}{V_m} x_{m-1} + \frac{W}{V_m} x_W$$

We assume that the flows of the species are approximately constant across each section of the column:

$$y_n = \frac{L}{V} x_{n.-1} + \frac{D}{V} x_D$$

$$y_m = \frac{L'}{V'} x_{m-1} + \frac{W}{V'} x_W$$

The feed to the column is partially liquid. Thus we define the liquid and vapor fraction of the feed as Φ_L and Φ_V, respectively:

$$\Phi_L = \frac{L' - L}{F} \quad \Phi_V = \frac{V - V'}{F}$$

Thus the q line can be written as follows:

$$q = \frac{\Phi_V - 1}{\Phi_V} x_{n.} + \frac{1}{\Phi_V} x_f$$

We now present the mass balances. See Fig. 3E8.1 for the flow across the low-pressure column.

1. Global:

$$F = D_1 + R_1$$
$$Fx_f = D_1 x_D + R_1 x_R$$

2. Low-pressure column:

$$F = D_2 + R_2$$
$$Fx_f = D_2 x_D + R_2 x_R$$

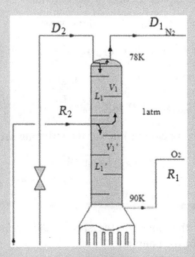

FIGURE 3E8.1

Flows across the low-pressure column.

Low-pressure column:

Let f be the liquefied fraction from stream D_2 and g the liquefied fraction from stream R_2:

$$L_1 = f \cdot D_2$$

$$L_1' = f \cdot D_2 + g \cdot R_2$$

$$V_1 + (1-f)D_2 = D_1$$

$$V_1' = L_1' - R_1$$

$$V_1' = V_1 - (1-g) \cdot R_2$$

Thus the operating lines at the rectifying and stripping sections, as well as the q line, are given as follows:

Rectifying section:

$$y_n = \frac{L_1}{V_1} x_{n.-1} + \frac{D_1 x_{D,1} - D_2 x_{D,2}}{V_1} = \frac{f \cdot D_2}{D_1 - (1-f)D_2} x + \frac{D_1 x_{D,1} - D_2 x_{D,2}}{D_1 - (1-f)D_2}$$

q line:

$$y_n = \frac{g}{1-g_1} x_{n.-1} + \frac{1}{1-g_1} x_f$$

Stripping line:

$$y_n = \frac{L_1'}{V_1'} x_{n.-1} + \frac{R_1}{V_1'} x_R = \frac{f \cdot D_2 + gR_2}{D_1 - (1-f)D_2 - (1-g)R_2} x + \frac{R_1 x_R}{D_1 - (1-f)D_2 - (1-g)R_2}$$

f corresponds to a flowrate of pure saturated nitrogen at 5 atm and $-178°C$ that is expanded to 1 atm. In the Hauser diagram we can see the result of that expansion (see Fig. 3E8.2). Using the lever rule we can compute the liquefied fractions. We assume that both streams are air, and thus the Hauser diagram can be used to compute both:

$$f = (68 - 31)/(68 - 22) = 0.8$$

$$g = (31 - 68)/(22 - 68) = 0.8$$

Solving the mass balances (see Table 3E8.1), we compute the operating lines. Next, we use the McCabe method to determine the number of trays needed in the low-pressure column (see Fig. 3E8.3).

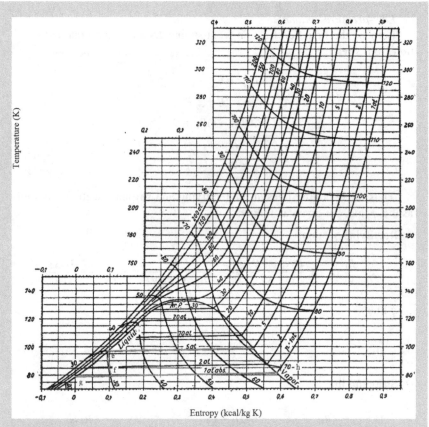

FIGURE 3E8.2

Liquefied fraction after expansion.

Table 3E8.1 Results of Low-Pressure Column

	F	D	R			
Feed	100	80.4347836	19.652174	−1E-06		
Feed N$_2$	79	78.0217401	0.97826087	79.000001		
	y	*L/V*		*D/V*	*L*1	34.9999959
Rectifying		0.4882486		0.49639886	*L*1′	80.244566
	y	*L′/V*		*R/V*	*V*1	71.6847847
Stripping		1.32243618		0.01612181	*V*1′	60.6793487

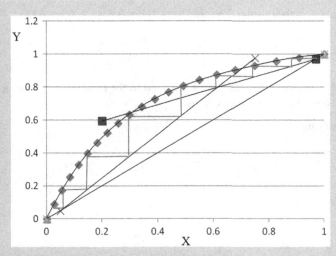

FIGURE 3E8.3

McCabe diagram for the low-pressure column.

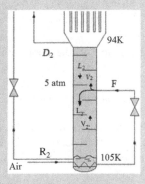

FIGURE 3E8.4

Flows across the high-pressure column.

High-pressure column:
   The energy balances to the condenser–reboiler couple both columns and determine the flows at the high-pressure column (Fig. 3E8.4):

$$V_2 \lambda_2 = L_1' \lambda_o$$

$$V_2 = \frac{L_1' \lambda_o}{\lambda_2}$$

$$\lambda_2 = f(N_2)$$
$$\lambda_o = f(O_2)$$

A mass balance to the feed tray is as follows:

$$L_2' = L_2 + h \cdot F$$

$$V_2' + R_2 = L_2'$$

where h is the liquefied fraction in the feed to the column. Thus the operating lines and the q line are as follows:

Rectifying section:

$$y_n = \frac{L_2}{V_2} x_{n-1} + \frac{D_2}{V_2} x_{D,2} = \frac{V_2 - D_2}{V_2} x + \frac{D_2}{V_2} x_{D,2}$$

q line:

$$y_n = \frac{h}{1-h} x_{n-1} + \frac{1}{1-h} x_f$$

Stripping line:

$$y_n = \frac{L_2'}{V_2'} x_{n-1} + \frac{R_2}{V_2'} x_{R,2} = \frac{L_2 + h \cdot F}{(L_2 + h \cdot F) - R_2} x + \frac{R_2 \cdot x_{R,2}}{(L_2 + h \cdot F) - R_2}$$

To compute the liquid fraction, h, the cool air is expanded from 120K and 200 atm to 5 atm:

$$h = (40 - 71)/(32 - 71)$$

In Table 3E8.2 we present the results of the mass balances and the operating lines. Next, using the McCabe method, the number of trays needed at the high-pressure column is six (Fig. 3E8.5).

Table 3E8.2 Results of High-Pressure Column

	F	D_2	R_2			
Feed	100	43.75	56.25	−1E-06		
Feed N_2	79	42.44	36.56			
	y	L/V		D/V	L2	54.2672357
Rectifying		0.55364996		0.43295954	L2'	133.754415
	y	L'/V'		R/V	V2	98.0172305
Stripping		1.72576524		0.25401783	V2'	77.504409

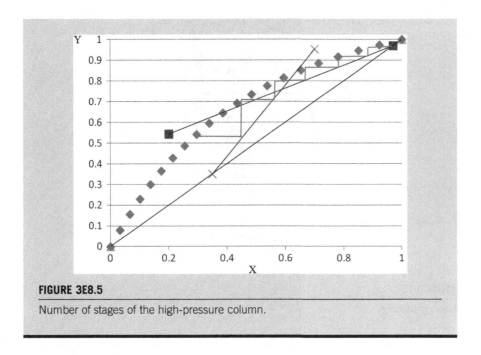

FIGURE 3E8.5

Number of stages of the high-pressure column.

Argon separation and purification

Argon is a valuable product that is in relatively high quantities in the air. It is possible to modify the double column to obtain argon from air (see Fig. 3.15). A stream from the low-pressure column is extracted at the level where Ar concentration is maximum, around 12%. This vapor is rectified in an auxiliary column that uses a partial condenser. As refrigerant of the second column, a fraction of the oxygen stream from the bottoms of the high-pressure column is used, after expansion. The bottoms of the auxiliary column are returned to the low-pressure column, while the distillate is the Ar produced at 98% purity. Oxygen is the main impurity of the Ar and can be eliminated using chemical separation, for instance, deoxoreactors where water is produced from the oxygen impurity.

Neon can also be separated. It is more volatile than nitrogen and can be obtained by rectification of the liquid product from the top of the high-pressure column. Neon comes with nitrogen and helium as impurities.

The evolution of liquefaction and separation processes has resulted in the current liquefaction processes (Linde, 2015). Fig. 3.16 shows the flowsheet of the typical layout of one of those plants. In the first stage air is compressed. Next, it is purified using molecular sieves to remove moisture, dust, and other impurities such as CO_2 or hydrocarbons. Then, the compressed air is cooled down using nonliquid oxygen and nitrogen. Subsequently, a double distillation column (Linde's design) with an auxiliary column for Ar recovery is used. The products obtained are not only gaseous oxygen and nitrogen, but also liquid oxygen and nitrogen. The oxygen comes from the bottoms of the low-pressure column, and the nitrogen from the top of the low-pressure column.

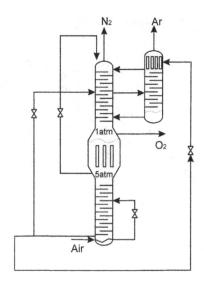

FIGURE 3.15

Argon production.

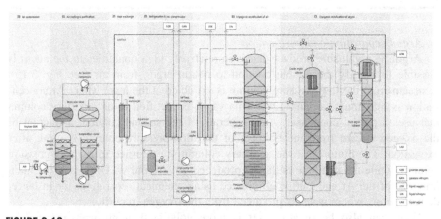

FIGURE 3.16

Typical liquefaction process. (© Linde AG, extract from "History and technological progress. Cryogenic air separation." p. 14–15.)

Storage. Two storage devices can be highlighted: vacuum insulated tanks and dewars. Vacuum insulated tanks consist of a double-walled vessel. The internal wall is made of stainless steel and the outer wall is made of carbon steel. Between them the chamber is under vacuum, filled with perlite as insulation. The size of these tanks range from 1 m^3 on, and they are typically designed for 14 atmg. It is also possible to have an evaporator to control the pressure inside

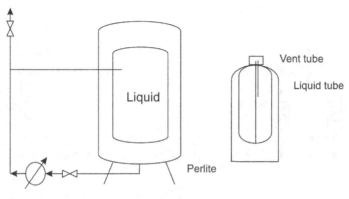

FIGURE 3.17

Vacuum-insulated tanks and dewars.

the tank and another one to serve gas products. Dewars are cryogenic mobile devices. They also have the same double-wall structure as the tanks. Their size ranges from liters to 1 m^3, and the designed pressure is 1.4 atmg. The discharge is carried out from the bottoms, and they have an internal evaporator for pressure control. Fig. 3.17 shows schemes for both.

3.2.2.1.2 Noncryogenic air separation

From an industrial perspective, liquid gases are expensive, but sometimes they are regasified for further use. Thus, a number of technologies have been developed to separate the air components without need for liquefaction. In this section, we discuss two alternatives, the use of pressure or temperature cycles known as Pressure Swing Adsorption (PSA) and Temperature Swing Adsorption (TSA), and the use of membranes. Note that these technologies are not fully developed, but they are gaining attention in the market.

Pressure Swing Adsorption cycles The use of PSA cycles to produce oxygen or nitrogen is based on the adsorption of one of the two on a bed of particular material. Typically, since air is free, in industry there is no recovery of the purge or the adsorbed species. The operation of these cycles consists of four stages: (a) pressurization, (b) adsorption, (c) depressurization, and (d) desorption. In order to ensure continuous operation of the plant, two beds operate in parallel at any one time. Fig. 3.18 shows the operation of the cycle and the pressure at the bed. An excess in the purge increases the purity of the effluent. A defect results in higher contamination.

Unlike TSA, PSA systems do not degrade the adsorbent and can operate longer without substitution. However, the presence of moisture and drops must be avoided. Filters are used to remove them before the air is fed to the PSA system.

The *production of nitrogen* above 99.5% purity consists of adsorbing the oxygen out of the air. The bed for this process consists of molecular sieves, either Zeolite 4 A or active carbon Bergbau—Foschung operating at 6—8 atmg. Desorption occurs

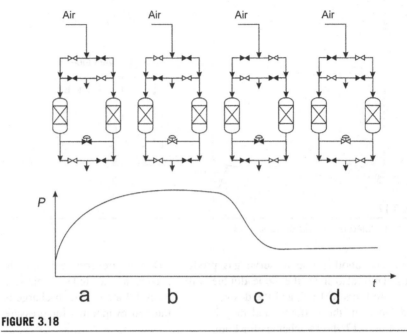

FIGURE 3.18

PSA cycle operation. Black valves are closed.

at atmospheric pressure. The purge stream consists of oxygen at 35–50% by volume. The energy consumption in the production of nitrogen is around 0.1 kWh/Nm³, and presents a lower investment than required for cryogenic units.

The *production of oxygen* uses a PSA system where the adsorbent is either Zeolite 13X or 5 A to adsorb the nitrogen. The working conditions are 1–3 atmg for adsorption, but the oxygen production cannot surpass 1500 Nm³/h. Furthermore, the oxygen produced has a low purity, lower than 90%, and contains Ar and N.

Apart from PSA systems for the production of oxygen, Vacuum Pressure Swing Adsorption can be used. In this case, gas desorption occurs below atmospheric pressure, 0.5 atmg. These units can produce up to 50 t/d, with more or less the same purity as the PSA systems (Pressure Swing Adsorption Picks Up Steam, 1988).

Membranes Gas separation using membranes is based on the different diffusion rates of the species. It is a process governed by Henry's and Fick's Laws. The permeability, a measure of the capacity of the membrane to transport gases, is the product between the Henry's Law constant and the diffusivity coefficient. The permeate is concentrated in the quicker-diffusing gases while the reject or purge comprises the slower-diffusing gases (Membranes, 1990).

Fick's Law dictates the diffusion rate:

$$\frac{\partial c_i}{\partial t} = D_i \frac{\partial^2 c_i}{\partial z^2}$$

(3.18)

where D_i is the diffusivity (m^2/s) of species i across the membrane, c_i is the concentration across the membrane, and z the distance. At steady state, we can compute the flux of component i across the membrane as follows:

$$N_i = D_i \frac{c_{i,1} - c_{i,2}}{\delta} \quad (3.19)$$

where δ is the thickness of the membrane. Thus, assuming that two gas phases are in contact through a membrane, to the left we have a total pressure P_1 and a partial pressure of each component given by:

$$p_{i,1} = x_{i,1} \cdot P_1 \quad (3.20)$$

To the right, we have P_2. For component i to diffuse, $P_1 > P_2$. Fig. 3.19 shows the scheme of the process. Gas solubility in polymers, the membrane, is assumed to follow Henry's Law (Perry and Green, 1997).

Henry's Law states:

$$c_i = p_i \cdot (1/H_i) = p_i \cdot S_i \quad (3.21)$$

Thus

$$N_i = D_i \frac{c_{i,1} - c_{i,2}}{\delta} = S_i D_i \frac{p_{i,1} - p_{i,2}}{\delta} = K_i \frac{p_{i,1} - p_{i,2}}{\delta} \quad (3.22)$$

where permeability is computed as $\frac{K_i}{\delta}$. The relative permeability of one gas compared to another depends on the molecule size and the interactions with the polymeric membrane. We can establish a relative scale of diffusion rate of the main gases involved in chemical processes:

Quicker $H_2 > He > H_2S > CO_2 > O_2 > Ar > CO > CH_4 > N_2$ Slower

Nitrogen and oxygen diffuse slowly across membranes while CO_2 is five times quicker than oxygen. Water vapor can go through polar membranes easily, but not as easily across nonpolar polymers. This relative diffusion velocity results

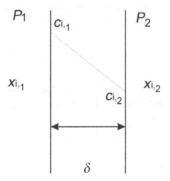

FIGURE 3.19

Membrane gas separation.

in the composition of the resulting permeate and reject phases. Based on the velocity, membranes are typically used to obtain nitrogen at 93–99%.

$$x_{i,2} = \frac{N_i}{\sum_i N_i} \tag{3.23}$$

The separation factor provided by a membrane, $S_{i,j}$, is defined as the ratio in concentrations of two species at both sides of the membrane:

$$S_{i,j} = \frac{\dfrac{x_{i,2}}{x_{j,2}}}{\dfrac{x_{i,1}}{x_{j,1}}} = \frac{\left.\dfrac{N_i}{N_j}\right|_2}{\left.\dfrac{N_i}{N_j}\right|_1} = \frac{K_i}{K_j} \frac{1 - \dfrac{x_{i,2}}{x_{i,1}} \dfrac{P_2}{P_1}}{1 - \dfrac{x_{j,2}}{x_{j,1}} \dfrac{P_2}{P_1}} = \alpha_{i,j} \frac{1 - \dfrac{x_{i,2}}{x_{i,1}} \dfrac{P_2}{P_1}}{1 - \dfrac{x_{j,2}}{x_{j,1}} \dfrac{P_2}{P_1}} \tag{3.24}$$

The selectivity of the membrane to both gases, $\alpha_{i,j}$, is the limit of the separation fraction for a certain pressure difference.

3.2.2.2 Chemical separation

These systems are based on the reactivity of the different gases in the mixture. Oxygen is much more reactive, generating some oxide and liberating the nitrogen. The deoxo process can be used to remove the traces of oxygen as water using hydrogen as reactant. Alternatively, we can have processes like $2BaO + O_2 \leftrightarrow 2BaO_2$ that go forward at 500°C and backward to recover the oxygen at 700°C.

Selection of the methods depends on two factors: production capacity and purity of the gas required. Membranes are typically used for $1-1000$ Nm^3/h of nitrogen and purities up to 99%. PSA can be used when slightly higher purity is needed, up to 10,000 Nm^3/h. Higher purity and production capacity requires the use of cryogenic fractionation. In the case of oxygen production, PSA is recommended for $1-1000$ Nm^3/h and up to 95% purity. Any other purity requires cryogenic fractionation.

Table 3.4 shows a brief comparison of the technologies discussed. It presents the status of the maturity of the technology, the relative economics, the purity of the products, and operating limitations (Smith and Klosek, 2001).

Table 3.4 Comparison of Air Separation Technologies

Process	Status	Economics	Byproduct Capability	Purity Limit (% volume)	Startup/ Shutdown Time
PSA	Semimature	$$$$$$	Poor	95	Minutes
Chemical	Developing	Unknown	Poor	>99	Hours
Cryogenic	Mature	$$	Excellent	>99	Hours
Membrane	Semimature	$	Poor	40	Hours

3.3 ATMOSPHERIC POLLUTION

There are two main sources of atmospheric pollution: *natural*, due to processes such as fire, volcanoes, fogs, etc., and *artificial*, due to human activities. Among them we can find phenomena like city smog, industrial smog, and the presence of gaseous species such as NO_x, SO_2, CO, CO_2, smoke, or odors.

3.4 HUMID AIR

Air can have a certain amount of moisture as a function of the conditions. Assuming ideal behavior, the total pressure results from the contribution of both gases, the air and the vapor:

$$P_T = P_{air} + P_{vapor} \tag{3.25}$$

Vapor pressure is a function of the temperature. Thus the higher the temperature the larger the amount of water vapor that the air can hold. If the air cannot handle a certain amount of water, then it condenses. This occurs for vapor pressures above the saturation at a certain temperature. Bear in mind that the presence of water vapor in air is an industrial problem for several processes. For instance, when dealing with solids, humidity can cause agglomeration, reducing the quality of the product (eg, detergents). In other cases, such as in the production of sulfuric acid, the presence of vapor generates the acid over time, representing a safety problem due to the corrosion of the pipes and reactor.

Here, the concepts of humid air are presented before they are applied to some simple examples.

Specific molar moisture:

$$y_m = \frac{P_v}{P_T - P_v} = \frac{\varphi P_v^{sat}}{P_T - \varphi P_v^{sat}} \tag{3.26}$$

Specific moisture:

$$y = \frac{Mw_{cond}}{Mw_{dry\,air}} \frac{P_v}{P_T - P_v} = \frac{Mw_{cond}}{Mw_{dry\,air}} \frac{\varphi P_v^{sat}}{P_T - \varphi P_v^{sat}} \tag{3.27}$$

Relative moisture:

$$\varphi = \frac{P_v}{P_v^{sat}} \tag{3.28}$$

We can use Antoine correlations to compute the saturation vapor pressure. In the case of water, we can use the following equations:

$$\ln(P_v(mmHg)) = 18.3036 - \frac{3816.44}{227.02 + T(°C)}$$

$$P_v(mmHg) = 2.12413 \cdot 10^{-10}(T(°C))^6 - 1.44107 \cdot 10^{-8}(T(°C))^5 + 4.2057 \cdot 10^{-6}(T(°C))^4$$
$$+ 1.41219 \cdot 10^{-4}(T(°C))^3 + 1.14812 \cdot 10^{-2}(T(°C))^2 + 3.29301 \cdot 10^{-1}T(°C) + 4.58643$$

$$(3.29)$$

Similarly, we can proceed for other vapor species such as ethanol or ammonia.

Specific volume of humid air:

$$v = \frac{RT}{P}\left(\frac{1}{Mw_g} + \frac{y}{Mw_v}\right) \tag{3.30}$$

Heat capacity (per kg of dry air):

$$c_{p,h} = c_{p,g} + c_{p,v}y \tag{3.31}$$

Enthalpy of humid air:

$$(T_{ref} = 0°C)\ H = (c_{p,g} + c_{p,v}y)T + \lambda_{T=0°C} \cdot y \tag{3.32}$$

EXAMPLE 3.9

Evaluate the process that transforms stream 1 from its initial conditions to the final ones given below. Determine whether water condenses, and compute the heat load for cooling this stream.

  Stream 1, initial conditions: $P = 750$ mmHg, $T = 40°C$, $\varphi = 0.7$
  Stream 1, final conditions: $P = 2550$ mmHg, $T = 22°C$

Solution

If the specific moisture at the beginning is larger than that at the end, water must condense since the air cannot hold the difference. Thus, using the concepts presented above, both specific moisture contents at the beginning and at the end, assuming saturation, are as follows:

$$y_{ini} = \frac{M_{water}}{M_{dry\ air}}\frac{\varphi P_v^{sat}}{P_T - \varphi P_v^{sat}} = 0.62\frac{0.7 \cdot 55.32}{750 - 0.7 \cdot 55.32} = 0.03375\frac{kg_{vap}}{kg_{dry\ air}}$$

$$y_{final} = 0.62\frac{19.83}{2550 - 19.83} = 0.0048\frac{kg_{vap}}{kg_{dry\ air}}$$

Since $y_{final} < y_{ini}$, water must condense. For 100 kg of dry air, initially 103.37 kg of gas are compressed ($100 \cdot (1 + y_{ini})$) and cooled. The gas phase at the end of the process consists of 100.48 kg, ($100 \cdot (1 + y_{final})$);

therefore the difference, 2.89 kg, has to condense. We now perform an energy balance:

$$\Delta H = Q$$

$$[(c_{p,\,g} + c_{p,\,v}y_f)(T_f)] + \lambda y_f + c_{p,\,\text{water}}\text{COND} \cdot T_f - [(c_{p,\,g} + c_{p,\,v}y_i)(T_i)] - \lambda y_i$$

$$= [(0.24 + 0.46y_f)(T_f)] + 597.2y_f + 1 \cdot 0.0289T_f - [(0.24 + 0.46y_i)(T_i)] - 597.2y_i$$

$$= 8.07 + 0.636 - 30.3765 = -22\ \text{kcal/kg}_{\text{dry air}}$$

EXAMPLE 3.10

Determine if mixing of the following two streams results in water condensation. The working pressure is 760 mmHg.

Stream 1	100 kg$_{da}$/s, $T = 54°C$, $\varphi = 0.95$
Stream 2	100 kg$_{da}$/s, $T = 9.6°C$, $\varphi = 0$

Solution
Stream 1

$$y_1 = 0.62 \frac{0.95 \cdot 112.5}{760 - 0.95 \cdot 112.5} = 0.1015 \frac{\text{kg}_{\text{vap}}}{\text{kg}_{\text{dry air}}}$$

$$H_1 = (0.24 + 0.46y_1) \cdot 54 + 597.2y_1 = 76.1 \frac{\text{kcal}}{\text{kg}_{\text{dry air}}}$$

Stream 2
Similarly:

$$y_2 = 0, \quad H_2 = 2.3 \frac{\text{kcal}}{\text{kg}_{\text{dry air}}}$$

Mass balance to water (assuming no condensation):

$$m_1 \cdot y_1 + m_2 \cdot y_2 = m_{\text{mix}} \cdot y_{\text{mix}}$$

$$y_{\text{mix}} = 0.05075 \frac{\text{kg}_{\text{vap}}}{\text{kg}_{\text{dry air}}}$$

For energy balance, we compute the temperature of the mixture:

$$m_1 \cdot H_1 + m_2 \cdot H_2 = m_m H_m$$

$$H_m = 39.2 \frac{\text{kcal}}{\text{kg}_{\text{dry air}}}$$

$$H_m = (0.24 + 0.46 \cdot 0.05075) \cdot T + 597.2 \cdot 0.05075$$

$$T = 33.83°C$$

At this temperature, with the computed specific moisture, the relative humidity becomes:

$$y_m = 0.62 \frac{\varphi P_v^{sat}(33.8)}{P_T - \varphi P_v^{sat}(33.8)}$$

$\varphi = 1.46 > 1 \rightarrow$ Water condenses

Therefore the mass and energy balances must consider the condensed water, COND:

Energy balance:

$$m_1 \cdot H_1 + m_2 \cdot H_2 = m_m H_m + COND \cdot h$$

$$100 \cdot 76.1 \frac{kcal}{kg_{dry\ air}} + 100 \cdot 2.3 \frac{kcal}{kg_{dry\ air}}$$

$$= 200[(0.24 + 0.46 y_f)(T_m)] + 597.2 y_f + COND \cdot 1 \frac{kcal}{kg} T_m$$

$$= 200 \left[\left(0.24 + 0.46 \left(0.62 \frac{P_v(T_m)}{760 - P_v(T_m)} \right) \right) (T_m) \right] + 597.2 \left(0.62 \frac{P_v(T_m)}{760 - P_v(T_m)} \right)$$

$$+ COND \cdot 1 \frac{kcal}{kg} T_m; \quad \varphi = 1$$

Mass balance to water:

$$m_1 \cdot y_1 + m_2 \cdot y_2 = m_m y_{sat} + COND$$

This is a complex problem to solve. We present here two methods: EXCEL and GAMS. For further details on the use of these packages, the reader can find the basics in Martín (2014).

EXCEL

We define the variables in different cells and use "solver" as shown in Fig. 3E10.1.

m_1	100
m_2	100
H_1	76.1
H_2	2.3
y_1	0.1045
y_2	0
Tm (G14)	39.7656916
Pv	54.5073231
Hm	39.0271005
Ym	0.04790204
COND (G18)	0.86959147
Balance(G19)	2.1368E-09

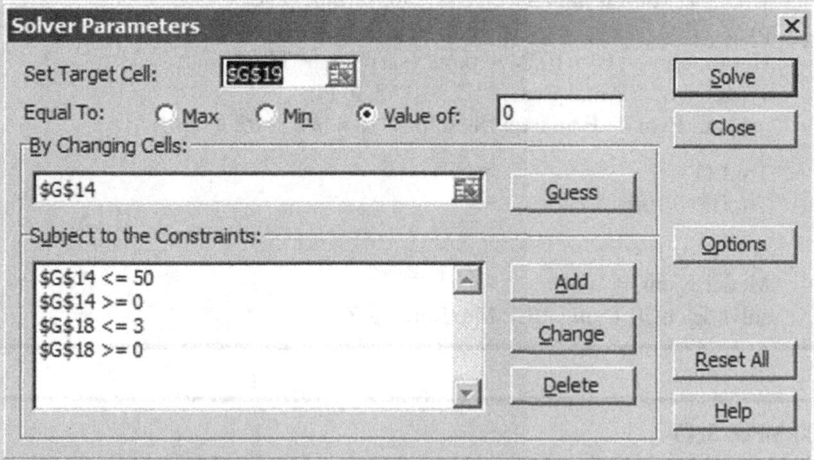

FIGURE 3E10.1

Example 3.10 in "Solver."

GAMS

In this case we define the data as scalar, the variables to be computed as positive variables, and write the equations of the problem by hand:

Scalar

mass1 /100/

mass2 /100/

T1 /54/

T2 /9.6/

phi1 /0.95/

phi2 /0/

Ptotal /760/;

positive variables

y1, y2, H1, H2, Hm, COND, ysat, Pv1, Pv2, Pvm, Tm;

variable

Z;

equations

BM, BE, y_1, y_2, H_1, H_2, pv_1, pv_2, H_sat, y_sat, pv_sat, obj;

BM.. mass1*y1 + mass2*y2 = E = (mass1 + mass2)*ysat + COND;

BE.. mass1*H1 + mass2*H2 = E = (mass1 + mass2)*Hm + COND*Tm;

H_1.. H1 = E = (0.24 + 0.46*y1)*T1 + 597.2*y1;

y_1.. y1 *(Ptotal-phi1*Pv1) = E = 0.62*phi1*Pv1;

pv_1.. Pv1 = E = exp(18.3036-3816.44/(227.02 + T1));

H_2.. H2 = E = (0.24 + 0.46*y2)*T2 + 597.2*y2;

y_2.. y2 *(Ptotal-phi2*Pv2) = E = 0.62*phi2*Pv2;
pv_2.. Pv2 = E = exp(18.3036-3816.44/(227.02 + T2));
H_sat.. Hm = E = (0.24 + 0.46*ysat)*Tm + 597.2*ysat;
y_sat.. ysat *(Ptotal-Pvm) = E = 0.62*Pvm;
pv_sat.. Pvm = E = exp(18.3036-3816.44/(227.02 + Tm));

Tm.LO = 0;
Tm.UP = 100;

obj.. Z = E = 2;
Model Ejem10 /all/;
solve Ejem10 Using NLP Maximizing Z

EXAMPLE 3.11

Compute the temperature that a stream of air at 800 mmHg, that at 20°C
has a relative moisture of 0.85, has to be heated to so that when mixed
with a second stream of air, whose dew point is 50°C, does not cause water
to condense.

Solution

We draw the psychometric diagram for 800 mmHg. To do that, for each
temperature, compute the vapor pressure using the Antoine equation, and
with it, the specific humidity at saturation using Eq. (3.26). See Fig. 3E11.1.

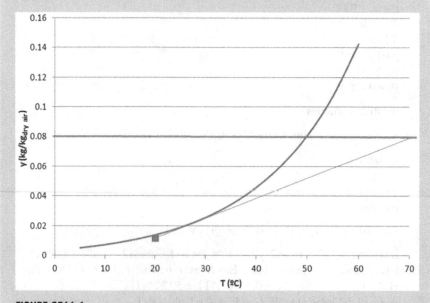

FIGURE 3E11.1

Graphic solution of example 3.11.

Next, the specific moisture of the second stream is computed. Since it is saturated at 50°C, y_2 is equal to 0.08. We draw a horizontal line at y equal to 0.08. Subsequently, we allocate the initial point representing steam 1. At 20°C, φ is equal to 0.85. Thus y_1 is equal to 0.012. See the square in Fig. 3E11.1.

In order for both streams to mix without condensation, we draw a line that, starting at the original conditions of stream 1, is tangent to the saturation line at 800 mmHg. The cross between this line and the horizontal one, corresponding to the specific moisture of the second stream, gives the temperature at which stream 1 has to be heated to in order to avoid condensation independently of the proportion in which both streams are mixed.

It turns out that 70°C is the solution to the problem.

EXAMPLE 3.12

Determine in what proportion the following two air streams, when mixed, condense:

Stream 1 $T = 10°C$, $\varphi = 0.55$
Stream 2 $T = 45°C$, $\varphi = 0.70$
The atmospheric pressure is 700 mmHg.

Solution

First, the saturation line is computed. For a number of temperatures we compute the specific moisture at saturation. Next, both mixtures are plotted as squares (see Fig. 3E12.1). Finally, mass and energy balances are formulated. It is assumed that the resulting mixture is saturated; then $\varphi = 1$.

If we draw a straight line between the two points, we see that there may be two points where saturation occurs. This is only an approximation to realize the possibility of two solutions.

Mass balance to water:

$$m_1 \cdot y_1 + m_2 \cdot y_2 = (m_1 + m_2)y_{m=sat}$$

$$100 \cdot 0.0044 + m_2 \cdot 0.048 = (100 + m_2)y_{sat}$$

Enthalpy balance:

$$m_1 \cdot H_1 + m_2 \cdot H_2 = (m_1 + m_2)H_{m=sat}$$

$$H_1 = (0.24 + 0.46 \cdot 0.0044) \cdot 10 + 597.2 \cdot 0.0044 = 5.07 \frac{kcal}{kg}$$

$$H_2 = (0.24 + 0.46 \cdot 0.0509) \cdot 46 + 597.2 \cdot 0.0509 = 40.43 \frac{kcal}{kg}$$

Assuming 100 kg for stream 1

$$100 \cdot 5.07 + m_2 \cdot 40.43 = (100 + m_2)H_{sat} = (100 + m_2) \cdot (0.24 + 0.46 \cdot y_{sat}) \cdot T_{sat} + 597.2 \cdot y_{sat}$$

$$\frac{100\cdot 0.0044 - 100 y_{sat}}{y_{sat} - 0.048} = m_2 = \frac{100\cdot 5.07 - 100((0.24 + 0.46\cdot y_{sat})\cdot T_{sat} + 597.2\cdot y_{sat})}{((0.24 + 0.46\cdot y_{sat})\cdot T_{sat} + 597.2\cdot y_{sat}) - 40.43}$$

This is a nonlinear system of equations, and there are actually two solutions:

Sol 1	18.23°C; $m_2 = 28.4$ kg
Sol 2	27.85°C; $m_2 = 96$ kg

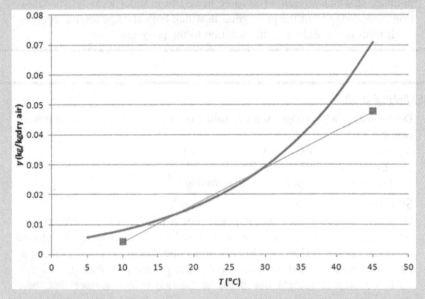

FIGURE 3E12.1

Graphic solution of example 3.12.

EXAMPLE 3.13

Atmospheric air at 25°C and $\varphi = 0.85$ is processed using an adsorbent bed to maintain the relative humidity of the product at 0.25. The total gas flow is 45 m³/min. The desiccant bed is made of silica gel and its dimensions are 15 inches (height) by 5 feet (radius). The adsorption capacity can be seen in Fig. 3E13.1. Compute the fraction of air bypassed as a function of time, and how long the bed can be in operation before regeneration.

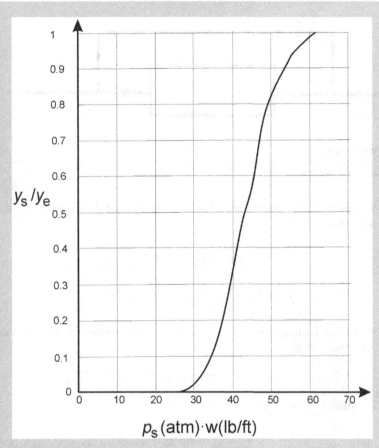

FIGURE 3E13.1

Adsorption capacity of the silica gel bed.

Solution

Fig. 3E13.2 presents a scheme of the operation. We used the following notation:

x: Bypassed fraction
y_s: Air specific moisture exiting the bed (kg/kg$_{dry\ air}$)
y_e: Air specific moisture fed to the bed (kg/kg$_{dry\ air}$)
y_f: Air specific moisture required (kg/kg$_{dry\ air}$)
P_{sat}: Saturation pressure (ata)
m: Mass of dry air processed (lb/ft^2)

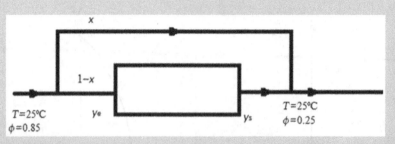

FIGURE 3E13.2

Dehydration system.

We compute the specific humidity entering, y_e, and that required for the product air, y_f:

$$y_e = 0.62 \frac{0.85 \cdot 21}{760 - 0.85 \cdot 21} = 0.015 \frac{kg_{vap}}{kg_{dry\ air}}$$

$$y_f = 0.62 \frac{0.25 \cdot 21}{760 - 0.25 \cdot 21} = 0.0043 \frac{kg_{vap}}{kg_{dry\ air}}$$

We perform a mass balance to water where x is the bypassed fraction of the initial stream:

$$x \cdot y_e + (1 - x) \cdot y_s = y_f \Rightarrow x = \frac{y_f - y_s}{y_e - y_s}$$

The mass flow is given as follows:

$$45 \frac{m^3}{min} \cdot v \cdot 0.454 \frac{kg}{lb} = \frac{45 \cdot \frac{m^3}{min} \cdot 0.454 \frac{kg}{lb}}{\left(\frac{1}{29} + \frac{y_e}{18}\right) \frac{0.082 \frac{atm\ L}{mol\ K} 298\ K}{1\ atm}} = \frac{45}{0.863 \cdot 0.454} = 115 \frac{lb}{min}$$

The fraction bypassed changes with time, and therefore not only x, but also the humidity of the air exiting the bed is as follows:

$$\frac{dw}{dt} = \frac{mass_{total}(1 - x(t))}{A} \rightarrow \frac{A \cdot dw}{(1 - x)} = mass_{total} dt$$

$$\frac{\pi \cdot r^2 \cdot dw}{1 - x} = 115 \frac{lb}{min} \cdot dt$$

$$\frac{dw}{1 - x} = \frac{115 lb}{\pi \cdot (5 ft)^2} \cdot dt = 1.46 \frac{lb}{ft^2} \cdot dt$$

Thus for pressures at the bed, the ratio y_s/y_e is computed. With that, at different time intervals we average the humidity and compute the bypassed fraction. Discretizing the differential equation, the time is determined (see Table 3E13.1).

Table 3E13.1 Results of the Discretized Operation of the Adsorbent Bed

P_{satw}	w	y_s/y_e	y_s	DW	y_{med}	x	Dt	t	y	x
0	0	0	0	1008.1691	0	0.42222222	1194	0	0	0.42222222
28	1008	0	0	72.0120789	0.0001499	0.41737011	85	1193.92208	0	0.42222222
30	1080	0.02	0.00029981	144.024158	0.00089942	0.39183354	162	1278.49202	0.00029981	0.41737011
34	1224	0.1	0.00149903	108.018118	0.00224855	0.33974346	112	1440.52981	0.00149903	0.39183354
37	1332	0.2	0.00299806	108.018118	0.00374758	0.27029946	101	1552.47033	0.00299806	0.33974346
40	1440	0.3	0.00449709				0	1653.75772	0.00449709	0.27029946
							0			
							1654			

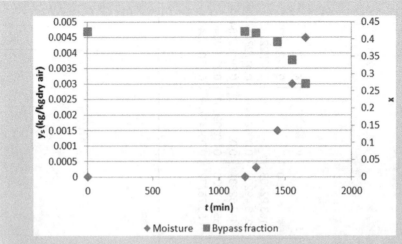

FIGURE 3E13.3

Profile of the operation of the dehydration system.

Fig. 3E13.3 shows the profile of the evolution of the bypass fraction and the specific moisture exiting the bed:

$$\frac{\Delta m}{1-x} = 1.46\Delta t \Rightarrow t = \sum \Delta t = 1653 \text{ min}$$

3.5 PROBLEMS

P3.1. In a Linde-type liquefaction process, air is compressed to 100 atm and expanded to 1 atm. The liquid fraction is 0.0385 kg/kg. Assuming that the heat loss is 0.313 kcal/kg and the temperature difference in the hot end of the heat exchanger is 2.8°C, compute the temperature of the compressed air before expansion.

P3.2. Air is liquefied using a modified Linde-type cycle. Atmospheric air at 80.6°F is compressed up to 100 atm. It is cooled down using the nonliquefied fraction and expanded isentropically to 1 atm, separating the gas and liquid phases. The liquefied fraction is 0.28 and the flowrate of air 0.83 ft³/s, measured at 1 atm, and 60°F, determine:
1. The temperature of the cold extreme of the heat exchanger assuming that the energy losses of the system are 40 BTU per kmol of fed air. The ΔT in the hot stream is 5°F.
2. The energy required by the compressor, as well as that produced in the expansion.

P3.3. Atmospheric air is liquefied using a process as in Fig. P3.3. Air at 27°C and 1 atm is compressed using 538.7 kJ/kg of initial air in a compressor whose efficiency is 80%. The compressed air is cooled down at constant pressure with the nonliquefied air. Next, it is expanded isenthalpically to obtain a liquid fraction of 0.05. Assume that all the heat losses take place in the heat exchanger. Compute the pressure of the air at the hot extreme of the heat exchanger (P_1), the temperature of the nonliquefied air (T_4), and the energy loss per kg of atmospheric air.

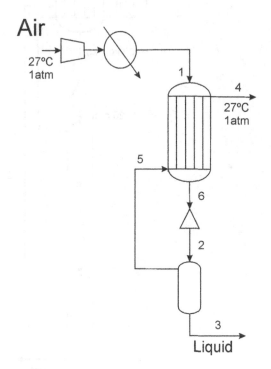

FIGURE P3.3

Flowsheet.

P3.4. Determine the liquid fraction of a process such as in Fig. P3.4, considering two isenthalpic expansions. The first one expands the air to 20 atm at 60 kcal/kg of enthalpy, and then the liquid fraction is further expanded to 1 atm.

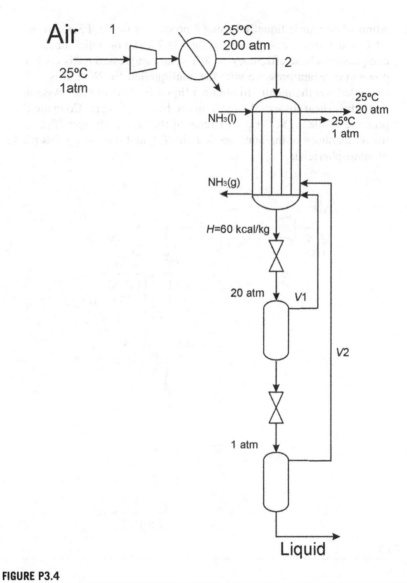

FIGURE P3.4

Flowsheet.

P3.5.   A flowrate of 1000 ft³/min of air at 70°F and 80% moisture is to be
        processed and fed to a facility with only 10% moisture. Silica gel is used
        as the absorbent bed (see Fig. P3.5). Part of the initial feed is bypassed
        dynamically so that the final moisture is controlled. The vessel diameter
        can be up to 5 ft and the operating time before regeneration must be 3 h.

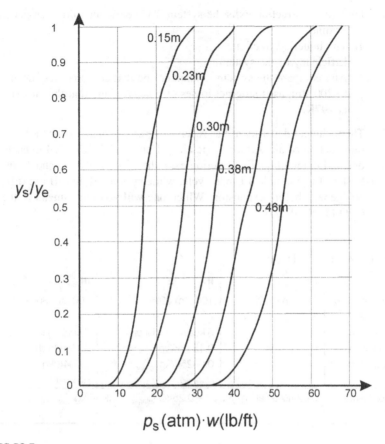

Adsorption capacity.

Assuming isothermal operation, determine the bed diameter and its depth. The absorption capacity can be seen in the figure.

P3.6. A Claude-type cycle is used to liquefy air originally at 1 atm and 300K. It is expected that the system recovers 10% of the compression energy that has compressed the air up to 80 atm. Fifty percent (50%) of the initial stream is sent to the expansion machine. Assume that the heat loss of the system is 6 kcal/kg of the initial processes air and that the nonliquefied fraction exits the system 3°C below the initial temperature. Determine:

1. The liquefied fraction,
2. Efficiency of the system,
3. And compare the system to another one that compress the air up to 200 atm, assuming net losses of 6 kcal/kg and that the air exits at 297K.

P3.7. The technical department has received three alternatives for the liquefaction of air. Atmospheric air at 1 atm and 27°C is fed to each one and the nonliquefied fraction exits the system at 24°C and 1 atm. System I is a nonideal Linde cycle, while systems II and III use different refrigerants before expansion. Which one will you recommend to the CEO (Table P3.7)?

Table P3.7 Process Data

	I	II	III
Compressed air	200 atm 300K	100 atm 300K	100 atm 300K
Heat losses	2 kcal/kg initial air	–	–
Refrigeration		5 kcal/kg initial air	8 kcal/kg initial air
Electricity cost	0.15 €/kWh	0.15 €/kWh	0.15 €/kWh
Refrigerant costs	–	0.0025€/kg liq. air produced	0.004€/kg liq. air produced
Environmental impact	$0.02 \cdot (W\ (kcal/kg_{liquid\ air})) + 0.25 \cdot (Q_{Refrigeration}\ (kcal/kg_{liquid\ air}))$		

P3.8. Air is liquefied using a modified Linde cycle with two consecutive compression stages, the first one from 1 to 20 atm, and the second one from 20 to 150 atm (see Fig. P3.8). The compressed air is cooled down with nonliquefied fractions. The cold air is expanded from 150 atm to 20 atm, and after being used to cool down compressed air, recycled before the second compression stage. The mixture of both streams is at 25°C. The liquid air at 20 atm is expanded to 1 atm.

Determine the liquid fraction and the work required at each stage per kg of initial air.

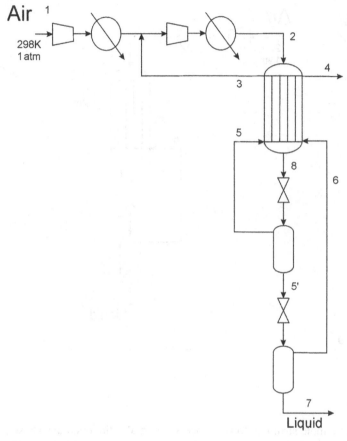

FIGURE P3.8

Flowsheet for the problem.

P3.9. The technical department receives the task of suggesting a system to liquefy air using only one expansion stage so that if the heat losses are 3 kcal/kg of initial air, it allows production of a liquid fraction of at least 15%. Initial air is at 27°C and 1 atm, and the nonliquefied fraction exits the system at 24°C and 1 atm (see Fig. P3.9). Determine the temperature at the inlet of the gas–liquid separator.

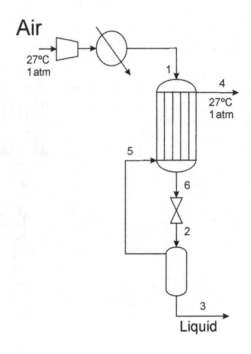

FIGURE P3.9

Flowsheet.

P3.10. A simple stripping tower processes partially liquefied air whose liquid fraction is the same as that obtained from a simple Linde system. The Linde system processes atmospheric air at 27°C and 1 atm that is compressed up to 200 atm and operates ideally with no heat loss. The column produces a residue that is 5% molar in nitrogen and obtains 95% of the maximum ideal composition in the distillate. Determine the number of trays and the distillate composition.

P3.11. A Linde cycle with prerefrigeration processes atmospheric air at 27°C and 1 atm. It compresses up to 200 atm. It is expanded to 1 atm and the nonliquefied fraction exits at 24°C. The heat losses in the system are 3 kcal/kg. Three different refrigerants are available. The aim is to obtain a liquid fraction of 0.15 kg per kg of initial air (at least). Each refrigerant has a cost and generates a certain environmental burden. The emission cost for the facility is 50 €/t of CO_2. Select the most appropriate refrigerant (Table P3.11).

Table P3.11 Process Data

	I	II	III
Q (kcal/kg initial air)	15	20	25
Coste (€/kcal refrigerant)	4	5	6
Impact (kg CO_2 produced/kcal refrigerant)	45	40	20

P3.12. A Linde-type air liquefaction system is used in a certain facility. The energy consumption of the plant is 109 kcal/kg of fed air. The liquefied fraction turns out to be 20% when 15 kcal/kg of refrigeration is used, and 29% when using 25 kcal/kg of refrigeration. Assume that the loss of energy is a constant fraction of the refrigeration. The nonliquid air exits at 290K and 1 atm. Determine the operating pressure and the design pressure of the compressor, and also its efficiency.

P3.13. A simple Linde column is used to produce rich oxygen and nitrogen. Over the top vapor with 85% in nitrogen is obtained. It represents 95% of the maximum concentration possible. From the bottom, a stream with 5% nitrogen is produced. Determine the liquefied fraction of the air fed to the column, a feasible air liquefaction cycle, and the number of trays for the column.

P3.14. Determine the refrigeration needed in a cycle (like the one in Fig. P3.14) to produce 0.3 kg of liquid air per kg of initial feed, assuming that the first expansion is down to 20 atm and the second one to 1 atm.

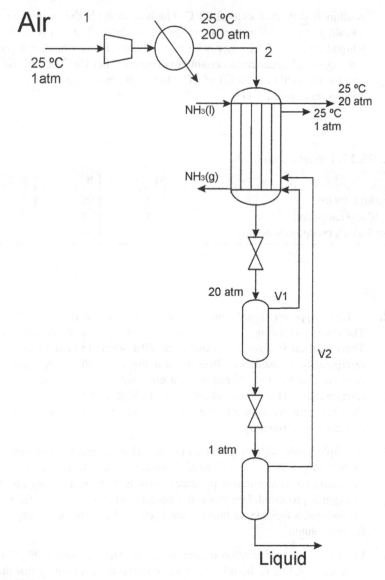

FIGURE P3.14

Flowsheet for the problem.

P3.15. In a Linde simple column system (see Fig. 3.11), air is fed at 300K. It is
compressed up to 200 atm. It is cooled down with the gas products N_2
and O_2, which exit the column at their boiling points of 77K and 90K,
respectively, with the system at 127K. Compute the number of trays

assuming that the nitrogen purity is 95% of the one corresponding to the thermodynamic equilibrium, and that the oxygen contains 5% nitrogen. Assume that the Hauser diagram (Fig. 3.1) holds for all enthalpy computations.

P3.16. Determine the maximum pressure at which a stream at 10°C and a specific humidity of 0.0045 kg_s/kg_{dryair}, and a second stream at 45°C and 0.048 kg_s/kg_{dryair}, can be mixed in any proportion with no condensation.

REFERENCES

Agrawal, R., Rowles, H.C., Kinard, G.E., 1993. Cryogenics. Kirk-Othmer Encyclopedia of Chemical Technology. Wiley.

American Chemical Society International Historic Chemical Landmarks, 1994. Discovery of Oxygen by Joseph Priestley. <http://www.acs.org/content/acs/en/education/whatischemistry/landmarks/josephpriestleyoxygen.html> (accessed 01.03.15.).

Hansel, J.G., 1993. Oxygen. Kirk-Othmer Encyclopedia of Chemical Technology. Wiley.

Hausen, H., 1926. Der Thomson-Joule-Effekt und die zustandsgrossen der Luft. C. Forsch. Arb. Ing.-Wes. Heft 274.

Hardenburger, T.L., Ennis, M., 1993. Nitrogen. Kirk-Othmer Encyclopedia of Chemical Technology. Wiley.

Häussinger, P., Leitgeb, P., Schmücker, B., 1998. Nitrogen. ULLMANN'S "Encyclopedia of Industrial Chemistry". Wiley-VCH.

Kerry, F.G., 2007. Industrial Gas Handbook: Gas Separation and Purification. CRC Press, Boca Raton, FL.

Kirschner, M.J., 1998. Oxygen. ULLMANN'S "Encyclopedia of Industrial Chemistry". Wiley-VCH.

Latimer, R.E., 1967. Distillation of air. Chem. Eng. Prog. 63 (2), 35−59.

Linde, 2015. engineering.com/internet.global.lindeengineering.global/en/images/AS.B1EN%201113%20-%20%26AA_History_.layout19_4353.pdf © Linde AG, extract from 'History and technological progress. Cryogenic air separation.', p. 14−15.

Martín, M., 2014. Introduction to Software for Chemical Engineers. CRC Press, Boca Raton, FL.

Membranes, Shoot for the big time. Chemical Engineering/April 1990, 37−43.

Perry, R.H., Green, D.W., 1997. Perry's Chemical Engineer's Handbook. McGraw-Hill, New York, NY.

Pressure Swing Adsorption Picks Up Steam. 1988. Chemical Engineering/September 26.

Smith, A.R., Klosek, J., 2001. A review of air separation technologies and their integration with energy conversion processes. Fuel Process. Technol. 70, 115−134.

Timmerhaus, K.D., Flynn, T.M., 1989. Cryogenics Process Engineering. Springer, New York.

Vancini, C.A., 1961. La sintesi dell'ammoniaca. Hoepli, Milano.

Vian Ortuño, A., 1999. Introducción a la Química Industrial. Reverté, Barcelona.

Water

4.1 INTRODUCTION

Water is a precious resource in our planet. Its relative availability—71% of the Earth's surface is water—has been responsible for a lack of further efforts towards water consumption efficiency. The total amount of water is $1.38 \times 10^{18}\,\mathrm{m}^3$. Only 2.6% is freshwater, and out of that, 10% alone is accessible, around $3.7 \times 10^{15}\,\mathrm{m}^3$. Table 4.1 shows this information in detail (Ortuño, 1999).

Recently, numerous reports have pointed out the fact that by 2025 two-thirds of the population will suffer water scarcity. Fig. 4.1 presents the distribution of the world's water and its scarcity. We can distinguish between physical scarcity, places where 75% of the available water is withdrawn for use, and economic scarcity, regions where there is enough water, but it is not affordable. Regions where less than 25% of the available water is withdrawn are classified as suffering little or no scarcity, while in those regions approaching scarcity, a limit of 60% withdrawn has already been reached. As a result, the current situation of water availability and use, as well as wastewater treatment technologies and reuse, has led society to a more conscious use of water, at least in developed countries Agthe and Billings (2003).

Water use is crucial for mankind. It represents 70% of the human body, and can be used as an industrialization index. Fig. 4.2 shows water consumption per country.

We must distinguish between water withdrawal and consumption because it represents a common mistake in the nonspecialized literature. Withdrawal refers to the water diverted from the source for its use. Consumption is the amount that does not return, not even as waste. Fig. 4.3 shows the comparison between both concepts across the world. We see that developed countries show efficient ratios between withdrawal and consumption, while developing countries are not that efficient.

The water withdrawal to water consumption ratio can be evaluated by sectors. While industrial and domestic sectors show large withdrawal but small consumption, agriculture shows large water consumption too. Only fresh water—that coming from rivers, lakes, streams, and below-surface water—is directly used,

Industrial Chemical Process Analysis and Design. DOI: http://dx.doi.org/10.1016/B978-0-08-101093-8.00004-5

Table 4.1 Water Distribution on Earth

Region	Volume (10^6 km^3)	(%)
Oceans	1348	97.39
Ice	27.82	2.01
Below surface	8.06	0.58
Rivers and lakes	0.225	0.02
Atmosphere	0.013	0.001
TOTAL	1384	100

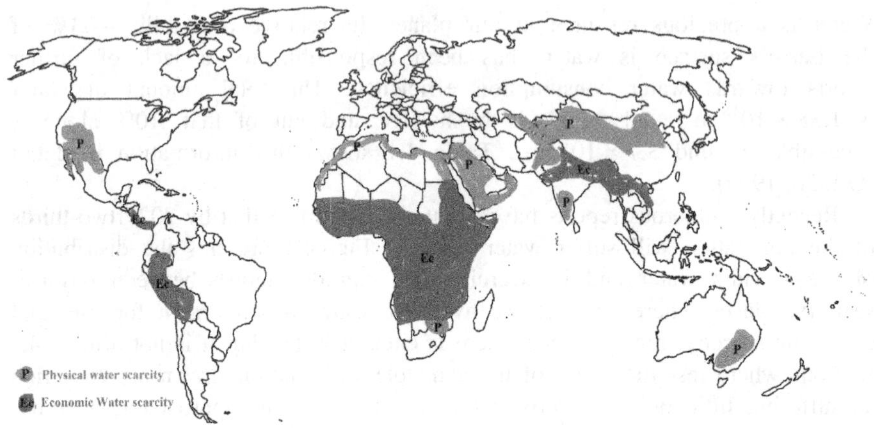

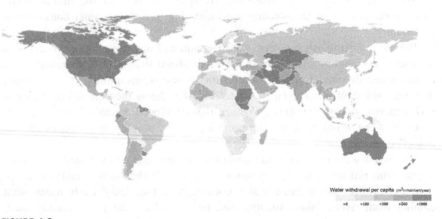

FIGURE 4.1

Water scarcity.

Based on the information of International Water Management Institute (2007)

FIGURE 4.2

Water consumption per capita.

With permission from: http://brittany-harris.com/projects/water-use.html.

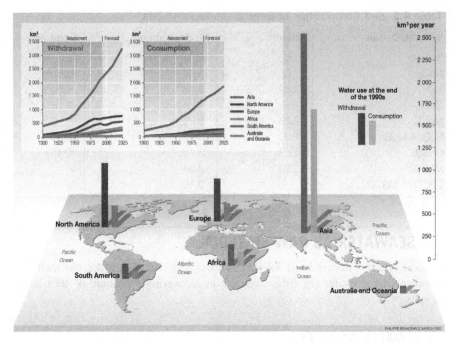

FIGURE 4.3

Water withdrawal and consumption across the world.

Igor A. Shiklomanov. State Hydrological Institute (SHI, St. Petersburg) and United Nations Educational, Scientific and Cultural Organisation (UNESCO, Paris), 1999. World Resources 2000–01, People and Ecosystems: The Fraying Web of Life, *World Resources Institute (WRI). Washington DC, 2000, Paul Harrison and Fred Pearce,* AAAS Atles of population 2001, *American Association for the Advancement of Science, University of California Press, Berkeley.*

otherwise it must be treated. There are only a few cases where seawater has a direct use, such as the growth of certain algae strains. The details of the impurities of water are out of the scope of this chapter, and we encourage the reader to look for specialized literature on the topic; however, it is interesting to highlight the common presence of a number of species:

Suspended species, organic, or inorganic.
Dissolved salts: CO_3^{2-}, NO_3^-, SO_4^{2-}, Cl^-
Dissolved gases: O_2, CO_2

In industry, water is used as a cooling agent, separation medium, raw material for chemical production, and as a mean to transmit pressure.

Table 4.2 Salt Composition of Several Oceans and Seas

% Salts/Ocean	Atlantic	Mediterranean	Dead Sea
% Salts	3.63	3.87	22.3
NaCl	77.03	77.07	36.55
KCl	3.89	2.48	4.57
$CaCl_2$	–	–	12.38
$MgCl_2$	7.86	8.76	45.20
$NaBr = MgBr_2$	1.30	0.49	0.85
$CaSO_4$	4.63	2.76	0.45
$MgSO_4$	5.29	8.34	–
$CaCO_3 + MgCO_3$	–	0.10	–

4.2 SEAWATER AS RAW MATERIAL

Seawater is a raw material and source for water and salts. Therefore, the methods for processing seawater are either focused on recovering the liquid or separating the salts.

4.2.1 WATER SALINITY

The salinity of water comprises the amount of solid matter in grams per kilogram of seawater when the bromine and iodine have been replaced by the equivalent chlorine, and when the carbonate has become oxide and the organic matter has been oxidized completely. In Table 4.2 the main composition of the salts of various oceans is presented. Note that the Dead Sea not only has a large salt content, but also that the salt composition is different due to the large presence of $MgCl_2$.

EXAMPLE 4.1

Compute the osmotic pressure of seawater assuming:

* The salt is NaCl.
* Van't Hoff's equation holds.
* NaCl dissociation degree can be found in Table 4E1.1.

Table 4E1.1 Dissociation Degree of NaCl

[NaCl] (mol/L)	α NaCl	[NaCl] (mol/L)	α NaCl
0.001	0.966	1	0.660
0.01	0.904	2	0.670
0.1	0.780	4	0.680

Solution

In 1885 Van't Hoff stated that for a diluted solution, the osmotic pressure can be computed using the ideal gas equation of state where n_{ef} is the effective concentration due to the actual dissociation degree of NaCl. Let n be the moles of NaCl:

$$NaCl \Rightarrow Na + Cl$$
$$(1 - \alpha) \quad \alpha \quad \alpha$$
$$n_{ef} = n(1 + \alpha)$$

Thus for the *Atlantic Ocean*:

$$C_{NaCl} \cong C_{salts} = 3.63 \frac{g}{10^2 cm^3} \left(\frac{10^3 cm^3}{L} \right) = 36.3 \frac{g}{L} \left(\frac{mol}{58.5g} \right) = 0.6205M$$

$$\alpha_{NaCl} = 0.780 + \frac{0.660 - 0.780}{1.0 - 0.1}(C_{NaCl} - 0.1) = 0.711$$

$$n_{ef} = n(1 + \alpha) = C_{NaCl}(1 + \alpha) = 1.063M$$

$$\pi_{Atl} = \frac{1.063 \cdot 0.082 \cdot 298}{1} = 25.94 \text{ atm}$$

Similarly, for the *Mediterranean Sea*, where the solute concentration is 0.00386 g/cm^3 (see Table 4.2):

$$C_{NaCl} \cong C_{salts} = 0.65983M$$

$$\alpha_{NaCl} = 0.705$$

$$n_{ef} = n(1 + \alpha) = C_{NaCl}(1 + \alpha) = 1.129M$$

$$\pi_{Med} = \frac{1.129 \cdot 0.082 \cdot 298}{1} = 27.54 \text{ atm}$$

4.2.2 SEAWATER DESALINATION

From a historical perspective, the first references to water treatment for purification date back to 2000 BC. It was already known by that time that seawater might be purified, in terms of its taste. For instance, Aristotle stated that saltwater, when turned into vapor, becomes sweet, and that vapor does not form saltwater upon condensation. In China, the use of bamboo and earthenware filters was reported. By 1500 BC in ancient Egypt, coagulation was developed to settle particles using alum. It was not until 500 BC that Hippocrates discovered the healing powers of water and invented bag filters to trap sediments responsible for bad odors. Another ancient method described by Tales and Democritus was the use of sand and gravel filtration. This was reported after analyzing the water that entered through wax walls in a vessel located in the ocean for some time. During the first century, Pliny the Elder (AD 23–79) gathered the information available on natural history, commenting on the methods he was aware of that related to water purification. In China, they used leaves to concentrate the wine at that time. Later in the second century, Alexander of Aphrodisia designed equipment for evaporation and condensation of water—in other words, distillation. During Roman times, civil engineering for water transportation was the main contribution to the history of water processing. Another important discovery was the fact that the ice generated out of seawater was sweet, a discovery made by Thomas Bartholin, Robert Boyle, Samuel Beyher by the 17th century. Over the next two centuries, the development of the already-known solar evaporation, distillation, and freezing was the main contribution. During the 20th century, technology allowed reverse osmosis and ion exchange. However, industrial development has been slow.

The cost of water desalination is determined by a number of factors, such as the process technology, the cost of energy, the unit capacity installed, and the value of the byproducts. Among the methods, two types can be distinguished, those aiming at water separation and those devoted to salt separation. The volume of water is around 30 times that of the salts, thus it would be expected that, from a thermodynamic point of view, it should be cheaper to recover the salts. However, the technologies required for that purpose are more expensive. From an industrial perspective, evaporation, freezing, and reverse osmosis are the most common methods. Table 4.3 presents the energy required to produce 1 kg of water, assumed to be at 20°C. For the sake of simplicity, no ebullioscopic or cryoscopic effects are shown in these calculations. From a purely energy balance point of view, evaporation is by far the most energy-intense method, while reverse osmosis is the most promising. However, this procedure presents some complications related to the membranes and their material. Comparing freezing with evaporation, the latter requires more than six times as much energy, which somehow points to freezing as a better alternative. Again, there are factors beyond energy consumption. For instance, the ice crystals obtained during freezing require a large amount of freshwater to be cleaned, to remove the rest of the salt. Therefore, evaporation and distillation are nowadays the most-used methods. Needless to say, these methods benefit from a knowledge of water thermodynamics.

Table 4.3 Energy Balance for Three Typical Desalination Processes

Method	$Q_{Sensible}$ (kcal/kg)	Q_{latent} (kcal/kg)	Q_{total}(kcal/kg)
Evaporation	80	540	620
Freezing	20	80	100
Reverse osmosis	–	–	3

Table 4.4 Reference Cost for Some Technologies

Technology	Multiple Evaporation	Steam Compression	Reverse Osmosis
Investment ($€/m^3/day$)	800–1000	950–1000	700–900
Cost ($€/m^3$)	75–85	87–95	45–92

Table 4.4 reports some values of the cost of production using different methods. The purer the water, the more expensive, but because of the distillation-based method, the cost is almost independent of the salt content.

4.2.2.1 Technologies based on water separation

There are five core technologies based on water separation: evaporation, freezing, hydrate production, extraction using solvents, and reverse osmosis.

4.2.2.1.1 Evaporation

It is the most widely used technology, and therefore the data available is considerable Billet (1998). Its advantages versus other technologies are as follows:

- High production capacities.
- Low investment cost.
- High heat transfer yield.
- Seawater, properly processed, is not corrosive.

Evaporation-type technologies are based on steam economy. It consists of reusing the steam generated at higher pressure to provide the energy for further evaporation stages so that only one of the effects requires utilities. Thus, there should be a negative gradient of pressure from one evaporation chamber to the next as we progress downstream. This is to maintain the driving force, ie, the temperature difference between the evaporation chamber and the condensation one.

The typical systems are multieffect evaporation systems. Seawater is fed to the first effect, where utility steam is used for the evaporation. The steam produced in the first effect is used to evaporate water out of the brine from the previous effect. As the number of effects increases, the energy integration is better, but the investment increases. Thus the optimal number of effects is typically 12. This gives a steam economy of 10, due to ebullioscopic effects, which is 10 kg of steam generated out of 1 kg of steam utility purchased. Before seawater is processed, bicarbonates and CO_2 must be removed. Fig. 4.4 shows a scheme of a three-effect system. Apart from these, the use of open boilers (directly heated) is interesting due to the possibility of producing salt in the form of big crystals (McCabe et al., 2001).

One particular feature of the evaporation of water from salt solutions is the effect of the salt concentration on the boiling point of the solution. Saline water does not boil at the saturation temperature corresponding to the working pressure,

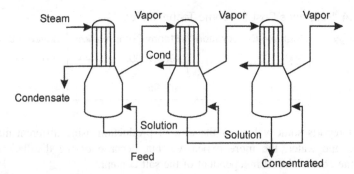

FIGURE 4.4

Three-effect evaporation system.

but above that. A Dühring diagram is used to compute the increase in the boiling point of the solution. Fig. 4.5 shows the Dühring diagram for NaCl solutions from Earle and Earle (2004).

Evaporator design. There are two main types of evaporator design: single-effect evaporators and multieffect evaporators Ocon and Tojo (1967).

Single-Effect Evaporators The aim is to compute the contact area required. We proceed by formulating a mass and energy balance, and apply the design equation for the energy transferred. Fig. 4.6 can be used as reference.

Let F, S, E, and W be the flow rates of the feed, the concentrated solution, the evaporated solvent, and the steam used as heating utility, respectively. H is used for vapor phase enthalpies, and h for liquid phase ones. T is used for temperature, and x_i is the mass fraction of the solute in stream i.

Global mass balance:

$$F = S + E \tag{4.1}$$

Mass balance to the solute:

$$Fx_f = S \cdot x_s \tag{4.2}$$

Enthalpy balance to the evaporator:

$$\sum f_{in} H_{in} = \sum f_{out} \cdot H_{out}$$
$$W \cdot H_w + F \cdot h_f = W \cdot h_w + E \cdot H_e + S \cdot h_s + q \tag{4.3}$$
$$W(H_w - h_w) = E \cdot H_e + S \cdot h_s - F \cdot h_f$$

Although the mass balance is formulated in a generic way, computing the enthalpies of the different streams is a function of the properties of the solutions.

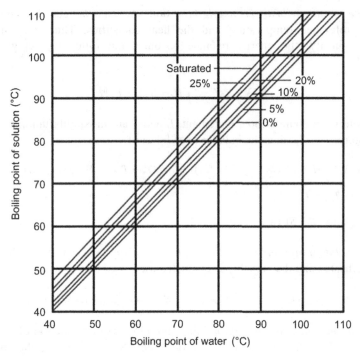

FIGURE 4.5

Dühring diagram for NaCl solutions.

Reproduced with permission from: Earle, R.L., Earle, M.D., 2004. Evaporation. Unit Operations in Food Processing, The New Zealand Institute of Food Science & Technology (Inc.).

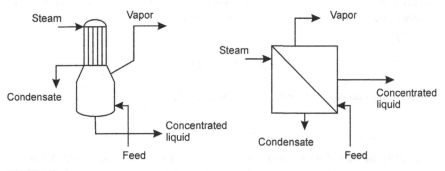

FIGURE 4.6

Scheme of single-effect evaporator.

If the solution heat is negligible, both h_f and h_s can be computed using a reference temperature and the heat capacities. Thus, for saturated steam as utility, the energy balance becomes the following, as given by Eq. (4.4):

$$W(\lambda_w) = (F - S) \cdot H_e + S \cdot h_s - F \cdot h_f \qquad (4.4)$$

If there is no ebullioscopic increment, E and S are in equilibrium at the same temperature, $T_s = T_e$; thus the balance is as follows:

$$W(\lambda_w) = (F - S) \cdot (\lambda_s + c_p(T_s - T_{ref})) + S \cdot c_p(T_s - T_{ref}) - F \cdot c_p(T_f - T_{ref})$$

$$\text{for} \quad T_{ref} = T_s \qquad (4.5)$$

$$W(\lambda_w) = (F - S) \cdot (\lambda_s) - F \cdot c_p(T_f - T_s)$$

If the solution presents an ebullioscopic increment, the steam leaves the chamber superheated and the enthalpy is computed as:

$$H_e = c_{p,vap}(T_s - T_{sat,water}(\text{Pressure})) + \lambda_s + c_p(T_{sat,water}(\text{Pressure}) - T_{ref}) \qquad (4.6)$$

If solution heat is not negligible, then enthalpy is not a function of temperature alone, but also of the concentration. For NaCl solutions, the increase in the boiling point of the solutions is presented in the Dühring diagram in Fig. 4.5. For NaOH solutions, Fig. 4.7 shows the correspondent diagram. Furthermore, the enthalpy of the NaOH liquid solutions can be determined using Fig. 4.8.

For an useful temperature interval, the driving force for the heat transfer in the evaporator is the temperature gradient between chambers. In the evaporation chamber, the temperature is that of the water boiling point plus the ebullioscopic increment due to the solutes dissolved. In the condensation chamber, there is saturated steam that condenses. Thus, the driving force is given by:

$$\Delta T_{useful} = T_{cond} - T_{eb} - \Delta e \qquad (4.7)$$

The design equation for the evaporator is as follows:

$$Q = U \cdot A \cdot \Delta T \qquad (4.8)$$

where U is the global heat transfer coefficient, A, the contact area, and ΔT is the temperature gradient. See example 4.2 for the design of a single-effect evaporator.

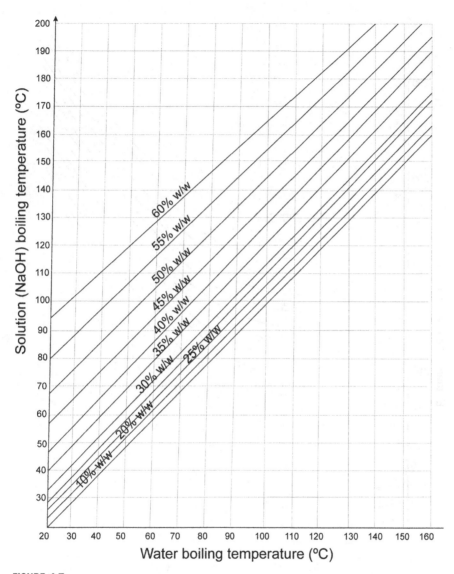

FIGURE 4.7

Dühring diagram for NaOH.

Reproduced with permission from: McCabe, W.L., 1935. The enthalpy-concentration chart: a useful device
for chemical engineering calculations. Trans. AIChE 31, 129–169; McCabe (1935).

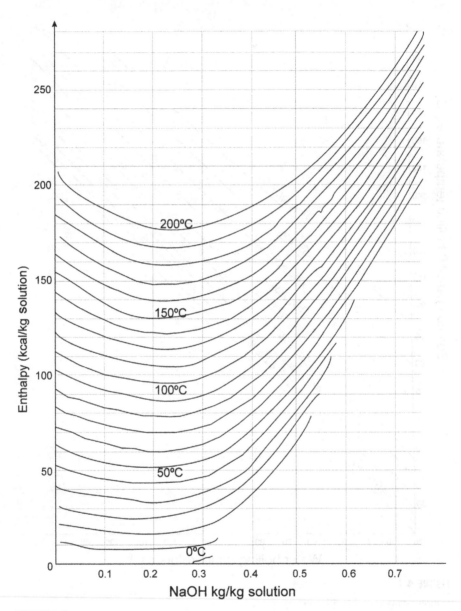

FIGURE 4.8

Enthalpy of NaOH solutions.

Reproduced with permission from: McCabe, W.L., 1935. The enthalpy-concentration chart: a useful device
for chemical engineering calculations. Trans. AIChE 31, 129–169; McCabe (1935).

EXAMPLE 4.2

A simple evaporator processes 10,000 kg/h of a NaOH solution, 20% by weight. A final concentration of 50% is required. The available heating utility is saturated steam at 5 atm, and the evaporation chamber operates at 150 mmHg. The global heat transfer coefficient is 2500 kcal/m²h°C. The feed to the system is at 20°C. Compute:

a. The flowrate of heating utility.
b. The area of the evaporator.
c. The steam economy of the system.

Solution

a. Fig. 4E2.1 shows a scheme of the operation.
   We formulate a mass balance to the solids to determine the concentrated liquid flow rate, L, and the evaporated water, E:

$$F \cdot x_f = L \cdot x_L$$

$$L = \frac{10000\,\frac{kg}{h} \cdot 0.2}{0.5} = 4000\,\frac{kg}{h}$$

$$F = L + E$$

$$E = 6000\,\frac{kg}{h}$$

Next, an energy balance to the system is formulated as follows:

$$W \cdot H_w + F \cdot h_f = W \cdot h_w + E \cdot H_e + L \cdot h_L$$

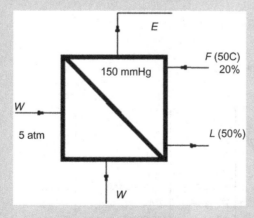

FIGURE 4E2.1

Simple evaporator.

The enthalpies of the liquid streams—the feed and the concentrated liquid, h_f and h_l, respectively—are shown in Fig. 4.8 as a function of their temperature and composition. H_e is the superheated steam enthalpy, which is a function of the ebullioscopic increment in the boiling point. Fig. 4.7 can be used to determine the difference in the boiling point of the product liquid phase, L, taking into account that at 150 mmHg the boiling point should be 60°C. The properties of the saturated steam can be taken from tables. Thus, the energy balance yields:

$$W \cdot \lambda_{w,5atm} + F \cdot h_f(20°C, 20\%) =$$
$$E \cdot H_e(H_{vsat,Teb_{150mmHg}} + 0.46 \cdot \Delta e_{50\%})$$
$$+ L \cdot h_L(Teb_{150mmHg} + \Delta e_{50\%}, 50\%)$$
$$W \cdot 503 \, kcal/kg + 10000 \cdot 14.2 \, kcal/kg = 6000 \cdot 643.4 \, kcal/kg + 4000 \cdot 136 \, kcal/kg$$
$$W = 8400 \, kg/h$$

See Table 4E2.1 for the results.
b. Thus, the heat transfer area is determined as follows:

$$Q = W \cdot \lambda_w = U \cdot A \cdot \Delta T_{util} \Rightarrow A = \frac{W \cdot \lambda_w}{U \cdot \Delta T_{util}} = \frac{4.23 \cdot 10^6 kcal/h}{2500 \dfrac{kcal}{m^2 h °C}(152.3 - 104.7)°C} = 36m^2$$

c. The vapor economy, Ec, of the process is given by the following equation:

$$Ec = \frac{E}{W} = \frac{6000}{8400} = 0.71$$

Table 4E2.1 Results

Heating Utility				
Pressure (mmHg)	3800			
T (°C)	152.315955			
λ (kcal/kg)	503.115296			

Solution				
Pressure (mmHg)	F	L	E (vapor rec)	
T_{ini} (°C)	20	104.7	104.7	
C (%w/w)	20	50		
H_{dis} (kcal/kg)	14.22852	136.374159	643.6	
	P (mmHg)	T (boiling, water)	C (%)	$T + \Delta e$
Outlet	150	60.08°C	50	104.7

Multieffect Evaporators This is the typical configuration for the operation of evaporators. The vapor phase generated in the first effect is used to evaporate water from the solution fed to the second effect. In order for the system to operate, as we go along the evaporator train, the evaporation chamber pressure decreases to maintain a driving force between chambers. There are a number of alternative operating conditions, depending on the direction of the flows and the feed to the evaporators.

There are four types of *feed systems*:

a. Direct Feed. The direction of the vapor and that of the liquid to be concentrated is the same along the evaporator train. Fig. 4.4 is an example of this operation. The pressure at the evaporator chamber decreases in the flow's direction. The mass and energy balances to this system are given by the following system of equations:

$$F - L_n = \sum_n E_i$$
$$W \cdot \lambda_w + F \cdot h_f = E_1 \cdot H_{e1} + L_1 \cdot h_{L1}$$
$$E_1 \cdot (\lambda_{e1} + 0.46\Delta e_1) + L_1 \cdot h_{L1} = E_2 \cdot H_{e2} + L_2 \cdot h_{L2}$$
$$E_2 \cdot (\lambda_{e2} + 0.46\Delta e_2) + L_2 \cdot h_{L2} = E_3 \cdot H_{e3} + L_3 \cdot h_{L3}$$
$$H_{ei} = H_{\text{sat,boilingwater}} + c_p \Delta e_i$$
$$h_{L,i} = h_{\text{dis}}(\%_i, T_{b,\text{water}} + \Delta e_i)$$

(4.9)

b. Countercurrent feed. The direction of the heating vapor and that of the liquid to be concentrated are opposite. The liquid is fed to the last effect. See Fig. 4.9A.
c. Mixed feed. Part of the system is fed following a direct scheme and the rest as countercurrent. See Fig. 4.9B for a scheme.
d. Feed in parallel. The liquid to be concentrated is simultaneously fed to all effects.

There is a *useful temperature gradient* in a multieffect system. The actual driving force is computed as the difference between the condensing temperature at the steam chamber and the boiling point of water at the last of the effects. From this difference we have to deduct the ebullioscopic increments in the various evaporators of the system. Thus, the useful gradient of temperature decreases with the number of effects, reducing the number of effects that can be used. The advantage of the multieffect evaporator system is a lower energy consumption, while its individual efficiency is lower than a single-effect system.

$$\Delta T_{\text{useful}} = T_o - T_n - \sum \Delta e_i \tag{4.10}$$

There are two approaches that can be used for the *design of a multieffect evaporator system*. There is the computational-based one, depicted in

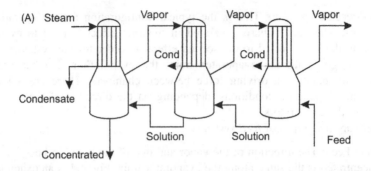

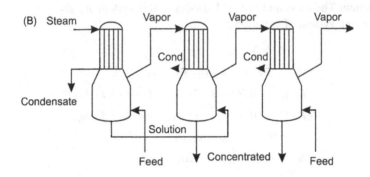

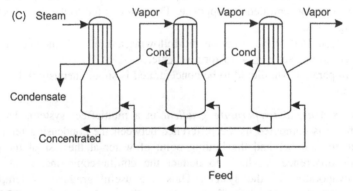

FIGURE 4.9

Alternative feed structures: (A) Counter current feed, (B) Mixed feed, and
(C) Parallel feed.

Martín (2014), where advanced modeling and simulation techniques were
required to include the thermodynamic models of the mixtures; or the traditional iterative approach, based on the schemes in Figs. 4.7 and 4.8. For the
sake of argument, here we only discuss the second one and leave it to

the reader to investigate the modeling-based approach. The iterative approach consists of six stages:

1. Perform a mass balance to the system to compute the total vapor generated:

$$F = L + E \tag{4.11}$$

2. Assume that all the effects are equal, including their size and the energy exchanged. Thus we distribute the useful driving force among the effects inversely proportional to the global heat transfer coefficient.
3. Formulate the energy balances to all effects.
4. Solve the system of equations evaluating the evaporated flowrate at each one of them.
5. Compute the contact area of each effect. If they are the same, we are done. If not, we must proceed to stage 6.
6. Redistribute the driving force as follows:

$$\Delta'T_1 = \Delta T \frac{q_1/U_1}{\sum q/U}; \quad \Delta'T_2 = \Delta T \frac{q_2/U_2}{\sum q/U} \tag{4.12}$$

EXAMPLE 4.3

In the production of NaOH from Na_2CO_3, a 12% w/w solution is obtained. Its concentration for further use must be 38%. Thus a three-effect evaporation system is employed. The initial feed consists of a flowrate of 13,500 kg/h at 25°C, and it is fed to the first effect in direct-feed operating mode. Saturated steam at 4 kg/cm² is available as heating utility. It leaves the first effect as saturated liquid. The absolute pressure at the evaporation chamber in the first effect is 0.1 kg/cm² and the global heat transfer coefficients are 1460, 1250, and 1050 kcal/m² h°C, respectively.

Assuming that all the effects are similar, and that we have as fuel a gas mixture that is 90% methane and 10% nitrogen (by volume) at 25°C and 1 atm, determine the flowrate of fuel (m³/h) required for the operation of the evaporation system if the combustion efficiency is 80%.

Solution

Fig. 4E3.1 shows the scheme of the process for concentrating NaOH using a three-effect evaporator system. We follow the method proposed above. Using a mass balance we compute the product, $L = 4263$ kg/h, and the total vapor, $E = 9237$ kg/h. Assuming that all the effects work similarly, the evaporated water at each effect is $E_i = 3079$ kg. With this and the mass balances to each of the effects:

$$L_{i-1} = L_i + E_i$$

$$L_{i-1} \cdot x_{Li-1} = L_i \cdot x_{Li}$$

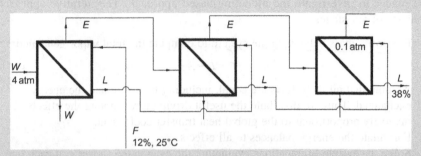

FIGURE 4E3.1

Three effect evaporation system.

The composition of the liquid flows is as follows:

$$x_1 = 0.155, \ x_2 = 0.22, \ x_3 = 0.38$$

Using Fig. 4.7, the ebullioscopic increments for the three-liquid composition are 3°C, 5°C, and 23°C, respectively. The boiling point at the third effect pressure, 0.1 atm, is 46°C.

Thus, the useful temperature gradient is as follows:

$$\Delta T_{\text{useful}} = 144 - 46 - 3 - 5 - 23 = 67°C$$

Redistributing the temperature gradient inversely proportional to U we have the following:

$$\Delta T_1 = 18.7°C, \ \ \Delta T_2 = 21.8°C, \ \ \Delta T_3 = 25.97°C$$

We now build Table 4E3.1. Starting from the highest temperature, that of the saturated steam, we go down bit-by-bit, computing the temperatures at the different effects and their boiling temperatures. Note that most of the energy transfer from the supersaturated vapor to the next effect takes place at the boiling point, and thus that temperature is used to compute the ΔT for area calculation purposes. With these values in the second column of Table 4E3.1, and the composition of the liquid phase, we compute the enthalpies. For instance, the latent heats are evaluated at the boiling temperatures (Tb), while the enthalpies of the vapor phases are those of the superheated vapor at the chamber temperature (T).

We now can solve the energy balances to each of the effects to compute W, E_1, E_2, and E_3:

$$W \cdot 510 + 13500 \cdot 20 = (13500 - E_1) \cdot 111 + E_1 \cdot 646$$

$$E_1 \cdot 527.5 + (13500 - E_1) \cdot 111 = 637E_2 + (13500 - E_1 - E_2) \cdot 89$$

$$E_2 \cdot 545 + (13500 - E_1 - E_2) \cdot 75 = 617E_3 + 4263 \cdot 75$$

$$E_1 + E_2 + E_3 = 9237$$

Table 4E3.1 Enthalpy Data for Example 4.3

	T (°C)	λ (kcal/kg)	Reheat	Total (kcal/kg)	H$_{dis}$ (kcal/kg)	H$_{vap}$ (kcal/kg)
Vapor 1	144	510	0	510		
ΔT_1	18.7					
Tl	125.2				111	
Δe_1	3					
Tb I	122.2	526	1.5	527.5		646
ΔT_2	21.8					
T II	100.4				89	
Δe_2	5					
Tb II	95.4	543	2.5	545		637
ΔT_3	26					
T III	69.4	559			75	
Δe_3	23					
Tb III	46.4	571	11.5	583		617

Table 4E3.2 Results for Example 4.3

1st Effect			2nd Effect			3rd Effect		
Steam/Vapor								
P (mmHg)	3040	3041.25626		1278.69079				571.573237
T (ºC)	144.1		T (ºC)	115.261369		T (ºC)	92.2082559	
λ	509.22488		λ	529.211242		λ	543.869365	

Solutions	F	L	E (vapor rec)	F	L	E	F	L	E
T (ºC)	25	119.751286	119.751286	119.751286	99.3782865	99.3782865	99.3782865	69.3534171	69.3534171
C	0.12	0.15129444		0.15129444	0.21393333		0.21393333	0.38	
hi (kcal/kg)	20.162124	109.122095	646.669776	109.122095	88.8793955	639.347591	88.8793955	75.8350406	627.853927
Flow (kg/h)	13500	10707.5974	2792.40261	10707.5974	7572.45245	3135.14494	7572.45245	4263.15789	3309.29456
U (kcal/m2 h)	1460			1250			1050		
A (m2)	76.0077598			76.0077598			76.0077598		

Chamber	P(mmHg)	Tboiling (ºC)	C(%)	T+Δe (ºC)	P(mmHg)	Tboiling (ºC)	C(%)	T+Δe (ºC)	P(mmHg)	Tboiling (ºC)	C(%)	T+Δe (ºC)
Out		115.261369	15.1294445	119.7513		92.2082559	21.3933334	99.37829		76	46.2	38	69.3534171
	1278.69079				571.573237				76.3419712			

Dtutil	24.3487139			15.8830825			22.8548388	

Energy Balance 1st effect		Energy Balance 2nd Effect		Energy Balance 3rd Effect	
W=	5306.12163	Entra	2677482.36	Entra	2401050.33
Entra	2974197.83	Sale	2677482.36	Sale	2401050.33
Sale	2974197.83	Q (Wlamb)	1509046.9	Q (Elamda)	1824002.35
Q (Wlamb)	2702009.15	Q(UADT)	1509046.9	Q(UADT)	1824002.35
Q(UADT)	2702009.15	Economy	1.74078974		

Thus we obtain:

$$W = 5100 \text{ kg/h}; \quad E_1 = 2566 \text{ kg/h}; \quad E_2 = 2921 \text{ kg/h}; \quad E_3 = 3751 \text{ kg/h}$$

We see that the evaporated flowrates are not the same, and neither are the areas. Thus, we redistribute the temperatures again and go back. Table 4E3.2 presents the results of the operation of the system.

The energy to be provided:

$$W\lambda_w = 2,683,000 \text{ kcal}/h.$$

This energy is obtained from natural gas. The stoichiometry is as follows:

$$CH_4 + 0.11N_2 + 2(O_2 + 3.76N_2) \rightarrow CO_2 + 2H_2O + (0.11 + 1.5 \cdot 3.76)N_2$$

Assuming liquid water, the energy produced in the combustion of the gas is computed as follows:

$$\Delta H_r = \sum \Delta H_f \Big|_p - \sum \Delta H_f \Big|_r = -94.052 + 2(-57.8) + 17.889 = -191.76 \text{ kcal/mol}$$

Assuming an efficiency of 80%, the flowrate of gas required is given by the following expression:

$$n = \frac{2.68 \cdot 10^6}{0.8 \cdot 191.76} = 17500 \frac{\text{mol}_{CH_4}}{h} \equiv 19700 \frac{\text{mol}_{gas}}{h} \equiv 480 \text{ Nm}^3/h$$

Sometimes in the evaporation the liquid is saturated and crystals are formed. Therefore, apart from evaporators, the units act as crystallizers. *Crystallization* is a separation based on the formation of a pure solid. In the case of penicillin, it is from a solution and it occurs by supersaturation. In this way the solubility of the solid in the solution is overcome. There is a phase equilibrium between the solid and liquid phases. Three regions are found: undersaturated, metastable, and labile solutions. Only in this last region are new crystals formed spontaneously, by nucleation. In the metastable region crystals can grow, while in the undersaturated region any crystals dissolve. Supersaturation can be accomplished by cooling/heating, via evaporation of the solvent at constant temperature, or by using a mixed process, adiabatic evaporation. Alternatively, supersaturation can be achieved via precipitation using a chemical or salting-out by adding a second solvent.

The analysis of crystallization processes is carried out using population balances, accounting for the generation, growth, and death of the crystals. This state can be modeled using a mass transport balance of the species from the bulk to the solid. Assuming that the volume and the surface area are proportional to a characteristic length, we obtain:

$$\rho_c \alpha L^2 \frac{dL}{dt} = \frac{d\rho_c V}{dt} = \frac{dM}{dt} = A \cdot N_A = A \cdot K_C \cdot \Delta C^n = \beta L^2 \cdot K_C \cdot \Delta C^n = G \qquad (4.13)$$

This holds if there is uniform supersaturation of the media and there is no crystal breakage (death); thus the growth rate is independent of the size, ie, McCabe's ΔL law. It is widely used in crystallization modeling. By performing a population balance to a perfectly mixed tank, the concentration of particles of size L, assuming mixed product removal, $n_o = n$, no seeding and no breakage of crystals, and that McCabe's ΔL law holds:

$$\frac{\partial Vn}{\partial t} = \frac{\partial n}{\partial t} + n\frac{\partial \ln V}{\partial t} = \frac{n_i}{\tau_i} - \frac{n_o}{\tau_o} - \frac{\partial Gn}{\partial L} + B - D \Rightarrow \frac{dn}{n} = -\frac{dL}{G\tau} \rightarrow n = n_o e^{-L/(G\tau)} \quad (4.14)$$

where G is the growth rate and τ the residence time (V/Q). The design of these crystallizers is carried out by solving:

- Mass balance to the solute.
- Global mass balance. Accounting for evaporation rate.
- Residence time. Compute the size of the tank.
- Nucleation rate.
- Growth rate.
- Population balance. Compute total mass of crystals, $M = 6\alpha\rho_p n_o (G\tau)^4$.
- Modal size of the crystal. The dominant crystal size, L_D, is $3\ G\tau$.

EXAMPLE 4.4

A three-effect evaporator system (see Fig. 4E4.1) is fed with 30 t of brine, 15% by weight of NaCl. From the first effect, 12 t of water are evaporated. From the second, 8 t are evaporated, and 6 t from the last one. The salt crystals are recovered in a conic hopper. During this operation, the level inside the evaporator decreased. The cross-sectional area of the evaporators is 15 m^3/m. Table 4E4.1 shows the operating data of the system. Compute the mass of NaCl obtained at each evaporator and the concentrated solution obtained from evaporator III, if its final composition has to be 35%.

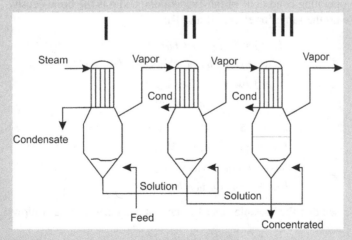

FIGURE 4E4.1

Scheme of three-effect evaporators with NaCl crystallization.

Table 4E4.1 Operating Parameters of the Multieffect Evaporator System

	Evaporator		
	I	II	III
Feed (t)	30		
Feed composition	0.15		
Level decrease (m)	0.04	0.05	0.03
Evaporated water (t)	12	8	6
Solubility of NaCl % weight	30	32	35
Concentrated compositions % weight	18	22	35
Density (t/m3)	1.140	1.145	1.150

Solution

We formulate the mass balances to the solid (a), the liquid (b), the brine (c), and the crystals (d), for each effect as follows:

$$F \cdot x_f + B_{C,I} = C_I + L_I x_{s,I} \qquad (a)$$

$$F \cdot (1 - x_f) + B_{Wa,I} = Vap_I + L_I(1 - x_{s,I}) \quad (b)$$

$$B_{C,I} + B_{Wa,I} = h_I \cdot A \cdot \rho_{Brine} \qquad (c)$$

$$B_{C,I} = sol_I \cdot (B_{C,I} + B_{Wa,I}) \qquad (d)$$

where B_C is the solids in the brine and B_{Wa} is the water in the brine; C is the crystals generated; F, L, and Vap are the flows of liquid fed, concentrated solution, and vapor; x is the salt concentration; h is the level of brine descended; A is the cross-sectional area; sol is the solubility of the salt in the operating conditions of the evaporators; and ρ is the brine density.

We do the same for effects II and III:

$$L_I \cdot x_{s,I} + B_{C,II} = C_{II} + L_{II} x_{s,II}$$

$$L_I \cdot (1 - x_{s,I}) + B_{Wa,II} = Vap_{II} + L_{II}(1 - x_{s,II})$$

$$B_{C,II} + B_{Wa,II} = h_{II} \cdot A \cdot \rho_{Brine}$$

$$B_{C,II} = sol_{II} \cdot (B_{C,II} + B_{Wa,II})$$

$$L_{II} x_{s,II} + B_{C,III} = C_{III} + L_{III} x_{s,III}$$

$$L_{II}(1 - x_{s,II}) + B_{Wa,III} = Vap_{III} + L_{III}(1 - x_{s,III})$$

$$B_{C,III} + B_{Wa,III} = h_{III} \cdot A \cdot \rho_{Brine}$$

$$B_{C,III} = sol_{III} \cdot (B_{C,III} + B_{Wa,III})$$

We see that the variables that we have at each effect are as follows:

$$B_{C,i}, C_i, B_{Wa,i}, L_i$$

Table 4E4.2 Operating Parameters of the Multieffect Evaporator System

	I		II		III
$F =$	30	$L_I =$	20.06	$L_{II} =$	11.59
$x_f =$	0.15	$x_{s,I} =$	0.18	$x_{s,II} =$	0.22
$B_{C,I} =$	0.2052	$B_{C,II} =$	0.27	$B_{C,III} =$	0.1811
$B_{a1} =$	0.4788	$B_{a2} =$	0.5839	$B_{a3} =$	0.3364
$C_I =$	1.43	$C_{II} =$	1.06	$C_{III} =$	0.73
$L_I =$	17.07	$L_{II} =$	11.59	$L_{III} =$	5.19
$x_{s,I} =$	0.18	$x_{s,II} =$	0.22	$x_{s,III} =$	0.35
Vap1 =	12	Vap2 =	8	Vap3 =	6
h_I (m)	0.04	h_{II} (m)	0.05	h_{III} (m)	0.03
$\rho 1$ (t/m^3)	1.14	$\rho 2$ (t/m^3)	1.145	$\rho 3$ (t/m^3)	1.15
Area (m^3/m)	15	Area (m^3/m)	15	Area (m^3/m)	15
Sol	0.3	Sol	0.32	Sol	0.35

We can actually solve each evaporator one after the other, or the three simultaneously. See Table 4E4.2 for the results.

Therefore we obtain 3.23 t of crystals and a concentrated solution of 5.19 t of water 35% of NaCl.

Vapor recompression. This modification involves the compression of the vapor generated in the evaporation. The compressed vapor is used as heating utility.

Solar evaporators: The operation of this units is based on gas—liquid separation with no boiling. The system consists of pools of water covered with plastic or glass. The water evaporates to a dry air so that the distance to saturation is the driving force for the process. Solar energy is the source for heating up the water, and the evaporated water condenses on the surface of the films or the glasses. Proper design allows collecting the condensed drops. Currently, work is being carried out to overcome losses in efficiency due to the reflection of the transparent film, the drops on the surface, the water surfaces and that of the ground, and the radiation from the water bulk. Fig. 4.10 shows schemes of several designs.

4.2.2.1.2 Freezing

Water desalination via freezing is based on the generation of a liquid—solid interphase that is permeable to water only. The solid is obtained by freezing a layer of water. The ice crystals remain in the water and must be recovered. The drawbacks of this technology are that on the one hand there is salt trapped within the crystal, and on the other hand the salt also remains on the crystal surface and freshwater is needed for washing.

Taking a look at Fig. 4.11, the water phase diagram, we realize that by reducing the pressure and removing the heat, it is possible to freeze water.

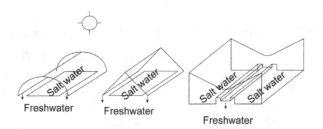

Freshwater   Freshwater   Freshwater

FIGURE 4.10

Solar evaporators.

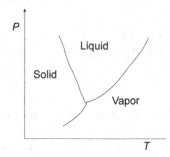

FIGURE 4.11

Water phase diagram.

Furthermore, by compressing a vapor phase at constant temperature, it is also possible to obtain crystals. There are two main methods for doing this:

a. Water expansion. Vacuum freezing. Seawater is frozen at $-4°C$ and 3 mmHg. In the evaporation it cools down, generating the ice crystals. To maintain the vacuum, the vapor generated in the expansion must be removed either using a compressor or by absorbing it into a hydroscopic solution. The large amount of vapor generated is a technical challenge.

b. Use of a refrigerant. We use a refrigerant whose vapor pressure is above that of water and that is not miscible with it; for instance, butane. When it is expanded, the cooling process freezes the water, generating ice crystals that are easily separated. Fig. 4.12 shows the cycle. The ice is melted, cooling down the refrigerant after recompression, and the water is either used to wash the crystals or as product.

4.2.2.1.3 Hydrate production

This method consists of the production of hydrates by combining water and halogenated organic chemicals. These species can be separated from the brine and later decomposed to recycle the organic species.

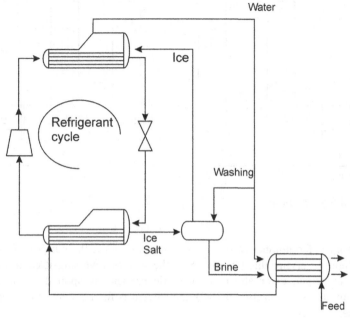

FIGURE 4.12

Freezing desalination.

4.2.2.1.4 Extraction using solvents

An organic solvent, immiscible with the brine and partially miscible with fresh water, is used for extraction.

4.2.2.1.5 Reverse osmosis

Reverse osmosis is the process by which a semipermeable membrane is used to separate or purify a liquid by permeation King (1980). The concentrated solution cannot permeate it. This technology allows the highest level of separation between salts and water, and it is based on the natural operation of living organisms. The osmosis process was the one observed. When two solutions of different concentration are put into contact across a membrane, water flows from the less concentrated chamber to dilute the second solution (Fig. 4.13A). This flow only stops when the pressure at both sides reaches an equilibrium (Fig. 4.13B). This phenomenon is known as osmosis. Osmotic pressure is so-called because the increase in the volume, and thus in the liquid height, generates a difference in hydrostatic pressure.

The osmotic pressure for two components (the solute and the liquid) in the equilibrium across a membrane can be estimated as follows:

$$\Delta P_{eq} = -(RT/VA)\ln \beta A \tag{4.15}$$

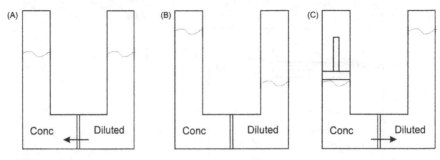

FIGURE 4.13

Reverse osmosis.

For a dilute solution:

$$\ln \beta A = -(n_B/n_A) \tag{4.16}$$

where n_A is the number of moles of solvent and n_B is the number of moles of solute. Upon observing this phenomenon, the idea of reversing it came to mind. If a pressure equal to or above the osmotic pressure is applied to the side of the concentrated solution (Fig. 4.13C), the solvent is forced across the membrane, ie, reverse osmosis.

Separation factors. In reverse osmosis there are two driving forces. On the one hand, there is the pressure difference. On the other hand, there is the gradient in solute concentration. Thus, the flow of each of the species, the solvent (A) and the solute (B), can be calculated as follows:

$$N_A = K_A(\Delta P - \Delta \pi) \tag{4.17}$$

$$N_B = K_B(C_{B1} - C_{B2}) \tag{4.18}$$

where N_A and N_B are the flows of both components across the membrane (A represents water and B represents the salt).

ΔP: Pressure difference across the membrane.
$\Delta \pi$: Osmotic pressure difference across the membrane.
C_{B1}, C_{B2}: Salt concentration at both sides of the membrane.
K_A, K_B: Empiric constants as a function of the membrane and the salts.

In the equilibrium the ratio $\dfrac{C_{A2}}{C_{B2}}$ should be proportional to $\dfrac{N_A}{N_B}$. Thus dividing both expressions we have:

$$\frac{C_{A2}}{C_{B2}} = \frac{N_A}{N_B} = \frac{K_A(\Delta P - \Delta \pi)}{K_B(C_{B1} - C_{B2})} \tag{4.19}$$

Assuming that the membrane does not allow the salt to cross, the following relationships hold:

$$C_{B2} \to 0 \text{ and } C_{B2} \ll C_{B1}$$

Thus, Eq. (4.19) becomes:

$$\alpha_{A-B} = \frac{C_{A2}}{C_{B2}} = \frac{N_A}{N_B} = \frac{K_A(\Delta P - \Delta \pi)}{K_B(C_{B1})} \qquad (4.20)$$

Rearranging the variables we have Eq. (4.21):

$$\frac{C_{A2}C_{B1}}{C_{B2}} = \frac{K_A(\Delta P - \Delta \pi)}{K_B} \qquad (4.21)$$

Thus, the separation factor f becomes:

$$\frac{C_{A2}C_{B1}}{C_{A1}C_{B2}} = \frac{(C_B/C_A)_1}{(C_B/C_A)_2} = \frac{K_A(\Delta P - \Delta \pi)}{K_B C_{A1}} = f \qquad (4.22)$$

where $1/C_{A1}$ is equal to $1/\rho_A$, water density, which is constant.

Membrane characteristics. Membranes are the key element in reverse osmosis. They allow the flow of solvents and reject salts. Their filtration capacity depends on the chemical composition of the fluid to be processed and the interaction with the solute. Furthermore, the material behaves differently when the process takes place at different temperatures and pressures. Finally, the amount of solids to be removed also affects the purification process. Therefore, there are some general characteristics that a material must have in order to be used as a membrane:

- It must be highly permeability to freshwater and must reject salts.
- It should be resistant to chemicals.
- It must have high stability over a wide range of pH.
- It must have high resistance to pressure.
- It must have high endurance.

Types of membranes by materials. *Inorganic membranes* are made of ceramic materials that present high chemical stability. However, they are fragile and expensive, and therefore their use is only recommended for high-temperature processing.

There are two types of *organic membranes* that can be distinguished as a function of their chemical composition: cellulosic membranes and aromatic polyamide membranes.

Cellulosic membranes are based on cellulose acetate, and are appropriate for processing large flow rates. They can be arranged as pipes, as plane sheets in spiral, or as hollow fibers.

Aromatic polyamide membranes typically process smaller flowrates, but their specific surface is 15 times larger than that provided by their cellulosic counterparts. Furthermore, they are more stable against chemical and biological agents.

Membrane configurations. There are four membrane configurations: modules with plane membranes, modules with tubular membranes, modules with spiral membranes, and modules of hollow fiber.

Modules with plane membranes: This configuration is no longer in use due to its high price. It typically provides $50-100$ m^3/m^3 and pressure drops of $3-6$ kg/cm^2.

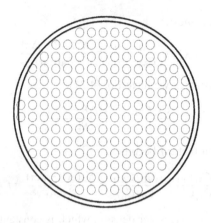

FIGURE 4.14

Tubular membranes.

A preliminary filtration is required to remove suspended solids, and the membrane must be supported. The regeneration requires high-pressure water or the use of chemicals. The product purity is high.

Modules with tubular membranes: These membranes are allocated inside porous tubes that provide support. The surface area that they provide is $50-70$ m^2/m^3. Pressure drops of around $2-3$ kg/cm^2 are also typical. These modules do not require previous filtration, and they can be regenerated chemically, mechanically, or using pressurized water. The cost is also typically high. Fig. 4.14 shows the packing of these membranes inside the tube.

Modules with spiral membranes (axial or radial flow): These modules are built by surrounding a permeable tube with the membranes separated by porous material. They allow good purification, and different structures and materials are employed such as spiral polyamide and spiral cellulose acetate and triacetate. The modules can be standard or for high rejection. In this last case the production capacity is lower in order to increase the purity of the product by rejecting $90-99\%$ of the NaCl. The surface area provided is $600-800$ m^2/m^3 for a pressure drop of $3-6$ kg/cm^2. They require preliminary filtration to remove particles from 10 to 20 μm. Membrane cleaning can be carried out with pressurized water or using chemicals. The cost is lower than the previous configurations, but they need support for the membrane. Fig. 4.15 from AMTA shows the scheme of the configuration, operation, and construction. The feed solution permeates across the layers of membrane material until it reaches the center tube that recovers the permeate. The rest remains in the membrane.

Modules of hollow fiber: The surface area they provide is large, $6000-8000$ m^2/m^3, with a low-pressure drop of $0.2-0.5$ kg/cm^2. These modules require preliminary filtration to remove particles from 5 to 10 μm. They can be cleaned chemically or with pressurized water. The cost is low and they do not need support for the membrane. Fig. 4.16 shows a scheme.

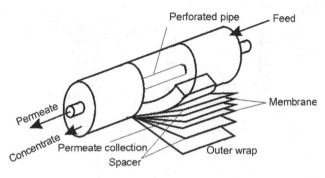

FIGURE 4.15

Spiral membranes.

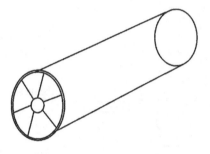

FIGURE 4.16

Hollow fiber membrane.

Operation of the membrane modules. Membrane modules can be operated in several configurations: parallel, series, and series production.

In *parallel* the units operate under the same pressure conditions, producing the same product quality. Typically after the filters, pH is adjusted and chemicals are added. Fig. 4.17 shows the scheme of the installation.

In *series* the material rejected from the previous unit is processed again to increase production capacity. Fig. 4.18 shows the scheme of the installation.

In *series production* they are meant to produce high-quality products by processing the permeate of the first system again. The concentrated solution for the second system is recycled to the first one. Fig. 4.19 shows the scheme of the installation.

The modules can operate at three different pressures, depending on the feed to process:

- Low-pressure (10–15 atm) modules: Appropriate for low salinity brines (800–1000 ppm). Ion exchange resins can compete with them.
- Medium-pressure modules (28–35 atm): For salt solutions of 4000–5000 ppm. These were the first on the market.
- High-pressure modules (35–80 atm): Designed to obtain freshwater from seawater since the osmotic pressure is typically from 22 to 27 atm.

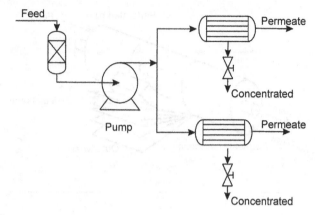

FIGURE 4.17

Parallel operation of membrane systems.

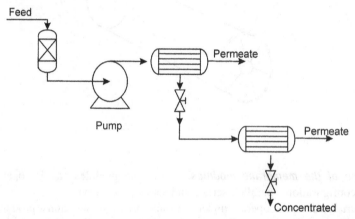

FIGURE 4.18

Series operation of membrane systems.

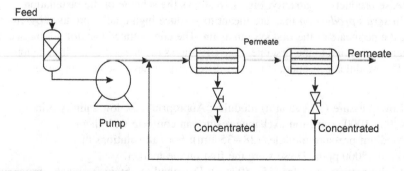

FIGURE 4.19

Series production configuration.

To design the membrane modules, we use example 4.5 below. The design equations are given by the flow of solvent and that of salts as presented above in the text.

EXAMPLE 4.5

Design of reverse osmosis systems. The system is fed with a flow rate of 2.5 kg/s at 8000 kPa. The permeate intended is 1 kg/s at 101 kPa, and the rejected is at a pressure of 7800 kPa.

Data:
Permeability to water: 2.095×10^{-6} kg/m^2 s kPa
Permeability to salt: 3.153×10^{-5} kg/m^2 s
Initial salinity: 42.000 ppm
Permeate salinity: 145 ppm

Solution

We formulate a global mass balance (M) where f, p, and R refer to feed, permeate, and retentate, respectively. Thus, the flow of retentate is as follows:

$$M_f = M_p + M_R$$

$$2.5 = 1 + M_R \rightarrow M_R = 1.5 \text{ kg}/s$$

Now, a mass balance to the solute is formulated to compute the concentration:

$$M_f \cdot x_f = M_p \cdot x_p + M_R \cdot x_R$$

$$x_R = 69903 \text{ ppm}$$

We assume that all the salts are NaCl. Thus the osmotic pressure of the three solutions is as follows:

$$\pi_f = (x_f / M_{NaCl}) \cdot 1000 \cdot 8.314 \cdot 298 = 1779 \text{ kPa}$$

$$\pi_R = 2960 \text{ kPa}$$

$$\pi_p = 0.006 \text{ kPa}$$

Bear in mind that in the side of the feed there is a gradient of concentrations since the flow is concentrated. Thus, an average osmotic pressure is calculated as follows:

$$\pi_f' = 0.5(\pi_f + \pi_R) = 2369 \text{ kPa}$$

And the gradient in osmotic pressure between the permeate and the feed region is as follows:

$$\Delta\pi = \pi_f' - \pi_p = 2369 \text{ kPa}$$

The pressure gradient applied to the membrane module is as follows, taking into account that at the feed side there is also a pressure gradient:

$$\Delta P = 0.5 \cdot (8000 + 7800) - 101 = 7799 \text{ kPa}$$

Thus, the permeate flow is as follows:

$$M_p = A \cdot kw \cdot (\Delta P - \Delta\pi)$$

That allows computation of the area required:

$$(1) = A(2.095 \times 10^{-6}) (7799 - 2369)$$
$$A = 87.9 \text{ m}^2$$

We can validate the design by using the salts flow. The mean salinity in the feed side is as follows:

$$x = (M_f \cdot x_f + M_R \cdot x_R)/(M_f + M_R)$$
$$= ((2.5)(42) + (1.5)(69.903))/(2.5 + 1.5)$$
$$= 52.46 \text{ kg/m}^3$$

Using the design equation for the solids, we have:

$$M_s = A \cdot ks \cdot (x - x_p)$$

We compute the area, and the results are the same as before:

$$(1)(0.145) = A(3.153 \times 10^{-5})(52.46 - 0.145)$$

$$A = 87.9 \text{m}^2$$

Reverse osmosis has a wide range of applications apart from seawater desalination; for instance, the cooling cycles in chemical plants and boiler operation both require pure water. In the operation of these systems, water is evaporated and thus salts are concentrated in the solution. Therefore, salt removal is required at some point.

4.2.2.2 Technologies based on the separation of salts

There are three technologies based on the separation of salts: electrodialysis, ion exchange, and chemical depuration.

4.2.2.2.1 Electrodialysis

This is used to transport salt ions from one solution through ion exchange membranes to another solution under the influence of an applied electric potential difference. The efficiency is computed using the following expression:

$$\xi = \frac{ZFQ(C^d_{inlet} - C^d_{outlet})}{I} \tag{4.23}$$

where Z is the charge of the ion, F is the Faraday constant, Q is the flow rate of the diluted solution (L/s), C is the concentration of the diluted solution at the inlet and at the outlet (mol/L), and I is the current in amperes.

4.2.2.2.2 Ion exchange

It is used to capture certain ions as a function of their charge. Thus, cationic and anionic resins are available. The exchange between the resin and the ions is as follows for both types of resins, respectively:

$$\text{Cationic} \quad R - H + M^+ \leftarrow \rightarrow R - M + H^+$$
$$\text{Anionic} \quad R - OH + A^- \leftarrow \rightarrow R - A + OH^-$$

The ion exchange beds cannot handle brines with concentrations above 3500 ppm. Therefore, the applicability is restricted to water cycles in boilers and low salinity aquifers with high content in calcium and magnesium.

4.2.2.2.3 Chemical depuration

In this process, seawater is treated with Cl_2 and $CuSO_4$ to precipitate the organic matter. Next, CaO and $CaCO_3$ are added to remove ions such as Cl^-, SO_4^{2-}, Mg^{2+}, and Ca^{2+} by precipitation. The water is next treated with NH_4HCO_3 to precipitate NaCl. Finally, active carbon allows one to achieve water with $200-300$ ppm of salts. The bed of active carbon is regenerated using NaOH and HCl. A flowsheet for the process is shown in Fig. 4.20.

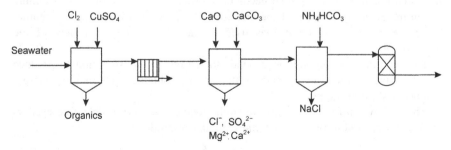

FIGURE 4.20

Chemical water purification.

4.2.3 WATER ELECTROLYSIS

Water decomposition leads to the production of hydrogen as the most valuable product. The use of electrolysis has been an alternative for quite some time. It has recently received more attention with the development of renewable technologies to produce power from solar to wind energy. However, the high cost (2.6–4.2 $/kg) makes it not as interesting from an industrial point of view, and thus natural gas steam reforming or hydrocarbon partial oxidation are the most widespread methods, as we will see in Chapter 5, Syngas. In spite of the reduced industrial use, less than 3% of hydrogen is produced using electrolysis; we need to bear in mind that the use of hydrocarbons as a source for hydrogen is not carbon-efficient since they have fixed CO_2 from the atmosphere and only H_2 is taken from them. Furthermore, electrolysis using renewable sources (Davis and Martín, 2014) has lately become an alternative to store solar (photovoltaic or PV solar, concentrated solar power) and wind energy in the form of chemicals (using CO_2 as a source of carbon). Thus, the process is twofold: renewable energy storage, and CO_2 capture and utilization.

Water decomposition into its components requires high energy consumption. Water formation energy, ΔH_f, is 68.3 kcal/mol at 25°C. However, only the fraction corresponding to the Gibbs free energy can be provided as work ($\Delta G = 56.7$ kcal/mol at 25°C), and the rest ($T\Delta S$) needs to be provided as heat.

$$\Delta G_f = \Delta H_f - T\Delta S \tag{4.24}$$

Solutions with free ions allow current conduction. Thus, the electromotive force that allows conduction is the decomposition potential. The theoretical minimum for the reversible water decomposition can be computed using the Nernst equation as follows:

$$\Delta G = nFE = 2 \times 96540 \times E$$
$$\Delta G \rightarrow E = 1.23 \text{ V at } 25°C \tag{4.25}$$
$$\Delta H \rightarrow E_{eq} = 1.48 \text{ V at } 25°C$$

The potential difference between both has to be provided by heat. In practice, the actual potential to be applied needs to be from 1.8 to 2.2 V due to the Ohmic loss in the electrolyte, those of the electrodes, and the polarization overpotential. There are a few basic definitions used to characterize the efficiency of the operation of a system.

Unit energy consumption corresponds to the energy consumption to produce a unit of the product. Its inverse is the product obtained per unit of energy, the energy yield of the process.

The *energy ratio* is the ratio between the theoretical work (Gibbs free energy) and the power used to produce a certain amount of product:

$$\eta = \frac{n_{electrons} \cdot F \cdot V}{E_e} \tag{4.26}$$

Energy efficiency is the ratio between the energy stored in the product, the low heating value, and that consumed to produce it. Commercial plants operate from 50% to 75% energy efficiency.

The *current yield* computes the ratio between the theoretical power and that used for the production of a certain amount of product:

$$\varepsilon = \frac{n \cdot F}{I \cdot t} \qquad (4.27)$$

where F is equal to 96,485 C/mol (1 F is the charge of a mol of electrons). The units are equivalent to J/V mol or A·s/mol since 1 Coulomb = 1 A·s and 1 Volt = Joule per Coulomb.

4.2.3.1 Commercial electrolyzers

There are two distinct designs: those that use electrolyte solutions and those that employ solid polymer electrolytes. The most used are the first class, and they present two configurations: tank-type and press filters.

1. *Tank-type electrolyzers* are also known as unipolar electrodes. Each cell consists of a cathode and anode and room for the electrolyte. The electrodes may be separated or even isolated by a diaphragm. The electrodes are connected in parallel and the number of cells needed to reach the production capacity represents the electrolytic system. The potential difference is the same for each cell. The H_2 and O_2 are collected through pipes connected to the anodic and cathodic places, respectively. This is the oldest configuration.

   Only a few parts are needed to build them, and they are relatively inexpensive. Tank-type electrolyzers optimize a lower thermal efficiency, and they are typically used for their lower power costs. Furthermore, in terms of maintenance, the individual cells can be isolated or replaced independently of the rest of the configuration. They provide low potential difference and high intensity, and the purity of the hydrogen produced is high.

2. *Filter press cells*, or *bipolar electrolyzers*, have their electrodes in series so that one face acts as the cathode and the other as the anode, which results in a compact configuration. Between each electrode there is a diaphragm. They are connected to the electrical grid through the terminal electrodes, simplifying the installation, so that the potential difference is the summation of that corresponding to each element. Thus they work with low current intensity and high potential difference.

In both cases, the electrolyte solution flows in order to be cooled down continuously, and is recycled back to the electrolyzer. This stage represents the major water consumption of the plant. Fig. 4.21A shows a scheme for tank-type electrolyzers, and Fig. 4.21B shows an example of a filter press cell.

Water electrolysis using solar or wind energy has focused on the production of hydrogen for the so-called hydrogen economy. Fig. 4.22 shows a flowsheet for that process. It can operate at atmospheric pressure or above (30 atm) at 60−80°C

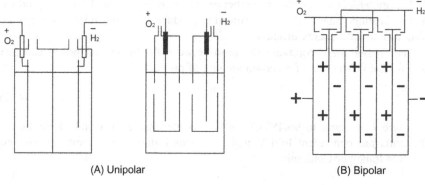

FIGURE 4.21

Schemes for electrolyzers: (A) Unipolar, and (B) Bipolar.

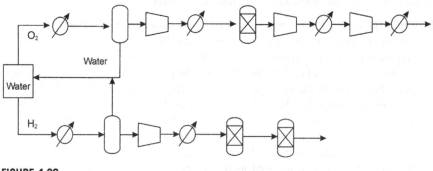

FIGURE 4.22

Production of hydrogen and oxygen from water electrolysis.

and with an electrolyte solution, KOH 20−40 wt%. Typically, 500 Nm^3/h of hydrogen are produced per unit, consuming 0.9 L of water per Nm^3 of hydrogen. The power required is 175,000 kJ/kgH_2 (NEL Hydrogen, 2012). Once the water is split, both gas streams must be purified. The oxygen stream contains water that is condensed, and is later compressed and dehydrated using silica gel or a similar adsorbent. For the hydrogen-rich stream, we do not only need to remove the water that saturates the stream, but more importantly, the traces of oxygen. Therefore, a deoxo reactor that uses a small amount of the hydrogen produced converts the traces of oxygen into water. Finally, a dehydration step using adsorbent beds is used before final compression.

Lately, this hydrogen has been used to hydrate CO_2 and to produce chemicals such as methane, methanol, and DME as a way to store energy. Table 4.5 shows the comparison of these processes based on the work by Martín (2016). See Chapter 5, Syngas, for synthesis details.

Table 4.5 Yield to Products and Consumption for CO_2/H_2-Based Processes

Main Product/Other Products and Raw Materials	CH_4	CH_3OH	DME
CO_2 captured (kg/kg)	3.0	1.4	1.9
H_2O consumption (kg/kg)	2.8	1.2	1.3
O_2 production(kg/kg)	3.3	1.5	2.1

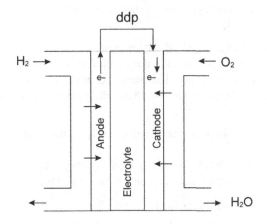

FIGURE 4.23

Fuel cell.

4.2.3.2 Fuel cells

Fuel cells perform the opposite chemical reaction—they transform hydrogen and oxygen into electrical energy. The advantage is the absence of mobile parts, so there is no need for recharge; they have been used in spaceships since the 1960s.

$$H_2 + \frac{1}{2}(O_2 + 3.76N_2) \rightarrow H_2O + Q + \text{Power}$$

The process is highly efficient, has no emissions, is quiet, and the byproduct is water. The main drawbacks of the technology are the storage of the hydrogen, the size and weight of the cells, and the high cost of hydrogen. There are a few car makers who have produced models running on hydrogen over the last few decades, but the supply chain of hydrogen is not developed to the point where making the cars is a commercial prospect.

Fig. 4.23 shows an acid fuel cell. The hydrogen is injected into the anode region. It crosses it and reacts with the oxygen at the electrolyte, producing water that exits with the excess of air used. The alkali kind of cell works the opposite, and the water is produced on the hydrogen side.

EXAMPLE 4.6

We would like to produce 2 kg/s of hydrogen at 30 bar via water electroly-sis. We have several sources of renewable energy, including solar, wind, and biomass. The oxygen produced has to be compressed up to 100 bar for its storage. Assume complete water breakup into hydrogen and oxygen, pure hydrogen from the cathode and oxygen from the anode. The compres-sor's efficiency is 85%, and the electrolysis occurs at 80°C. The electro-lytic cell has an energy ratio of 60%. Assuming that the polytropic coefficient is $k = 1.4$:

a. Compute the energy required for the process.
b. Select the technology using the information in Table 4E6.1.
c. Determine the price of the wind energy for it to be more profitable if the carbon tax for CO_2 emission is 40€/t and the ground is at a cost of 25€/m^2.

Solution

a. The energy required is that needed to split the water, and for gas compression.

$$\text{Energy} = E + W(\text{Oxygen line}) + W(\text{Hydrogen line})$$

For the electrolyzer we have the following reaction:

$$H_2O \rightarrow \tfrac{1}{2} O_2 + 2H^+ + 2e^- \quad E = -1.23 \text{ V}$$
$$2H^+ + 2e^- \rightarrow H_2 \quad\quad\quad E = 0 \text{ V}$$
$$E = 0 - (-1.23) = 1.23 \text{ V}$$

$$\eta = 0.6 = \frac{n_{H_2} \cdot e \cdot F \cdot V}{E_e} = \frac{\dfrac{2000 \ g/s}{2} \cdot 2 \cdot 96500 \cdot 1.23}{E_e} \Rightarrow E_e = 395650 \text{ kW}$$

The energy consumption of the compressors is computed using the next equation. Considering that the compression cannot be made in one stage (due to the pressure difference), but must be made in three stages, the work used is computed as follows:

$$W(\text{Comp}) = 3(F) \cdot \frac{8.314 \cdot k \cdot (T_{in} + 273.15)}{((MW) \cdot (k-1))} \frac{1}{\eta_s} \left(\left(\frac{P_{out}}{P_{in}} \right)^{\frac{1}{3}\frac{k-1}{k}} - 1 \right)$$

The consumption of the H_2 line is equal to 4568 kW.
The consumption of the O_2 line is equal to 3285 kW.

Table 4E6.1 Information on the Technologies

Technology	Cost (€/kW)	CO$_2$ Generated (kg/kW)	Area (m^2/kW)
Wind	1300	0.5	0
Solar	1150	1	5
Biomass	1000	0	10

b. Now we compare the three technologies in terms of cost and emission generation. In the table below, we see the cost of each technology considering not only the investment in the plant, but also the emission tax and the cost for the area that will be used. Biomass turns out to be the cheapest option.

				Cost (€)	CO_2 (kg)	Area (m²)	Emission (€/kg)	Ground (€/m²)	Environmental Impact (€)
Wind	1300	0.5	0	524,555,513	201,752	0	0.04	25	524,563,583
Solar	1150	1	5	464,029,877	403,504	2,017,521	0.04	25	514,484,047
Biomass	1000	0	10	403,504,241	0	4,035,042	0.04	25	504,380,301

c. In order for wind to be the cheapest option, its cost must be lower than 1250 €/kW.

4.2.4 THERMOCHEMICAL CYCLES FOR WATER SPLITTING

The energy for water splitting that can be provided as work, ΔG, becomes zero at 4000°C. Therefore at that temperature, only thermal energy would be necessary for that operation, reducing the yield loss due to the production of power. Note that it is infeasible from the practical point of view. In order to thermally split water, we should be able to work at 700−900°C, which can be achieved at nuclear reactors, for instance. To reach those temperatures, we can produce a series of chemical reactions whose global operation yields water splitting, recycling the other chemicals involved.

There are more than three hundred different cycles. To decide on a promising alternative, the characteristics of the cycle that must be considered are the number of reactions or steps; the number of elements involved; the cost and availability of chemicals; the types of materials needed for construction and processes; the corrosivity of the species involved; the safety, health, and environmental considerations of the chemicals and the operating conditions; the effect of temperature on costs; the maturity of the cycle; the thermal efficiency; the operating temperature; the similarity of reaction kinetics among the different stages (fast in general); the purification of the products; and the fact that ΔG must be zero or negative to reduce the energy requirements. We classify the cycles into four groups: halides, oxides, sulfur, and hybrids (Aporta et al. (2011)).

4.2.4.1 Family of the halides

The general chemical reaction is as follows:

$$3\,MeX_2 + 4\,H_2O \rightarrow Me_3O_4 + 6\,HX + H_2\,(Me = Mn\ or\ Fe\ y\ X = Cl, Br\ or\ I)$$

Since the 1970s the Euratom initiative (Ispra, Italy) has studied different cycles. For instance, the one named Mark I uses mercury and bromide. The reactions and their respective temperatures can be found below. The major risks are in the operations with Hg and Br, which are health hazards:

$$CaBr_2 + 2H_2O \Rightarrow Ca(OH)_2 + 2HBr \qquad [730°C]$$

$$Hg + 2HBr \Rightarrow HgBr_2 + H_2 \qquad [250°C]$$

$$HgBr_2 + Ca(OH)_2 \Rightarrow CaBr_2 + HgO + H_2O \quad [200°C]$$

$$HgO \Rightarrow Hg + \frac{1}{2}O_2 \qquad [600°C]$$

$$2H_2O \Rightarrow H_2O + H_2 + \frac{1}{2}O_2$$

Another cycle that avoided the use of such dangerous chemicals received the name Mark 9, and was also proposed by Euratom. The cycle is as follows:

$$6FeCl_2 + 8H_2O \rightarrow 2Fe_3O_4 + 12HCl + 2H_2 \qquad [650°C]$$

$$2Fe_3O_4 + 3Cl_2 + 12HCl \rightarrow 6FeCl_3 + 6H_2O + O_2 \quad [175°C]$$

$$6FeCl_3 \rightarrow 6FeCl_2 + 3Cl_2 \qquad [420°C]$$

4.2.4.2 Family of the oxides

The general reaction is as follows:

$$3 MeO + H_2O \rightarrow Me_3O_4 + H_2$$

$$Me_3O_4 + Q \rightarrow 3 MeO + (1/2)O_2$$

where the metal (Me) can be Mn, Fe, or Co.

4.2.4.3 Family of sulfur

In this family, we highlight two cycles. The first one is the iodine–sulfur cycle, given by the following reactions:

$$I_2 + SO_2 + 2H_2O \rightarrow 2HI + H_2SO_4 \quad < 100°C$$

$$2HI + Q \rightarrow H_2 + I_2 \qquad [700°C]$$

$$H_2SO_4 + Q \rightarrow H_2O + SO_2 + \frac{1}{2}O_2 \quad [850°C]$$

The second cycle is the sulfate one, given by the following set of reactions:

$$4FeSO_4 + Q \rightarrow 2Fe_2O_3 + 3SO_2 + SO_3 + \frac{1}{2}O_2 \quad [700°C]$$

$$2Fe_2O_3 + 3SO_2 + SO_3 + H_2O \rightarrow H_2 + 4FeSO_4$$

4.2.4.4 Hybrid cycles

In parallel to the efforts presented above, the Westinghouse Corporation proposed a hybrid cycle that combined an electrochemical stage (taking place at low temperature), and another thermal one. The optimum efficiency, 40%, is obtained for 65% sulfuric acid. However, it can be increased up to 46% if the electrolysis takes place in a series of stages.

$$SO_2 + 2H_2O \rightarrow H_2SO_4 + H_2 \quad [25°C, electrolysis]$$

$$H_2SO_4 \rightarrow H_2O + SO_2 + \frac{1}{2}O_2 \quad [850°C]$$

4.2.5 SODIUM CHLORIDE INDUSTRY

Sodium chloride (NaCl) is the most abundant salt in seawater, and is the main raw material for sodium and chloride chemical production Bertram (1993). The separation of NaCl had already been done in ancient times by solar evaporation of water. An evolution of this method, adding green dyes (solivap green), is still in operation today. Thus water is stored in low, deep pools exposed to high solar intensity. As solar and wind energy evaporates water from the pools, different precipitates remain in solution. In the beginning, $CaCO_3$ and $Fe(OH)_3$ precipitate. Later, as the density increases, other salts precipitate. Table 4.6 shows the salt and purity obtained.

Following this procedure, we can recover up to two-thirds of the NaCl. In the mother liquor there are still NaCl (up to 35%), $MgSO_4$, $MgCl_2$, KCl, and NaBr. Thus, the procedure can continue. Therefore, we can obtain $MgSO_4$ and KCl with a low level of purity. We can treat the mother liquor with $CaCO_3$ and/or $MgCO_3$ to produce $Mg(OH)_2$. Next, by bubbling Cl_2 gas, Br_2 is generated. There are two processes to carry out the production of Br_2 from the mother liquor. The first one (just described) is known as the Kubiersky method. The second is Dow's method. It is an electrolytic-based method to extract bromine from brine. It was patented in 1891. The brine is treated with sulfuric acid and bleaches to produce bromine by oxidation of bromide, which remains dissolved in the brine. Next, the solution is dripped onto burlap, and water is blown through it so that the bromine volatilizes. Bromine is thus trapped with iron turnings to produce a solution of ferric bromide. By using more iron metal, the ferric bromide is transformed into ferrous bromide. Thus free bromine can be obtained by thermal decomposition of the last species.

Table 4.6 Salts Precipitated by Solar Evaporation

Density (kg/L)	Species Precipitated	Use
1.21	$CaSO_4 \cdot H_2O$	
1.21−1.23	NaCl 96−98%	Food industry
1.23−1.25	NaCl 92−95%	Chemical industry
1.25−1.29	NaCl 92%	Brine

Various salts can be obtained from seawater, but the following sections focus on the use of NaCl as a raw material for Na_2CO_3. Until the 1800s it was believed that K_2CO_3 and Na_2CO_3 were the same species since they were both alkalis that were obtained by calcinations and lixiviation of plants. The Na_2CO_3, also known as washing soda, or soda ash, is of particular interest in the production of soaps, glasses, as well as paper, phosphates, oxalates, and borax. The two processes that have been used industrially are Leblanc's process and Solvay's process Solvay Alkali GmbH (1998), Rauh (1993).

4.2.5.1 Production and use of sodium carbonate (Na₂CO₃)

4.2.5.1.1 Production via Leblanc's process

Historical perspective: In 1775 the French Academy of Sciences offered an award of 2400 Livres for a method to produce soda ash from sea salt. Nicolas Leblanc (1742−1806), chemist and physician to Louis Phillip II, Duke of Orléans, proposed a method and obtained the patent in 1791. That same year he built a plant at Saint−Denis (Paris) worth 200,000 Louises to produce 320 tons of product a year. The facility only operated until the Duke's death by guillotine in 1793. That forced him to open the patent and he was denied the award. Napoleon returned the plant to Leblanc in 1801, but unable to upgrade it to compete with the new processes, he committed suicide in 1806. In 1818 the first plant was built in Germany, and later, in 1823, another was built in England.

Process: The process follows the chemical reactions below.

$$2NaCl + H_2SO_4 \rightarrow 2HCl + Na_2SO_4$$

$$Na_2SO_4 + 2C \rightarrow Na_2S + 2CO_2 \quad (Na_2SO_4 + 4C \rightarrow Na_2S + 4CO)$$

$$CaCO_3 + Na_2S \rightarrow Na_2CO_3 + CaS$$

Fig. 4.24 shows the scheme of the process. NaCl is mixed with sulfuric acid and then the solution is heated. The HCl is vented to the atmosphere. Next, the mass is exposed to direct flame to remove the remaining chloride. Coal and calcium carbonate are then used in a proportion of 2:2:1 for the salt, the carbonate, and the coal, respectively. The mixture is heated up to 1000°C in a furnace. The black ash produced needs to be lixiviated to avoid oxidation to sulfate. For that it is covered with water. The lixiviation takes place in stages. The final

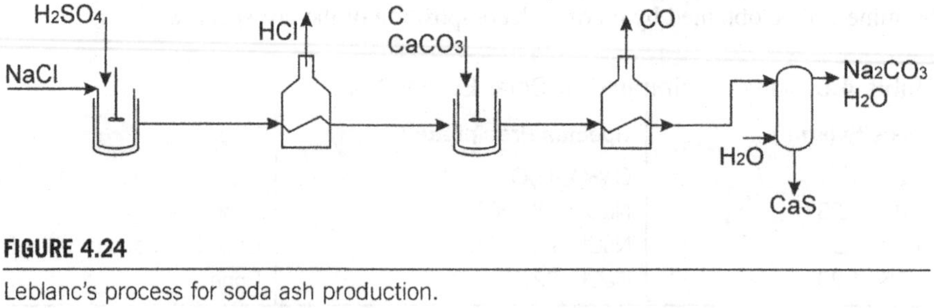

FIGURE 4.24

Leblanc's process for soda ash production.

liquor is processed to precipitate the carbonate after cooling down the mixture. Some of the drawbacks of the process are that all the reactions occur in solid phase, being slow, and consume large amounts of energy. Furthermore, the process presents several environmental concerns.

In those early days the HCl was useless and was vented. Furthermore, smelly solid CaS was also produced, and had no further value either. When piled up, it produced hydrogen sulfide. Because of these emissions, in 1839 the facilities received a formal complaint due to the effect of the process on the surroundings. As a result, the British Parliament in 1863 passed the first of several Alkali Acts, regarded as the first modern air pollution legislation. In this particular case, the Act did not allow venting of more than 5% of the HCl produced in alkali plants. Absorption beds using charcoal were installed in the plants and absorbed in water, producing hydrochloric acid. Later, the CaS was used to recover the S and produce sulfuric acid, and the HCl was used to obtain Cl_2 via oxidation.

EXAMPLE 4.7

We would like to produce 10 t/day of Na_2CO_3 using the Leblanc process, given by the following reactions:

$$NaCl + H_2SO_4 \rightarrow HCl + NaHSO_4$$
$$NaHSO_4 + NaCl \rightarrow Na_2SO_4 + HCl$$
$$Na_2SO_4 + 2C \rightarrow 2CO_2 + Na_2S$$
$$Na_2S + CaCO_3 \rightarrow CaS + Na_2CO_3$$

Compute the amount of raw material and byproducts, assume that the conversion of the reactions is 100%, and the volume of commercial sulfuric acid, considering that the concentration is $60°Be$.

Solution

$°Be$ is a density-based measurement. For liquids denser than water $10°Be$ is the density of a 10% solution of NaCl in water and $0°Be$ is the density of distilled water at 15.6°C.

$$°Be = 145 - \frac{145}{\rho_{relative,15.6°C}}$$

For liquids less dense than water:
$0°Be$ is the density value of a solution of 1 g of NaCl and 9 g of water; $10°Be$ is the value for distilled water at 15.6°C.

$$°Be = \frac{140}{\rho_{relative,15.6°C}} - 130$$

Table 4E7.1 Sulfuric Acid Densities (g/cm^3)

% peso	15°C	20°C	15.6°C
77	1.6976	1.6927	1.6970
78	1.7093	1.7043	1.7087

Since sulfuric acid is denser than water, and $°Be = 60$, we use the first of the two equations to compute the relative density to water, with a result of 1.706. Water density at 15.6 is 0.999007 g/cm^3 based on tables. Thus the actual density is 1.7042 g/cm^3. Using tables for sulfuric acid (see Perry and Green 1997), we can interpolate the weight percentage for the temperature of 15.6°C (Table 4E7.1).

Thus, for a density of 1.7042 g/cm^3, we interpolate a mass fraction of 77.62%. From the mechanisms we compute a global reaction for the entire process as follows:

$$2\,NaCl + H_2SO_4 + 2C + CaCO_3 \rightarrow 2HCl + 2\,CO_2 + CaS + Na_2CO_3$$

Based on the stoichiometry of the reaction, and using a day as the basis for the calculations, the moles of sodium carbonate needed in the process are computed as follows:

$$m_{Na_2CO_3} = 10000\ kg \equiv \frac{10,000\ kg}{106\ kg/kmol} = 94.34\ kmol = n_{Na_2CO_3}$$

The need for raw materials is as follows:

$$m_{NaCl} = 2 \cdot n_{Na_2CO_3} \cdot M_{NaCl} = 2 \cdot 94.34\ kmol \cdot 58.1\frac{kg}{kmol} = 11038\ kg$$

$$m_{H_2SO_4} = n_{Na_2CO_3} \cdot M_{H_2SO_4} = 94.34\ kmol \cdot 98\frac{kg}{kmol} = 9245.32\ kg$$

$$m_C = 2 \cdot n_{Na_2CO_3} \cdot M_C = 2 \cdot 94.34\ kmol \cdot 12\frac{kg}{kmol} = 2264.16\ kg$$

$$m_{CaCO_3} = n_{Na_2CO_3} \cdot M_{CaCO_3} = 94.34\ kmol \cdot 100\frac{kg}{kmol} = 9434\ kg$$

Thus, the commercial acid that we need is computed from the pure raw material as follows:

$$m_{H_2SO_4}\big|_{commercial} = \frac{m_{H_2SO_4}}{0.7762} = 11911\ kg$$

$$V_{H_2SO_4}\big|_{commercial} = \frac{m_{H_2SO_4}\big|_{commercial}}{1.7042\ kg/L} = 6989\ L$$

The products of the process are determined as follows:

$$m_{CaS} = n_{Na_2CO_3} \cdot M_{CaS} = 94.34 \text{ kmol} \cdot 72 \frac{\text{kg}}{\text{kmol}} = 6792 \text{ kg}$$

$$m_{CO_2} = 2 \cdot n_{Na_2CO_3} \cdot M_{CO_2} = 94.34 \text{ kmol} \cdot 44 \frac{\text{kg}}{\text{kmol}} = 8301.92 \text{ kg}$$

$$m_{HCl} = 2 \cdot n_{Na_2CO_3} \cdot M_{HCl} = 2 \cdot 94.34 \text{ kmol} \cdot 36.5 \frac{\text{kg}}{\text{kmol}} = 6887 \text{ kg}$$

4.2.5.1.2 Production via Solvay's process (ammonia soda process)

Chemical history of the process. In 1811 the French physicist Augustin Fresnel discovered that sodium bicarbonate precipitates when carbon dioxide is bubbled through ammonia brine.

$$NaCl + CO_2 + NH_3 + H_2O \rightarrow NaHCO_3 + NH_4Cl$$

This reaction takes place in a series of reactions as follows:

$$NH_3 + H_2O \rightarrow NH_4OH$$

$$2NH_4OH + CO_2 \rightarrow (NH_4)_2CO_3 + H_2O$$

$$(NH_4)_2CO_3 + CO_2 + H_2O \rightarrow 2NH_4HCO_3$$

$$NH_4HCO_3 + NaCl \rightarrow NH_4Cl + NaHCO_3$$

Ammonia acts as a buffer for high pH, at which sodium bicarbonate is not soluble and precipitates. The bicarbonate is decomposed by heating, as given by the following reaction:

$$2NaHCO_3 \rightarrow Na_2CO_3 + H_2O + CO_2$$

However, the reversibility of the reactions requires that one of the raw materials ($NaCl$ or NH_4HCO_3) be fed in excess instead of in stoichiometric proportions. The cheapest one, the $NaCl$, is the one fed in excess. To recover the precipitated bicarbonate, the stream must be cooled down to 30°C so that it crystallizes.

From an industrial point of view, this process did not require sulfuric acid, which was in high demand back then, and the reactions took place in solution. However, the drawback of the process was the consumption of ammonia. Before the Haber–Bosch process, ammonia was produced as a byproduct in charcoal production or from manure decomposition. The high demand for ammonia, and thus its cost, made the process difficult to scale up. The recovery of ammonia within the process changed this aspect. In 1822 the recovery of ammonia was already known. Using a strong alkali, $Ca(OH)_2$, the ammonium cation was transformed into ammonia that could be distilled. The $Ca(OH)_2$ was internally produced using calcium carbonate, which decomposed into calcium oxide. Finally, CaO reacted with water to produce the calcium hydroxide.

$$CaCO_3 \rightarrow CaO + CO_2$$

$$CaO + H_2O \rightarrow Ca(OH)_2$$

Thus NH_3 was recovered as follows:

$$2NH_4Cl + Ca(OH)_2 \rightarrow 2NH_3 + CaCl_2 + H_2O$$

$$CO_2 + NH_3 + H_2O \rightarrow NH_4HCO_3$$

The Solvay process can be summarized in the following global reaction:

$$2\,NaCl + CaCO_3 \rightarrow Na_2CO_3 + CaCl_2$$

Although some of the reactions had been known since 1811, there were problems in the process design. In 1861 the Belgian industrial chemist Ernest Solvay proposed the use of a 24 m tall absorption tower in which carbon dioxide was bubbled through a flow of ammonia containing brine. This unit incorporated an arrangement for cooling down the bicarbonate to separate it from the products. Finally, ammonia was also recovered as suggested by Augustin Fresnel. By 1864 Ernest and his brother Alfred had built a plant near Charleroi (because of the existing chemical industry in the region) to provide for coal and ammonia. The production capacity was 200–250 t/yr. By 1880 Solvay's process, which was more environmentally friendly than Leblanc's, completely displaced it.

Flowsheet. Fig. 4.25 shows the flowsheet for the production of sodium carbonate from NaCl, coal, and calcium carbonate as raw materials.

The solution feed is treated with $Ca(OH)_2$ and Na_2CO_3 to eliminate the cations Mg^{2+} and Ca^{2+} in the form of $Mg(OH)_2$ and $CaCO_3$, since otherwise they would precipitate with CO_2. Next, Tower T-01 is fed with a saturated solution of NaCl. This column receives the regenerated NH_3 from T-05. The ammoniacal brine is diluted with the water that accompanies the ammonia. To reconcentrate the solution, it is fed to T-02, a packed bed of solid NaCl. The ammonia brine is then fed to Solvay Tower T-03, where it is put in countercurrent contact with the CO_2 that rises across the tower. CO_2 streams with different concentrations are fed at various levels. Concentrated CO_2 from the decomposition of the sodium bicarbonate is fed into the bottoms of the columns. Diluted CO_2 from the decomposition of calcium carbonate is fed above the cooling stage of the column (see Fig. 4.25). The gas with a small content of CO_2 exits the tower from the top. Along the tower, the ammoniacal brine descends and is put into contact with gases with higher concentrations of CO_2 so that an equilibrium among the species $NaHCO_3/NH_4Cl/NaCl/NH_4HCO_3$ is reached. The Solvay Tower uses double trays. Furthermore, in the lower part there is a bundle of cooling tubes for reducing the temperature to 30°C, allowing precipitation of the sodium bicarbonate and its separation. A suspension is obtained from the bottom of the tower.

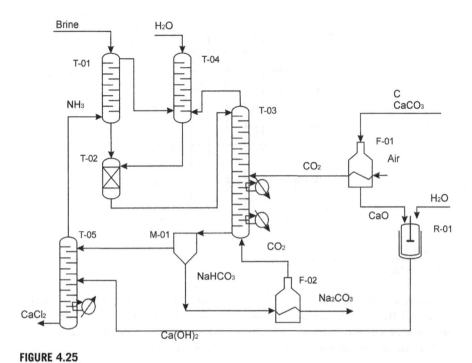

FIGURE 4.25

Solvay's flowsheet.

The equilibrium of the species is studied using the Jaenecke diagram (see Fig. 4.26), which represents the pairs of species. From top to bottom the anions are represented (Cl^- and HCO_3^-), and from left to right the cations (NH_4^+ and Na^+). The feed to the column is at opposite vertexes, NaCl and NH_4HCO_3. The lever rule can be used to determine the feed composition as a function of the ratio between those two. The product, $NaHCO_3$, is at the right bottom vertex. A line connecting this point to the feed point is extrapolated to the equilibrium lines. Next, this point is connected to the NaCl vertex. We draw a line parallel to this one by the feed point and to the right vertical axis and we read the amount of product obtained.

A solution of NH_4Cl with the suspended $NaHCO_3$ exits from the bottom of the Solvay Tower. A centrifuge filter is used to recover the crystals that are sent to Furnace F-02 to be decomposed. The CO_2 produced is a concentrated stream that is injected at the bottom of the Solvay Tower. The sodium carbonate is obtained from F-02. The NH_4Cl solution is recycled to recover the ammonia. In order to do this, a solution of calcium hydroxide, obtained from the decomposition of calcium carbonate and the reaction of the CaO with water, is used to produce ammonia out of NH_4Cl.

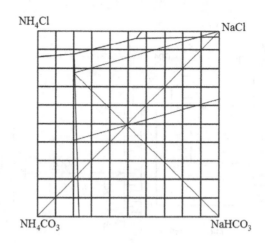

FIGURE 4.26

Jaenecke diagram.

The rest of the units involved are T-04, which uses water to absorb the ammonia from T-01 to T-03, and Furnace F-01, used for the production of calcium oxide, the raw material used to produce the strong alkali in R-01 to recover the ammonia.

We can summarize the process into the following steps:

1. Production of a saturated solution of NaCl in water.
2. Decomposition of calcium carbonate:

$$CaCO_3 \rightarrow CaO + CO_2$$

   The diluted CO_2 is fed to the Solvay Tower at a medium point.
3. Production of the ammoniacal brine:

$$NaCl + H_2O + NH_3$$

4. Precipitation of bicarbonate by reaction of the ammoniacal brine with CO_2:

$$NaCl + H_2O + NH_3 + CO_2 \rightarrow NH_4Cl + NaHCO_3$$

5. Filtration of the bicarbonate crystals.
6. Thermal decomposition of the sodium bicarbonate. The concentrated CO_2 is recycled to T-03:

$$2\,NaHCO_3 \rightarrow Na_2CO_3 + H_2O + CO_2$$

7. Production of calcium hydroxide in R-01:

$$CaO + H_2O \rightarrow Ca(OH)_2$$

8. Ammonia recovery by distillation from the mother liquor coming from the centrifuge using the calcium carbonate:

$$2NH_4Cl + Ca(OH)_2 \rightarrow 2NH_3 + CaCl_2 + 2\,H_2O$$

The process has a global chemical reaction given by the following equation:

$$2\,NaCl + CaCO_3 \rightarrow Na_2CO_3 + CaCl_2$$

Solvay's process consumes 0.8 t of coal per ton of soda ash produced, while Leblanc's process requires 3.5 t per ton, which represents a large savings in raw materials. The only byproduct is calcium chloride, and it may represent a hazard to the environment if not properly discarded.

EXAMPLE 4.8

A Solvay plant produces 10 kg/s of soda ash. The process is fed with a 10% molar excess of NaCl. Determine the amount of NaCl and NH_3 needed, as well as the calcium carbonate used for the production of the CO_2 needed. Finally, compute the composition of the mother liquor in terms of the ions in it.

Solution

We need to use the Jaenecke diagram to determine the equilibrium established in Solvay's Tower. In Fig. 4E8.1, the feed to the tower is plotted. Since we use an excess of NaCl, and its composition with respect to the ammonia is measured along the diagonal, the point is closer to the NaCl vertex. We draw a line from the bicarbonate vertex to the feed, and extend it to the equilibrium lines. Finally, a parallel to the line from this point to NaCl is drawn from the feed point in the diagonal. We obtain the $NaHCO_3$.

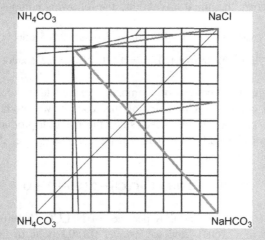

FIGURE 4E8.1

Equilibrium composition.

The composition of the mother liquor is shown in Fig. 4E8.1 in the contact between the line from the bicarbonate vertex to the equilibrium line. It is 12% HCO_3^-, and the rest Cl^-; and around 11% Na^+, and the rest ammonium.

Therefore, 40% sodium bicarbonate is obtained per mol of initial mixture. The rest is in the mother liquor. NaCl usage is 40/52. Therefore, per 100 mol of NaCl fed to the system, we produce 76 mol of $NaHCO_3$. The rest of the NaCl remains in the mother liquor.

Using the production capacity as a basis for our calculations, according to the stoichiometry of the reaction,

$$2\,NaHCO_3 \rightarrow H_2O + CO_2 + Na_2CO_3$$

The moles of bicarbonate needed are twice those of the final product obtained:

$$N_{Na_2CO_3} = 10/(23\cdot2 + 12 + 3\cdot16) = 0.094\ kmol/s$$

$$N_{NaHCO_3} = 2\cdot N_{Na_2CO_3} = 0.189\ kmol/s$$

On the other hand, in Solvay's Tower we have the reaction given by the following equation:

$$NaCl + CO_2 + NH_3 + H_2O \rightarrow NaHCO_3 + NH_4Cl$$

Per mol of initial mixture, we produce 0.4 mol of $NaHCO_3$. Therefore to produce 0.189 kmol of bicarbonate, we need $0.189/0.4 = 0.4717$ kmol with a composition of 52% NaCl and 48% $NaHCO_3$. With the yields determined above, the initial mixture has the following composition:

$$N(NaCl) = (100/77)\cdot0.189 = 0.4717\cdot0.52 = 0.245\ kmol/s$$

$$N(NH_3) = (0.189)/(0.4/0.48) = 0.4717\cdot0.48 = 0.226\ kmol/s$$

To evaluate the recovery of ammonia, we consider the NH_4Cl produced. Based on the stoichiometry of the reaction we produce as much NH_4Cl as the bicarbonate produced, since the reaction produces 1 mol of NH_4Cl per mol of bicarbonate. Thus, to treat it and recover the ammonia we use the following reaction:

$$Ca(OH)_2 + 2\,NH_4Cl \rightarrow CaCl_2 + 2\,NH_3 + 2\,H_2O$$

We need 0.5 mol of $Ca(OH)_2$ per mol of NH_4Cl. Therefore, we need 0.0943 kmol/s of $Ca(OH)_2$ and we recover 0.189 kmol/s of ammonia. To obtain the calcium hydroxide, we need as many moles of calcium carbonate:

$$CaCO_3 \rightarrow CaO + CO_2$$

$$CaO + H_2O \rightarrow Ca(OH)_2$$

Thus we need 0.0943 kmol/s of $CaCO_3$. That produces 0.0943 kmol/s of CO_2. The other source of CO_2 is the decomposition of the bicarbonate:

$$2\,NaHCO_3 \rightarrow H_2O + CO_2 + Na_2CO_3$$

From this reaction 0.0943 kmol/s of CO_2 are also produced. Thus the total CO_2 feed to the process is 0.189 kmol/s.

Typically the Cl^- goes to the mother liquor. As a result, and with the composition of the mother liquor, it is possible to compute the kmols of liquor.

There are a number of variants of the ammonia–soda process. Among them the one that has received the most attention is the dual process that combines soda ash production with ammonium chloride production. It is of particular interest in Japan due to the high cost of imported rock salt. Another one is based on the oversaturation of ammonia chloride with ammonia, which dissolves NaCl. When adding ammonia and CO_2, $NaNH_4CO_3$ precipitates. It is filtered and decomposed into Na_2CO_3 and NH_3. The mother liquor is heated up to 40–50°C, releasing ammonia so that NH_4Cl precipitates. The crystals are separated by filtration and the mother liquor is treated with ammonia and CO_2 to produce NH_4HCO_2. By adding NaCl, $NaHCO_3$ precipitates and we obtain NH_4Cl.

4.2.5.1.3 Usage of soda ash

The product is obtained as crystals of sodium carbonate decahydrated. It is used for the *production of glass*. This use represents more than half of the production of the soda ash worldwide since bottle and window glass are made by mixing Na_2CO_3, $CaCO_3$, and SiO_2. It is also used to *treat water* to remove Ca^{2+} and Mg^{2+} from it. Soda ash is used in the *production of soaps* since it is cheaper and easier to handle than NaOH. Instead of transporting NaOH, which is hydroscopic, it is produced in situ via the following reactions:

$$Na_2CO_3 + Ca(OH)_2 \Rightarrow CaCO_3 + 2NaOH$$
$$CaCO_3 \Rightarrow CaO + CO_2$$
$$CaO + H_2O \Rightarrow Ca(OH)_2$$

It is also used in the *production of paper* to separate the lignin from the cellulose via the sulfite method. Moreover, it can be used to *produce pigments and dyes*. Its anion belongs to a weak acid that allows easy release of the sodium. Furthermore, we can *produce sodium bicarbonate*. Even though the $NaHCO_3$ is produced in the Solvay process directly, it is actually cheaper to use the final product and react it with CO_2 to produce pure bicarbonate. The bicarbonate is also used in baking soda production and fire extinguishers. As an *alkali it is used to remove SO_2* from the flue gases in thermal plants.

EXAMPLE 4.9

Evaluate a process for the production of NaOH from a solution of sodium carbonate whose composition is given in Table 4E9.1.

This solution reacts with commercial calcium hydroxide (slaked lime) that contains CaO (burnt lime or quick lime), and as an impurity, $CaCO_3$, due to the production process of the calcium hydroxide:

$$CaCO_3 \rightarrow CaO + CO_2$$

$$CaO + H_2O \rightarrow Ca(OH)_2$$

The desired product is a solution with the composition given in Table 4E9.2.

Compute, per 100 kg of product:

1. The amount of alkali solution.
2. Composition and amount of commercial lime solution.
3. Reactant in excess and excess percentage.
4. Conversion.

Solution.

The main reactions that take place are the following:

$$Na_2CO_3 + Ca(OH)_2 \rightarrow CaCO_3 + 2NaOH$$

$$CaO + H_2O \rightarrow Ca(OH)_2$$

$$CaCO_3 + Q \rightarrow CaO + CO_2$$

Table 4E9.1 Alkali Solution

Chemical	(%)
NaOH	0.5
Na_2CO_3	14
H_2O	85.5

Table 4E9.2 Product Composition

Chemical	(%)
$CaCO_3$	12
$Ca(OH)_2$	0.5
Na_2CO_3	0.3
NaOH	10
H_2O	77.2

The process we evaluate mixes the alkali solution and the commercial lime as (Table 4E9.2):

$$m_{alkali} + \text{Commercial lime } (m_{CaO}, m_{Ca(OH)_2}, m_{CaCO_3}) \rightarrow \text{Desired product.}$$

For 100 kg of product we formulate a number of mass balances to the main atoms.

1. *To compute the amount of commercial lime we perform a mass balance to the sodium atom*:

$$Na_{in} = Na_{out}$$

$$m_{alkali}\left[0.005\cdot\left(\frac{23 \text{ kmol}}{40 \text{ kg}}\right) + 2\cdot0.14\cdot\left(\frac{23 \text{ kmol}}{106 \text{ kg}}\right)\right] = 2\cdot0.003\cdot100\left(\frac{23 \text{ kmol}}{106 \text{ kg}}\right)$$

$$+0.10\cdot100\left(\frac{23 \text{ kmol}}{40 \text{ kg}}\right)$$

$$m_{alkali} = 92.413 \text{ kg}$$

Next, by performing a global mass balance, we have the mass of commercial lime:

$$m_{alkali} + \text{lime} = \text{Product}$$

The amount of commercial lime is 7.587 kg. Thus

$$m_{CaO} + m_{Ca(OH)_2} + m_{CaCO_3} = 7.587 \text{ kg}$$

2. *Next, we perform a mass balance to the calcium atom*:

$$Ca_{in} = Ca_{out}$$

$$m_{CaO}\cdot\left(\frac{40 \text{ kmol}}{56.1 \text{ kg}}\right)_{CaO} + m_{Ca(OH)_2}\cdot\left(\frac{40 \text{ kmol}}{74.1 \text{ kg}}\right)_{Ca(OH)_2} + m_{CaCO_3}\cdot\left(\frac{40 \text{ kmol}}{100 \text{ kg}}\right)_{CaCO_3}$$

$$= 12\left(\frac{40 \text{ kmol}}{100 \text{ kg}}\right)_{CaCO_3} + 0.5\left(\frac{40 \text{ kmol}}{74.1 \text{ kg}}\right)_{Ca(OH)_2}$$

$$0.7148\cdot m_{CaO} + 0.5412\cdot m_{Ca(OH)_2} + 0.4006\cdot m_{CaCO_3} = 5.07 \text{ kg}$$

And a mass balance to carbon:

$$C_{in} = C_{out}$$

$$0.14\cdot m\left(\frac{12 \text{ kmol}}{106 \text{ kg}}\right)_{Na_2CO_3} + m_{CaCO_3}\cdot\left(\frac{12 \text{ kmol}}{100 \text{ kg}}\right)_{CaCO_3}$$

$$= 12\left(\frac{12 \text{ kmol}}{100 \text{ kg}}\right)_{CaCO_3} + 0.3\left(\frac{12 \text{ kmol}}{106 \text{ kg}}\right)_{Na_2CO_3}$$

$$1.465 + 0.1199 m_{CaCO_3} = 1.474 \text{ kg}$$

Table 4E9.3 Lime Composition

Chemical	(%)
CaO	75
Ca(OH)$_2$	24
CaCO$_3$	1

Therefore we have a system of three equations with three variables that we can solve:

$$m_{CaO} = 5.587 \text{ kg}$$

$$m_{Ca(OH)_2} = 1.822 \text{ kg}$$

$$m_{CaCO_3} = 0.078 \text{ kg}$$

The composition of the commercial lime is given in Table 4E9.3.

3. *To determine the raw material in excess we use the global reaction*:

$$Na_2CO_3 + Ca(OH)_2 \rightarrow CaCO_3 + 2NaOH$$

Bear in mind that CaO can be converted into Ca(OH)$_2$ using just water, which is in excess in the process. Therefore, if Ca(OH)$_2$ is consumed, the CaO will automatically react with water if needed for the reaction. Thus, the total amount of calcium hydroxide potentially available is given by both burnt and slaked lime as follows:

$$m_{Ca(OH)_2} + m_{Ca(OH)_2 \text{ as CaO}} = m_{Ca(OH)_2} + m_{CaO}\frac{M_{Ca(OH)_2}}{M_{CaO}} = 1.822 + 5.687 \left(\frac{74.1\frac{\text{kg Ca(OH)}_2}{\text{kmol}}}{56.1\frac{\text{kg CaO}}{\text{kmol}}}\right)$$

$$= 9.335 \text{ kg}$$

This is equivalent to 0.1259 kmol.

On the other hand, we have Na$_2$CO$_3$, $0.14 \cdot m_{alkali} = 12.938$ kg, equivalent to 0.1221 kmol. Since the stoichiometry requires a 1:1 relationship, we have an excess of calcium hydroxide. The excess is computed as follows:

$$\% = \frac{0.1260 - 0.1221}{0.1221} \cdot 100 = 3.21\%$$

4. The *conversion* of the process is determined with respect to the limiting reactant, the sodium carbonate. We feed 13.938 kg to the system and in the product we still have 0.3 kg. Therefore, the conversion becomes:

$$\text{Conver}(\%) = \frac{12.678}{12.938} \cdot 100 = 98\%$$

4.2.5.2 Electrochemical decomposition of melted NaCl

Although it could be considered simple to process NaCl and produce metal sodium and Cl_2 gas, it actually presents technical challenges since the melting point of NaCl is rather high ($806°C$), and there are corrosion problems in the electrolytic basin and in the electrodes. Furthermore, the boiling point of Na is $877°C$. Therefore, at the required operating conditions for the electrolysis of NaCl, there are already high losses of Na by evaporation. Finally, Ba is soluble in Na under these extreme conditions, which makes the final product impure.

To deal with the problem, a eutectic mixture consisting of 33% NaCl and 67% $CaCl_2$ allows reduction of the melting point to $505°C$. Another possibility is the addition of Na_2CO_3, which allows working at $600°C$; the potential difference used is $8-9$ V.

The reaction is carried out in steel tanks recovered by refractory bricks. The two semi-reactions are the following:

Graphite electrode (anode) $Cl^- - e- \rightarrow (1/2) Cl_2$	(clear gas)
Steel electrode (cathode) $Na^+ + e- \rightarrow Na$	(Metal sodium)

4.2.5.3 Electrolytic decomposition of NaCl in solution

This is currently the optimal process for chlorine production. NaOH and hydrogen are also produced as byproducts. The advantage of this process is the high value of the byproducts.

$$NaCl \rightarrow Na^+ + Cl^-$$

Anode semireaction:

$$2Cl^- - 2e^- \rightarrow Cl_2$$

Cathode semireaction:

$$2 \cdot H_2O + 2e^- \rightarrow H_2 + 2OH^-$$

EXAMPLE 4.10

Evaluate the production of hydrogen and chlorine from a solution of NaCl in an electrolytic cell. The initial brine of 50 kg has 30% NaCl by weight. Current yield for the production of hydrogen is 75%. Using a current of 2000 A and 5 V, determine the amount of Cl_2 and H_2, the effluent composition, and the energy ratio of the process.

Solution

The product mass flow rate is computed assuming that the moles of Na in the feed are equal to those in the product. The molecular weights of the species involved are as follows:

$$HCl = 36.5; \quad NaCl = 58.5; \quad Cl = 35.5; \quad NaOH = 40;$$

Hints:

$F = 96,485$ (J/V mol) (A·s/mol)
1 Coulomb $= 1$ A · s
$1 C = V \cdot F$
$F = C/mol$

(Table 4E10.1)
Since the current yield is 0.75, the production of Cl is as follows:

$$\varepsilon = \frac{n_{Cl} \cdot F}{I \cdot t} = 0.75$$

$$n_{Cl} = 55.96 \text{ mol}$$

By the stoichiometry of the reaction, the moles of hydrogen produced are the same as those of Cl_2 and twice as much as those of NaOH. The NaCl decomposed is at a 1:1 ratio to the NaOH produced (Table 4E10.2).

Table 4E10.1 Potential

SemiReaction	Standard Potential E° (V)
$H_2O + 1e- \rightarrow \frac{1}{2} H_2 + OH^-$	−0.8277
$Cl^- \rightarrow \frac{1}{2} Cl_2 + 1e-$	+1.3583
$Na^+ + Cl^- + H_2O \rightarrow NaOH + \frac{1}{2} H_2 + \frac{1}{2} Cl_2$	2.2860

Table 4E10.2 Mass Balance Results

Brine	50				
%	0.3				
	mol	g		mol	g
Cl				55.96	1986.53
H				55.96	55.96
NaOH				111.92	4476.68
NaCl	256,41	15,000		144.49	8452.85
H₂O	630,000	35,000		629,888.08	32,985.49

The energy ratio is given as follows:

$$\eta_{energy} = \frac{n_{NaCl} \cdot F \cdot V_{redoc}}{I \cdot t \cdot V} = \frac{111.92 \cdot 96500 \cdot 2.286}{2000 \cdot 3600 \cdot 5} = 0.668 \equiv 66.8\%$$

4.2.5.4 Production of HCl from NaCl (Mannheim process)

The set of reactions is given as:

$$NaCl + H_2SO_4 \rightarrow HCl + NaHSO_4 \quad [room]$$

$$NaCl + NaHSO_4 \rightarrow HCl + Na_2SO_4 \quad [300°C]$$

$$2NaCl + H_2SO_4 \rightarrow 2HCl + Na_2SO_4$$

The gas produced condenses at $-73°C$ as a colorless liquid and solidifies at $-112°C$ in the form of colorless crystals.

Alternatively, the HCl can also be obtained from the chlorination of alkanes as a byproduct:

$$CH_4 + Cl_2 \rightarrow HCl + CH_3Cl$$

$$CH_3Cl + Cl_2 \rightarrow HCl + CH_2Cl_2$$

$$CH_2Cl_2 + Cl_2 \rightarrow HCl + CHCl_3$$

$$CHCl_3 + Cl_2 \rightarrow HCl + CCl_4$$

4.2.5.5 Production of lime from calcium carbonate

In Solvay's process, calcium carbonate is used to produce a dilute stream of carbon dioxide for the carbonation process. The reaction is at equilibrium:

$$CaCO_3 \Leftrightarrow CaO + CO_2$$

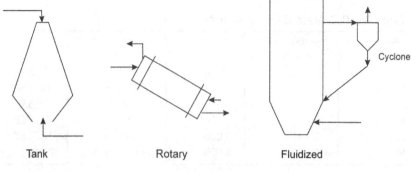

Tank   Rotary   Fluidized

FIGURE 4.27

Furnace designs for calcium carbonate decomposition.

To drive the equilibrium to products, the furnace operates at 1000–1200°C. There are a number of furnace designs (see also Fig. 4.27):

- Tank furnace: The product is low-purity CaO.
- Rotatory furnace: By feeding the gases and the calcium carbonate in countercurrent, the purity of the CaO produced is higher.
- Fluidized bed furnace: These are capable of producing the highest purity of CaO; the unit has the solids in suspension inside so that the rock decomposes.

EXAMPLE 4.11

In Solvay's process, calcium carbonate is used to produce CO_2 for the carbonation of the ammoniacal brine in a vertical continuous furnace. Flue gas is burned to generate the energy. The gas products are in countercurrent to the rock. The calcium carbonate is fed at 25°C and the lime exits the furnace at 900°C. The gas products leave the furnace at 250°C and the flue gas is fed at 725°C. Its composition is 9% CO_2, 2% O_2, 14% CO, and 75% N_2, and it is burned with a theoretical amount of air. The combustion is complete. Determine the flowrate of flue gas required (under standard conditions) to calculate 100 kg of calcium carbonate, assuming no energy loss and ignoring the moisture of the flue gas.

Solution

The problem is formulated as a mass and energy balance. The energy required for the decomposition of the calcium carbonate,

$$CaCO_3 \rightarrow CaO + CO_2,$$

is obtained by burning the CO in the flue gas:

$$CO + \tfrac{1}{2} O_2 \rightarrow CO_2$$

Furthermore, the flue gas is fed at a certain high temperature. The gas products leave at 250°C, but CaO and the air also need energy.

Thus the energy balance of the furnace is as follows:

$$\sum \Delta h_f \Big|_{r,ti} - \sum \Delta h_f \Big|_{p,tp} = 0$$

The energy required for the decomposition of the $CaCO_3$ is its formation heat. Since we process 100 kg of it, 1 kmol is our basis for the calculations:

$$\Delta H_R \Big|_{25°C} = \Delta H_{f,CaO} + \Delta H_{f,CO_2} - \Delta H_{f,CaCO_3} \Big|_{25°C} = (-151700) + (-94052) - (-289540)$$
$$= 43788 \frac{kcal}{kmol}$$

The reaction does not take place at 25°C, and the CaO exists at 900°C with no phase change:

$$c_{p,CaO}^{900°C} = 0.28 \frac{kcal}{kg°C}$$

$$c_{p,CaO}^{25°C} = 0.18 \frac{kcal}{kg°C} \Rightarrow c_{p,CaO} = 0.23 \frac{kcal}{kg°C}$$

$$Q_{CaO} = 1 \, kmol \frac{56 \, kg}{kmol} 0.23 \frac{kcal}{kg°C} (900°C - 25°C) = 11270 \, kcal$$

For the CO_2 we also need to adjust the enthalpy calculation since the CO_2 leaves at 250°C with the product gases:

$$Q_{CO_2} = 1 \, kmol \int_{25°C=298K}^{250°C=473K} 6.339 + 10.14 \cdot 10^{-3}T - 3.415 \cdot 10^{-6}T^2 = 2230.11 \, kcal$$

The total energy is the addition of the three terms:

$$Q_{requerido} = 43788 + 11270 + 2230.11 = 57288.11 \, kcal$$

The energy is provided by burning the CO. Using a basis of 100 kmol of flue gas, we have 14 kmol of CO, for which we need 7 kmol of O_2. We already have 2 kmol, and the rest (5 kmol) should come from the air. The composition of the air is 21% oxygen and 79% nitrogen. Thus:

$$N_2^{air} = \frac{79}{21} 5 = 18.8 \, kmol$$

The energy provided by the combustion of CO at 25°C and 1 atm is computed as:

$$\Delta H_R\Big|_{25°C} = n_{CO}\left(\Delta H_{f,CO_2} - \Delta H_{f,CO} - \frac{1}{2}\Delta H_{f,O_2}\Big|_{25°C}\right) = 14((-94052)-(-26416))$$

$$= -946904 \text{ kcal}$$

However, the nonreacting species exit at 250°C while they are fed at different temperatures. Air enters at 25°C and must be heated up to 250°C. The flue gas enters at 750°C and the unconverted gases leave at 250°C. Oxygen is assumed to be completely consumed:

$$Q_{\text{fluegas}} = \sum n_i \cdot \int_{298}^{997} c_{p,i}\, dT$$

where

$c_{p,CO_2} = 6.339 + 10.14 \cdot 10^{-3}T - 3.415 \cdot 10^{-6}T^2;\quad n_{CO_2} = 9 \text{ kmol}$

$c_{p,CO} = 6.350 + 1.811 \cdot 10^{-3}T - 0.2675 \cdot 10^{-6}T^2;\quad n_{CO} = 14 \text{ kmol}$

$c_{p,N_2} = 6.457 + 1.389 \cdot 10^{-3}T - 0.069 \cdot 10^{-6}T^2;\quad n_{N_2} = 75 \text{ kmol}$

$c_{p,O_2} = 6.117 + 3.167 \cdot 10^{-3}T - 1.005 \cdot 10^{-6}T^2;\quad n_{O_2} = 2 \text{ kmol}$

$Q_{\text{fluegas}} = 7922.36 \cdot 9 + 5172.31 \cdot 14 + 5119.92 \cdot 75 + 5386.05 \cdot 2 = 538480 \text{ kcal}$

The product gas, per consists of N_2 and CO_2. The CO_2 produced in the decomposition of the calcium carbonate is already accounted for:

$$Q_{\text{product}} = \sum n_i \cdot \int_{298}^{523} c_{p,i}\, dT$$

where

$c_{p,CO_2} = 6.339 + 10.14 \cdot 10^{-3}T - 3.415 \cdot 10^{-6}T^2;\quad n_{CO_2} = (10_{\text{fluegas}} + 14_{\text{comb}})$

$c_{p,N_2} = 6.457 + 1.389 \cdot 10^{-3}T - 0.069 \cdot 10^{-6}T^2;\quad n_{N_2} = (75_{\text{fluegas}} + 18.8_{\text{comb}})$

$Q_{\text{product}} = 2230.11 \cdot (10+14) + 1578.43 \cdot (75+18.8) = 201,579 \text{ kcal}/100 \text{ kmol of flue gas}$

The only variable is the mass of flue gas Ca_2CO_3:

$$Q_{\text{available}} = -n_{CO} \cdot \Delta H_{CO_2} + (n_{N_2} \cdot Q_{N_2gp} + n_{O_2} \cdot Q_{o_2gp} + n_{CO} \cdot Q_{COgp} +$$
$$n_{CO_2} \cdot Q_{CO_2gp}) - (n_{N_2} + n_{N_2air}) \cdot Q_{N_2,\text{product}} - (n_{CO_2} + n_{CO}) \cdot Q_{CO_2,\text{product}}$$

Thus the moles of flue gas are given as follows:

$$n_{\text{fluegas}} = \frac{Q_{\text{required}}}{Q_{\text{available}}} = \frac{57,288 \text{ kcal}}{12,860 \text{ kcal}} = 4.45 \frac{\text{kmol}_{\text{gas}}}{\text{kmol}_{CaCO_3}}$$

The energy balance can be formulated in a different form:

$$Q_{\text{in}} + Q_{\text{generated}} = Q_{\text{out}} + Q_{\text{absorbed}}$$

$$\sum_{j=\text{inlets}} \sum_i n_i \cdot \int_{T_{\text{ref}}}^{T_{\text{in},j}} c_{p,i} dT + Q_{\text{Combustion}} = \sum_{j=\text{outputs}} \sum_i n_i \cdot \int_{T_{\text{ref}}}^{T_{\text{out},j}} c_{p,i} dT + Q_{\text{Decomposition}}$$

Thus:

$$m_{\text{CaCO}_3} \cdot c_p \cdot (T_{\text{in}} - T_{\text{ref}}) + \sum_{i \in \text{air}} n_i \cdot \int_{T_{\text{ref}}}^{T_{\text{in}}} c_{p,i} dT + \sum_{i \in \text{fluegas}} n_i \cdot \int_{T_{\text{ref}}}^{T_{\text{in}}} c_{p,i} dT + n_{\text{CO}} \left| \Delta H_{\text{fCO} \to \text{CO}_2} \right|$$

$$= m_{\text{CaO}} \cdot c_p \cdot (T_{\text{out}} - T_{\text{ref}}) + \sum_{i \in \text{CO}_2(\text{from solids})} n_i \cdot \int_{T_{\text{ref}}}^{T_{\text{out}}} c_{p,i} dT + \sum_{i \in \text{fluegas}} n_i \cdot \int_{T_{\text{ref}}}^{T_{\text{in}}} c_{p,i} dT$$

$$+ n_{\text{CaCO}_3} \left| \Delta H_{\text{fCaCO}_3 \to \text{CaO}} \right|$$

Since air and $CaCO_3$ are at ambient temperature, the balance becomes:

$$n_{\text{fluegas}} \sum_{i \in \text{fluegas}} y_i \cdot \int_{T_{\text{ref}}}^{T_{\text{in}}} c_{p,i} dT + n_{\text{CO}} \left| \Delta H_{\text{fCO} \to \text{CO}_2} \right|$$

$$= m_{\text{CaO}} \cdot c_p \cdot (T_{\text{out}} - T_{\text{ref}}) + \sum_{i \in \text{CO}_2(\text{from solids})} n_i \cdot \int_{T_{\text{ref}}}^{T_{\text{out}}} c_{p,i} dT + \sum_{i \in \text{N}_2, \text{CO}_2} n_i \cdot \int_{T_{\text{ref}}}^{T_{\text{in}}} c_{p,i} dT$$

$$+ n_{\text{CaCO}_3} \left| \Delta H_{\text{fCaCO}_3 \to \text{CaO}} \right|$$

Now we compute the needs of the air and the gas products as a function of the feed of flue gas:

$$n_{\text{CO}_2} = n_{\text{CO}_2,\text{fluegas}} + n_{\text{fluegas}} \cdot y_{\text{CO}_2} = (y_{\text{CO}_2} + y_{\text{CO}}) \cdot n_{\text{fluegas}}$$

$$n_{\text{N}_2} = n_{\text{N}_2,\text{fluegas}} + n_{\text{N}_2,\text{air}} = n_{\text{fluegas}} \cdot y_{\text{N}_2} + n_{\text{N}_2,\text{air}}$$

$$n_{\text{N}_2,\text{air}} = 3.76 \cdot n_{\text{O}_2,\text{air}}$$

$$n_{\text{O}_2} = \frac{1}{2} n_{\text{CO},\text{fluegas}} = n_{\text{O}_2,\text{air}} + n_{\text{O}_2,\text{fluegas}} \Rightarrow n_{\text{O}_2,\text{air}} = n_{\text{fluegas}} \left(\frac{1}{2} y_{\text{CO},\text{fluegas}} - y_{\text{O}_2} \right)$$

Using these definitions in the energy balance, it becomes:

$$n_{\text{fluegas}} \sum_{i \in \text{fluegas}} y_i \cdot \int_{T_{\text{ref}}}^{T_{\text{in}}} c_{p,i} dT + n_{\text{fluegas}} \cdot y_{\text{CO}} \left| \Delta H_{f\text{CO} \to \text{CO}_2} \right| =$$

$$m_{\text{CaO}} \cdot c_p \cdot (T_{\text{out}} - T_{\text{ref}}) + n_{\text{CO}_2} \cdot \int_{T_{\text{ref}}}^{T_{\text{out}}} c_{p,i} dT + n_{\text{CaCO}_3} \left| \Delta H_{f\text{CaCO}_3 \to \text{CaO}} \right|$$

$$n_{\text{fluegas}} \left[(y_{\text{CO}} + y_{\text{CO}_2}) \int_{T_{\text{ref}}}^{T_{\text{out}}} c_{p,\text{CO}_2} dT + \left(y_{\text{N}_2} + 3.76(\frac{1}{2} y_{\text{CO}} - y_{\text{O}_2}) \right) \int_{T_{\text{ref}}}^{T_{\text{out}}} c_{p,\text{N}_2} dT \right]$$

Thus we can solve for the moles of flue gas:

$$n_{\text{fluegas}} = 4.45 \frac{\text{kmol}_{\text{gas}}}{\text{kmol}_{\text{CaCO}_3}}$$

The volume in standard conditions is calculated as follows:

$$V_{\text{fluegas}} = \frac{n_{\text{fluegas}} \cdot R \cdot T}{P} = \frac{4.45 \text{ kmol} \cdot 0.082 \frac{\text{atm} \cdot \text{m}^3}{\text{kmol} \cdot \text{K}} \cdot 273\text{K}}{1 \text{ atm}} = 99.7 \text{ m}^3$$

The *capture of CO$_2$* during the production of lime from calcium carbonate is an additional feature since the process is an equilibrium that can be displaced onwards and backwards under the proper operating conditions. Lately it has been used as a carbon capture technology. The equilibrium vapor pressure of CO_2 can be computed using Eq. (4.28):

$$P(\text{atm}) = \exp \left[7.079 - \frac{8308}{T(\text{K})} \right] \tag{4.28}$$

Using it we produce Fig. 4.28. Above the line, for high pressure, we have the carbonation reaction:

$$CaO + CO_2 \to CaCO_3$$

Below the line, calcination governs:

$$CaCO_3 \to CaO + CO_2$$

Therefore we can propose a flowsheet for the capture and regeneration of the bed as in Fig. 4.29. The flue gases containing the CO_2 are fed to the carbonator where a bed of CaO is made to react with the incoming CO_2. The operating

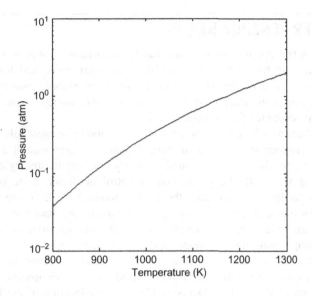

FIGURE 4.28

Vapor pressure of CO_2.

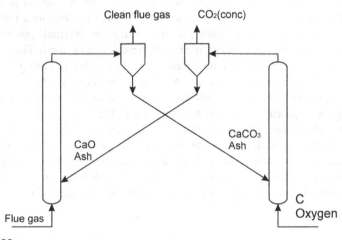

FIGURE 4.29

CO_2 capture system.

conditions are 650−700°C. This reaction is exothermic, generating 170 kJ/mol. The products of the reaction, $CaCO_3$ and CaO, are send to calcination. The decomposition of the calcium carbonate is endothermic, and coal and oxygen are fed to produce the energy to regenerate the CaO and produce a concentrated stream of CO_2.

4.3 WATER–ENERGY NEXUS

Water and energy are two interrelated natural resources. However, for decades water consumption has only been a problem in desert areas, and has therefore received little attention. Its price (low in most developed countries) is not important in economic analyses either. Therefore, only energy consumption and efficiency are accounted for in process design.

The production of energy requires a considerable amount of water. For instance, the production of power in thermal plants requires around 2 L/kWh. In this particular case, the water consumed is mostly that amount lost by evaporation in the cooling tower. In the production of petrol or diesel, there is a certain amount of water injected to extract the crude, around 2.5 L/L. These two examples show that although there is a strong link between both resources, each product and process determines the actual value, and most importantly, where it is possible to implement further water- and energy-saving technologies. When it comes to renewable resources, the use of water is more important since we need it to grow the crops in the case of biomass-based fuels and chemicals; it can be a limiting raw material in solar processes. The proper allocation of the facility has a large impact on water consumption. Native regions for crops do not require irrigation; for instance, no freshwater consumption is reported for the sugar cane produced in Brazil, which is the raw material of choice for the production of ethanol in that country. When it comes to solar energy, its availability is inversely correlated to water. Thus, concentrated solar power facilities require a fair amount of water if a wet cooling cycle is used, like in thermal power plants. Alternatively, dry cooling systems such as A-frames can be used. These systems consume 5–10% of the energy produced in the facilty for powering the fans—reducing the efficiency of the process—but they do not require water.

While on the one hand energy production (or fuels) requires energy, on the other hand energy is required to treat and transport water. The water used in any process comes out with increased levels of chemicals, particles, etc. This water needs to be reintegrated into the system to reduce the actual consumption, and for that the levels of contaminants must be reduced to the levels established by environmental regulations. Furthermore, water desalination is the only source of fresh water in many regions of the world, and it is an energy-intense process (Martín and Grossmann, 2015).

4.4 PROBLEMS

P4.1. To process a mixture of water and NaCl using reverse osmosis, a minimum theoretical pressure of 15 atm is required at room temperature. Determine the minimum pressure required to process a mixture with twice the NaCl concentration. The dissociation of NaCl is as shown in Table P4.1.

Table P4.1 Dissociation Degree

M	α
10^{-3}	0.966
10^{-2}	0.904
10^{-1}	0.780
1	0.660

P4.2. A solution of NaOH is to be produced from a solution of sodium carbonate with the mass composition given in Table P4.2.1.

Table P4.2.1 Sodium Carbonate Composition

	(%)
NaOH	0.6
Na_2CO_3	14.75
H_2O	84.65

The carbonate solution is processed with a mixture of CaO, $CaCO_3$, and calcium hydroxide to obtain a product with the composition shown in Table P4.2.2.

Table P4.2.2 Product Composition

Component	(%)
$CaCO_3$	14
$Ca(OH)_2$	0.65
Na_2CO_3	0.25
NaOH	10.25
H_2O	74.85

Per 100 kg/s of product:
1. Determine the amount of sodium carbonate solution.
2. Determine the amount of calcium mixture required.
3. Determine the limiting and excess reactants and the percentage in excess.
4. Compute the conversion of the process.

P4.3. In the production of CO_2 within the Solvay process, calcium carbonate with 100% purity is used. It is fed in countercurrent into a vertical furnace with a flue gas to be decomposed. The gases exit from the top of the furnace while the solids (CaO) are collected at the bottom. $CaCO_3$ is fed at 25°C and the CaO leaves at 900°C. Gas products leave the furnace at

250°C, and the flue gas is fed at 600°C with the following composition: 10% CO_2, 15% CO, and 75% N_2. Air is fed at 25°C in stoichiometric proportions to burn it. The combustion of the flue gas is complete. Compute the flowrate of flue gas in standard conditions to process 100 kg/s of $CaCO_3$. Assume that heat loss and gas moisture are negligible.

P4.4. A single-stage evaporator has a contact area of 30 m². It is used to concentrate 4000 kg/h of NaOH solution from 10% to 40% that is fed at 60°C. The hot utility is saturated steam that condenses at 115°C. In the evaporation chamber, the total pressure is 20 mmHg. A flow of 4000 kg/h of steam is required for the operation; see Fig. P4.4. Determine:
a. The heat loss as a ratio of the heat provided.
b. The global heat transfer coefficient.

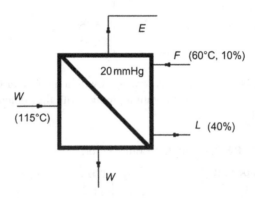

FIGURE P4.4

Scheme of an evaporator.

P4.5. A facility manager believes that the alkali solution they are buying does not meet specifications. It consists of NaOH, Na_2CO_3, and water. It reacts with a lime solution whose composition is given in Table P4.5.1.

Table P4.5.1 Lime Mass Composition

Limestone	(% w)
CaO	0.092
Ca(OH)$_2$	0.796
CaCO$_3$	0.111

The product has the composition given in Table P4.5.2.
Evaluate, for 100 kg of product:
1. Amount and composition of the alkali solution.
2. Amount of commercial lime required.
3. Reactant in excess and percentage in excess.
4. Conversion of the process.

Table P4.5.2 Product Mass Composition

Species	(% w)
$CaCO_3$	13.48
$Ca(OH)_2$	0.61
Na_2CO_3	0.36
NaOH	10.96
H_2O	74.59

P4.6. Calcium carbonate (100% purity) is fed to a vertical furnace in countercurrent with a flue gas to be decomposed for the production of CO_2 for the Solvay process. The resulting gases exit from the top of the furnace while the solids (CaO) are collected at the bottom. $CaCO_3$ is fed at 25°C and the CaO leaves at 875°C. Gas products leave the furnace at 225°C, and the flue gas is fed with the following composition: 10% CO_2, 15% CO, and 75% N_2. Air is fed at 25°C in a stoichiometric proportion to burn it. The combustion of the flue gas is complete. 5 kmol/s of flue gas are required to process 100 kg/s of $CaCO_3$. Assume that heat loss and gas moisture are negligible. Determine the inlet temperature of the flue gas.

P4.7. A solution of NaOH and $CaCO_3$ is to be produced by mixing an alkali solution (84.526% water and the rest NaOH and Na_2CO_3) with a commercial lime solution (($Ca(OH)_2$ + CaO) that contains $CaCO_3$ as an impurity). The product has the composition given in Table P4.7. Per 100 kg of product, and considering that we have an excess of $Ca(OH)_2$ of 4.45%, determine:
1. The amount and composition of the NaOH + $CaCO_3$ solution.
2. The amount and composition of the commercial lime solution.

Table P4.7 Product Composition

Species	(%)
$CaCO_3$	13.48
$Ca(OH)_2$	0.61
Na_2CO_3	0.28
NaOH	10.36
H_2O	75.27

P4.8. A single-stage evaporator processes 15,000 kg/s of an NaOH solution that is fed at 40°C. The solution concentration is 10%. Saturated steam at 3.5 atm (138.2°C and $\lambda = 513$ kcal/kg) is available as hot utility. It exits as saturated liquid. The economy of the steam is 0.785. The vacuum at the evaporation chamber is 620 mmHg with respect to 1 atm. The global heat transfer coefficient is 1600 kcal/m^2 h°C. Assuming no heat loss, compute: (a) hot utility flowrate, (b) evaporator surface area, and (c) composition of the concentrated solution.

P4.9. Compute the optimal ratio of NH_3 and NaCl to be fed into a Solvay Tower for maximum Na_2CO_3 production. Determine the amount of $NaHCO_3$ produced per ton of NaCl fed, the calcium carbonate consumed in the process, as well as the Na_2CO_3 produced.

P4.10. We would like to produce a flowrate of 2 kg/s of hydrogen at 30 bar via water electrolysis using different energy sources: solar, wind, or biomass. The oxygen produced must be compressed up to 100 bar for its storage. The compressors behave as polytropic, with an efficiency of 85% and $k = 1.4$, and the electrolysis takes place at 80°C, consuming 175.000 kJ/kgH$_2$. Assuming ideal water split so that we obtain pure hydrogen from the cathode and pure oxygen from the anode:
   a. Determine the energy required for the process.
   b. Suggest the best technology (see Table P4.10 for the processing parameters).
   c. Determine the cost of wind energy so that it becomes competitive, assuming a carbon tax of 40€/t of CO_2 and the fact that the ground required costs 25€/m^2.

Table P4.10 Technology Operating Data

Technology	Cost (€/kW)	CO$_2$ Produced (kg/kW)	Area (m^2/kW)
Wind	1300	0.5	0
Solar	1150	1	5
Biomass	1000	0	10

P4.11. A certain renewable source of energy produces 50,000 kW during June. We would like to use it to produce hydrogen. Hydrogen has to be compressed up to 30 bar and the oxygen is sold at 100 bar. Assuming perfect water splitting, pure hydrogen and oxygen streams from cathode and anode, respectively, a polytropic ($k = 1.4$) efficiency of 85% for the compressors, and the fact that the electrolysis takes place at 80°C and consumes 175.000 kJ/kgH$_2$:
   a. Compute the flow rate of hydrogen produced.
   b. Select the technology.
   c. Determine the cost of wind energy so that it becomes competitive, assuming a carbon tax of 40€/t of CO_2 and the fact that the ground required costs 25€/m^2 (Table P4.11).

Table P4.11 Technology Operating Data

Technology	Cost (€/kW)	CO$_2$ Produced (kg/kW)	Area (m^2/kW)
Wind	1300	0.5	0
Solar	1150	1	5
Biomass	1000	0	10

P4.12. Determine the ratio and amounts of NaCl and NH$_3$ to produce 10 kg/s of Na$_2$CO$_3$ using Solvay's process. The equilibrium in the tower yields 0.3 mol of NaHCO$_3$ per mol of initial mixture. Also compute the CO$_2$ required for the process and the calcium carbonate needed for the production of the Ca(OH)$_2$ used for ammonia recovery.

P4.13. A membrane module is available for the purification of a brine that contains 30,000 ppm of NaCl. We need to process 1000 kg/h, and the permeate represents 75% of the initial flow. The concentration of the permeate must be 200 ppm of NaCl. Determine the average pressure applied to the feed side. The discharge occurs at atmospheric pressure.
  Water permeability is 2.05×10^{-6} kg/m^2 s kPa.
  Salt permeability is 4.03×10^{-5} kg/m^2 s.

P4.14. The production of a solution of 35% HCl from hydrogen and chlorine is attempted. The gas product from the reaction has 95% HCl, and it is absorbed in water (see Fig. P4.14). Compute:
  a. The temperature of the converted gases and the energy to be removed from them.
  b. The energy removed so that the HCl is fed to the absorption tower at 25°C.

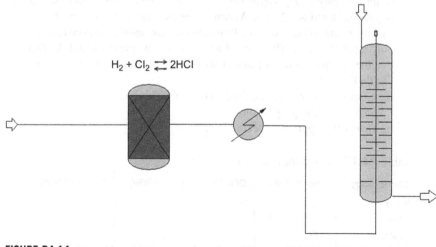

$$H_2 + Cl_2 \rightleftarrows 2HCl$$

FIGURE P4.14

Flowsheet for HCl production.

c. The water used in the process.

d. The heat removed for the absorption tower to operate isothermally.

$$\log(kp) = \frac{9586}{T(\text{K})} - 0.44 \log(T(\text{K})) - 2.16$$

P4.15. At what potential is Cl_2 released from an electrolytic cell that processes a solution of NaCl, 0.5 M?

P4.16. A certain renewable source of energy produces 75,000 kW during June. We would like to use it to produce hydrogen. Hydrogen has to be compressed up to 30 bar, and the oxygen is sold at 125 bar. Assuming perfect water splitting, pure hydrogen and oxygen streams from cathode and anode, respectively, a polytropic (k = 1.4) efficiency of 85% for the compressors, and the fact that the electrolysis takes place at 80°C with an energy ratio of 66%:

a. Compute the flowrate of hydrogen and oxygen produced.

b. Select a technology considering a carbon tax for emitting CO_2 of 40€/t and a cost for ground of 25€/m² (Table P4.16).

Table P4.16 Technology Operating Data

Technology	Cost (€/kW)	CO₂ Produced (kg/kW)	Area (m²/kW)
Wind	1300	0.5	0
Solar	1150	1	5
Biomass	1100	0	10

P4.17. An electrolytic cell is used to produce syngas for ammonia synthesis. Nitrogen comes from air separation while hydrogen is produced from water electrolysis. Hydrogen needs to be compressed up to 300 bar and the oxygen is stored at 125 bar. Assuming perfect water splitting, pure hydrogen and oxygen streams from cathode and anode, respectively, a polytropic (k = 1.4) efficiency of 85% for the compressors, and the fact that the electrolysis takes place at 80°C with a current efficiency of 66% (Table P4.17):

a. Compute the power required to obtain 50 kmol/s of syngas in stoichiometric proportions.

b. Select the appropiate technology.

Table P4.17 Technology Operating Data

Technology	Cost (€/kW)	CO₂ Produced (kg/kW)	Area (m²/kW)
Wind	1250	0.5	0
Solar	1150	1	5
Biomass	1100	0	10

P4.18. Syngas for the production of methanol is obtained from C and water in two processes in parallel.

— Using 10 kmol/s of C, the water gas cycle is used. Only 70% of the steam reacts with C. The WGSR equilibrium is established at 900K among the gases from the previous reaction. Assume the following equation for the equilibrium constant:

$$\log Kp = \left(\frac{2073}{T} - 2.029\right); T[=]K$$

— The proper H_2-to-CO ratio in the final mixture is obtained by mixing the resulting gas with hydrogen from electrolysis. The hydrogen must be compressed up to 50 bar and the oxygen stored at 125 bar. Assuming perfect water splitting, pure hydrogen and oxygen streams from cathode and anode, respectively, a polytropic (k = 1.4) efficiency of 85% for the compressors, and the fact that the electrolysis takes place at 80°C with a current efficiency of 66%, compute:

   a. The composition of the gases from the water gas cycle.
   b. The fraction of H_2 in the water cycle and that provided in electrolysis.
   c. The energy required for electrolysis and compression of the gases.

P4.19. A single-stage evaporator system is used to concentrate 10,000 kg/h of an NaOH solution from 10% to 35%. The feed is heated up using the residual heat from the condensed steam before entering the evaporator. The concentrated solution is cooled down using water in a second heat exchanger. Finally, the evaporated water is condensed in direct contact with water (see Fig. P4.19).

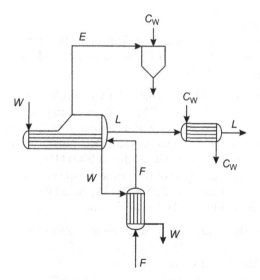

FIGURE P4.19

Evaporator scheme.

The hot utility is a saturated steam at 3.2 atm, which condenses at the evaporator and exits the heat exchanger at 40°C. The evaporation chamber is maintained at 150 mmHg vacuum with respect to 760 mmHg. The liquid from the direct contact cooling device is at dew point. The NaOH solution is fed to the system at 20°C and to the evaporator at 50°C. The cooling water used to cool down the concentrated solution enters at 20°C and leaves at 35°C. The global heat transfer coefficient is 2000 kcal/(m² h°C). Determine:
a. Solution boiling point.
b. Flowrate of hot utility.
c. Evaporator area.
d. Required cooling water.
e. Area of the heat exchangers. The global heat transfer coefficients are 800 and 400 kcal/ (m² h°C) for the feed heated and the solution cooler, respectively.

P4.20. A single-stage evaporator system is used to concentrate 10,000 kg/h of a NaOH solution from 10% to 45%. The feed is heated up using the residual heat from the condensed steam before entering the evaporator. The concentrated solution is cooled down using water in a second heat exchanger. The hot utility is a saturated steam at 3.5 atm, which condenses at the evaporator and exits the heat exchanger at 35°C. The evaporation chamber is maintained at 150 mmHg vacuum with respect to 760 mmHg. The NaOH solution is fed to the system at 20°C and to the evaporator at 45°C. A cooling water flowrate of 15,728 kg/h is used to cool down the concentrated solution, enters at 20°C, and leaves at 30°C. The global heat transfer coefficient is 1500 kcal/(m² h°C). Determine:
a. Solution boiling point.
b. Evaporator area.
c. Area of the heat exchangers. The global heat transfer coefficients are 650 and 125 kcal/(m² h°C) for the feed heated and the solution cooler, respectively.

P4.21. A membrane module is available for the purification of a flow of brine that contains 30,000 ppm of NaCl. We need to process 1000 kg/h, and the permeate represents 75% of the initial flow. The concentration of the permeate must be 200 ppm of NaCl. Determine the average pressure applied to the feed side. The feed is at 9000 kPa and the concentrate at 8500 kPa. The discharge of the permeate occurs at atmospheric pressure.
Water permeability is 2.05×10^{-6} kg/m² s kPa.
Salt permeability is 4.03×10^{-5} kg/m² s.

P4.22. Compute the needs of calcium carbonate and flue gas, with the composition 2% O_2, 78% N_2, and 20% CO required to operate a Solvay-based facility that produces 10 kg/s of sodium carbonate. The facility uses an excess of 10% molar NaCl. The furnace is fed with calcium carbonate, 100% at 25°C. The lime exits the furnace at 900°C. The gas products leave the unit at 250°C and the flue gas is fed at 600°C. The air used is at 25°C and assumed to be dry. Combustion is complete.

REFERENCES

Agthe, D.E., Billings, N.B., 2003. Managing Urban Water Supply. Springer science, Dordrecht.

Aporta, C.H., Martínez, P.E., Pasquevich, D.M., 2011. Estudio de los ciclos termoquímicos para la producción de hidrógeno nuclear Instituto de Energía y Desarrollo Sustentable – Comisión Nacional de Energía Atómica.

Bertram, B.M., 1993. Sodium Chloride. Kirk-Othmer Encyclopedia of Chemical Technology. Wiley.

Billet, R., 1998. Evaporation. ULLMANN'S "Encyclopedia of Industrial Chemistry". Wiley-VCH, Ruhr-Universität Bochum, Bochum, Federal Republic of Germany.

Davis, W., Martín, M., 2014. Optimal year-round operation for methane production from CO_2 and Water using wind energy. Energy 69, 497–505.

Earle, R.L., Earle, M.D., 2004. Evaporation. Unit Operations in Food Processing. The New Zealand Institute of Food Science & Technology (Inc.).

International Water Management Institute, 2007. Annual Report 2006/07;Colombo, Sri Lanka, 2006; http://www.iwmi.cgiar.org/About_IWMI/Strategic_Documents/Annual_Reports/2006_2007/theme1.html.

King, C.J., 1980. Separation Processes. McGraw-Hill, NewYork, NY.

Martín, M., 2014. Introduction to Software for Chemical Engineers. CRC Press, Boca Raton, FL.

Martín, M., Grossmann, I.E., 2015. Water—energy nexus in biofuels production and renewable based power. Sustainable Production and Consumption 2, 96–108. Available from: http://dx.doi.org/10.1016/j.spc.2015.06.005.

Martín, M., 2016. Optimal year-round production of DME from CO_2 and water using renewable energy. J. CO_2 Utilization 13, 105–113.

McCabe, W.L., 1935. The enthalpy-concentration chart: a useful device for chemical engineering calculations.. Trans. AIChE 31, 129–169.

McCabe, W.L., Smith, J.C., Harriot, P., 2001. Unit Operations. McGraw-Hill, New York, NY.

NEL Hydrogen, 2012. Technical data. http://www.nel-hydrogen.com/home/?pid = 75.

Ocon, J., Tojo, G., 1967. Problemas de ingeniería Química Tomo, first ed. Aguilar, Madrid.

Ortuño, A.V., 1999. Introducción a la Química Industrial. Reverté, Barcelona.

Perry, R.H., Green, D.W., 1997. Perry's Chemical Engineer's Handbook. McGraw-Hill, New York, NY.

Rauh, F., 1993. Sodium carbonate. Kirk-Othmer Encyclopedia of Chemical Technology. Wiley.

Solvay Alkali GmbH, 1998. Sodium carbonates: Christian thieme. ULLMANN'S "Encyclopedia of Industrial Chemistry". Wiley-VCH, Solingen, Federal Republic of Germany.

<http://brittany-harris.com/projects/water-use.html>

Syngas

5

5.1 INTRODUCTION

In this chapter, various carbonous raw materials, from biomass to fossil fuels (ie, coal, crude oil and natural gas) are evaluated for the production of syngas. Syngas refers either to the mixture of nitrogen and hydrogen, the kind of mixture needed for ammonia production, or carbon monoxide/hydrogen ($CO:H_2$) mixtures, the building blocks for the production of methanol, hydrocarbons, synthetic gasoline and diesel, or ethanol. Chapter 8, Biomass, is devoted to biomass processing, therefore here the technologies will be discussed, but applied to processing fossil resources, leaving further discussion of biomass-based fuels and chemicals to that chapter. Fig. 5.1 shows the relative availability of fossil resources worldwide. Coal is the major resource, and only in the Middle East oil is widely available. The picture was prepared before the industrial exploitation of shale gas, which, as it will be discussed, has changed the map. In the following paragraphs, the reader will find a brief description of the raw materials (covered throughout this chapter) that can be used to obtain syngas.

5.1.1 COAL

Coal has been widely used by mankind. Its consumption has increased steadily over the last 40 years, surpassing 7500 million tons (www.iea.org). The reserves are located in a few regions. Around 23% (122,000 million tons oil equivalent) is in the United States, 15% in Russia, 13% in China, 9% in Australia, 7% in India, 5% in Germany, 4% in Ukraine and Kazakhstan each, and 3.5% in Serbia. The production and consumption by region follow the same trend. The cost has been rather stable since 2000 (around $60/t), but peaked in 2008 at $140/t (http://www.infomine.com).

5.1.2 NATURAL GAS

Natural gas (NG) is one of the cleanest fuels. Its typical composition consists of 70–90% methane; 0–20% ethane, propane, and butane; 0–8% carbon dioxide (CO_2); 0–5% nitrogen; 0–5% hydrogen sulfide (H_2S); 0–0.2% oxygen; and traces of noble gases. The major proven reserves are in Russia (42,000 million tons of oil equivalent), followed by Iran and Qatar with half the amount of the Russian

Industrial Chemical Process Analysis and Design. DOI: http://dx.doi.org/10.1016/B978-0-08-101093-8.00005-7

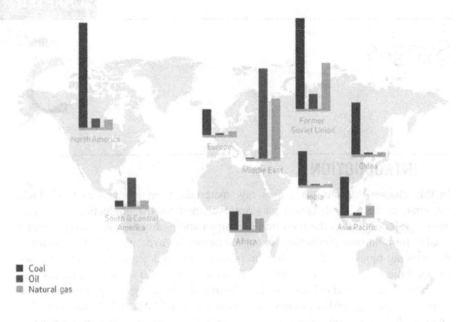

FIGURE 5.1

Fossil reserves.
http://www.worldcoal.org/bin/pdf/original_pdf_file/coal_matters_2_-_global_availability_of_coal
(16_05_2012).pdf "Coal Matters—Global Availability of Coal" (2012) The World Coal
Association, Copyright permission by the World Coal Association, London, UK.

reserves. Saudi Arabia, the United Arab Emirates, the United States, Nigeria, and
Algeria with 4000–6000 million tons oil equivalent are next. However, the natural
gas market has recently changed due to the development of technically feasible
solutions for exploiting unconventional reservoirs using horizontal drilling combined with fracking for the release of so-called shale gas. Shale gas is nothing
more than natural gas trapped in shale formations. The use of a high-pressure fluid
with chemicals to push the gas out has created new opportunities. The major
reserves of shale gas are located in the United States, Algeria, Canada, Mexico,
China, and Argentina, with over 500 trillion cubic feet available. As a result,
the current prices of natural gas have dropped to around $4/MMBTU (Henry
Hub prices). There are two qualities: dry and wet shale gas. The wet type contains
C2+ hydrocarbons in a certain proportion. It is more valuable since these chemicals
are used in industry for the production of high-value products, including plastics.

5.1.3 CRUDE OIL

Crude oil is a mixture of hydrocarbons. The standard unit is the barrel, which is
equal to 42 gal (1 gal = 3.785 L). Fig. 5.2 shows the typical composition of a

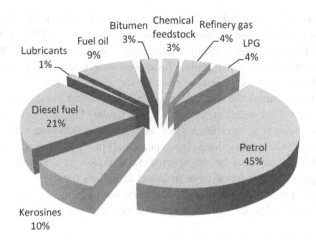

FIGURE 5.2

Typical composition of crude oil.

barrel. Petrol represents half of it, while diesel reaches around 25%. The major reserves of oil are located in the Middle East, representing 65% of the total worldwide (695,000 million barrels), with Saudi Arabia, Iraq, Kuwait, Iran, and the United Arab Emirates the most plentiful areas with 37.8%, 16.2%, 13.9%, 14.3%, and 14% shares, respectively. Venezuela with 76,000 million barrels, and Russia with 65,000 million barrels, are the main proven reserves on other continents. The price per barrel is quite volatile, and has varied from $90 in the 1980s down to $25 by 1997, increasing again to $100 by 2010. The EIA forecast covers a range from $60 to $200. By 2015 it was $30−$50 per barrel.

5.2 STAGE I: SYNGAS PRODUCTION

5.2.1 HYDROGEN AND H$_2$:CO MIXTURE PRODUCTION

Hydrogen, although one of the most abundant atoms, can only be found free in very small quantities in the atmosphere. It mostly appears in the form of water. It has two isotopes: deuterium (D) and tritium (T). Deuterium represents 0.015% of total hydrogen and is typically used for the production of heavy water (D$_2$O) to control nuclear reactions. Tritium is radioactive with a semidecay period of 12.5 years. The H−H bonding is strong, 436 kJ/mol. H$_2$ is slightly soluble in water, but reacts with most elements, generating hydrides with those less electronegative. Its small size allows its diffusion across steel. Furthermore, with carbon steel it generates methane, reducing the strength of the material. This quick diffusivity determines the design of the units that process it. With air it generates explosive mixtures; however, they result in a pressure decrease. The process generates implosions instead of explosions. It is not considered to be toxic. It is difficult to liquefy, and at 1 atm the melting and boiling points are $-259°C$ and $-253°C$, respectively. The critical temperature and pressure are 12.8 atm and $-240°C$. It has a peculiar characteristic—its Joule–Thompson coefficient is negative above 100K—so that it heats up upon expansion for a wide range of temperatures. The liquefaction requires the use of liquid nitrogen so that we can reach 80K for the expansion to allow entering the gas–liquid region. Fig. 5.3 shows the Hauser diagram. Compression is hazardous since leakages may generate explosive atmospheres. Hydrogen is transported by pipes, and in bottles under supercritical conditions, but it can also be transported as a liquid in cryogenic containers. The advantage is its high density. For instance, at 20K its density is 850 times higher than the gas phase, and therefore it has been used as fuel in rockets Reinmert et al. (1998). The typical uses of hydrogen as a raw material are the following:

- Production of ammonia with nitrogen.
- Production of hydrocarbons with CO and/or CO_2.
- Saturation of double bonds.
- Desulfurization of hydrocarbons, producing H_2S.
- As a reducing agent.

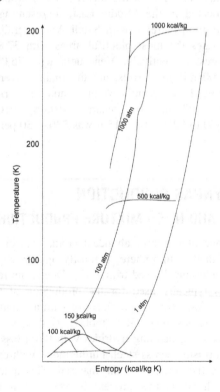

FIGURE 5.3

Hauser diagram hydrogen.

In the following pages the main methods for the production of hydrogen from hydrocarbons are presented. Remember that water electrolysis has already been covered in this book (see Chapter 4: Water). The method of choice depends on the economy of the raw material and its availability. Therefore, it is a dynamic problem.

5.2.1.1 Gas generator—water gas method

This method, developed by Bosch in 1920, is characterized by two distinct reactions for processing coal into syngas. On the one hand, the partial oxidation of the coal, and on the other hand, its processing with steam. Each of the main reactions is also affected by the chemical equilibrium among the species involved, Boudouard in the case of carbon oxidation, and water gas shift (WGS) in the case of steam processing.

5.2.1.1.1 Partial oxidation (gas generator)

This stage consists of flowing oxygen through a bed of incandescent coal. Limited air and a certain bed height are required for the production of CO and CO_2 together. In the first 15 cm of the bed C is combusted into CO and CO_2. When oxygen is no longer available, Boudouard equilibrium occurs so that the CO_2 reacts with the bed to produce CO (see the reactions below). The global process is exothermic:

$$2C_{(s)} + 1.5O_{2(g)} \rightarrow CO_{(g)} + CO_{2(g)}$$
$$C_{(s)} + CO_{2(g)} \leftrightarrow 2CO_{(g)} \quad \text{(Boudouard)}$$
$$3C_{(s)} + 1.5O_2 \rightarrow 3CO_{(g)}$$
$$C_{(s)} + \frac{1}{2}O_{2(g)} \rightarrow CO + 26.41 \text{ kcal/mol}$$

The Boudouard reaction is of particular interest. It is endothermic,

$$C_{(s)} + CO_{2(g)} \leftrightarrow 2CO_{(g)} \quad 41 \text{ kcal/mol}$$

with an equilibrium constant given by the following expression: Ortuño (1999)

$$K_p = \frac{P_{CO}^2}{P_{CO_2}} \Rightarrow \log(K_p) = 9.1106 - \frac{8841}{T}, \quad T[=]K \tag{5.1}$$

Pressure and temperature have an effect on the composition of the gas. High temperatures favor the production of CO (it is an endothermic reaction), while pressure reverts the equilibrium into CO_2; the number of moles is higher on the right-hand side. Let n_i be the moles of species i, X the conversion, and P_T the total pressure. Note that the solid coal does not take part in the equilibrium constant given by Eq. (5.2). Fig. 5.4 shows the effect of pressure and temperature in the equilibrium.

$$K_p = \frac{P_{CO}^2}{P_{CO_2}} = \frac{n_{CO}^2 \cdot P_T}{n_T \cdot n_{CO_2}} = \frac{(2X)^2 \cdot P_T}{(1+X) \cdot (1-X)} = \exp\left[9.1106 - \frac{8841}{T(K)}\right] \tag{5.2}$$

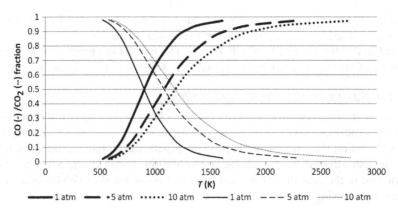

FIGURE 5.4

Boudouard equilibrium.

Thus, for the production of CO, we need limited availability of oxygen, a certain bed height, and high temperature (at the risk of melting the ash).

5.2.1.1.2 Steam processing (water gas)

The other stage in the process for syngas production from coal using this method is the use of steam. The main reaction produces CO and H_2. It is an endothermic reaction (31.35 kcal/mol). Thus, CO is favored at high temperature:

$$C_{(s)} + H_2O_{(g)} \rightarrow CO_{(g)} + H_{2(g)}$$

The reaction is an equilibrium whose constant can be written as follows:

$$\exp\left[\frac{-131,000}{(8.314 \cdot T(K)} + \frac{134}{8.314}\right] = \frac{(P_T \cdot n_{CO} \cdot n_{H_2})}{(n_{H_2O} \cdot (n_{Total}))} \quad (5.3)$$

The second reaction that occurs is the well-known *water gas shift reaction (WGSR)*, given by the following equation:

$$CO_{(g)} + H_2O_{(g)} \leftrightarrow CO_{2(g)} + H_{2(g)}$$

It is an exothermic reaction affected by temperature. The wide application of this reaction deserves more attention. Thus, a more detailed discussion is needed.

The equilibrium between the four species is as follows:

$$CO_{(g)} + H_2O_{(g)} \leftrightarrow CO_{2(g)} + H_{2(g)} \quad \Delta H = -41 \text{ kJ/mol}$$

It is widely used for modulating the concentration of CO and hydrogen in a stream, in the production of hydrogen from steam reforming of hydrocarbons, or in carbonous mass gasification. The equilibrium is reversible, and the reverse WGS can also be used if needed. The reaction is an equilibrium whose constant has been widely evaluated in the literature. The following are several correlations by Graaf et al. (1986), eq. (5.4) and by Susanto and Beenackers (1996), eqs. (5.5) and (5.6):

$$\log K_p = \log\left(\frac{P_{CO_2} P_{H_2}}{P_{H_2O} P_{CO}}\right) = \log\left(\frac{n_{CO_2} n_{H_2}}{n_{H_2O} n_{CO}}\right) = \left(\frac{2073}{T(K)} - 2.029\right) \qquad (5.4)$$

$$\log(K_p) = \frac{2232}{T(K)} + 0.0836\log(T(K)) + 0.00022T(K) + 2.508 \qquad (5.5)$$

$$K_p = 0.0265 e^{\left[\frac{33,010}{8.314 \cdot T(K)}\right]} \qquad (5.6)$$

There is some flexibility for locating the WGS reactor within the processes, depending on the presence of sulfur compounds in the gas. It can be located either before the sulfur removal step (sour shift) or after sulfur removal (sweet shift).

Sweet operation of the WGSR occurs in two steps at high and low temperature, respectively. They are known as high-temperature shift (HTS) and low-temperature shift (LTS).

A conventional *high-temperature sweet shifting* operates between 315°C and 530°C and uses chromium- or copper-promoted, iron-based catalysts. The catalysts are calcined over 500°C so that the Fe reaches α phase. Next, they are reduced at temperatures between 315°C and 460°C. Their half-time life is around three years. The difficulty in removing the energy generated in the reaction makes them work adiabatically. Thus, the feed is at around 310−360°C and 10−60 atm, and it is allowed to heat up. The conversion of CO is over 96%. The spatial velocity of the gas is typically 300−4000/h.

Low-temperature sweet shifting (up to 370°C) uses copper−zinc−aluminum catalysts. For instance, ICI catalyst contains 30% by weight of CuO, 45% of ZnO, and 13% of Al_2O_3. Furthermore, the use of supported Cu over chromium oxide (15−20% of CuO, 68−73% of ZnO, and 9−14% of Cr_2O_3) constitutes the Haldor Tøpsoe commercial catalyst that operates from 220°C to 320°C. The aim of this catalyst is to achieve 99% conversion of CO operating at 3−40 atm. The feed enters the reactor at 200−220°C and operates adiabatically. The spatial velocity of the gas is typically 300−4000/h. The half-life time of the catalyst ranges from two to four years.

A second type of WGSR is *sour WGSR*. For processing gases containing H_2S, a catalyst based on cobalt−molybdenum is used. It is possible to add Li, Na, Cs, or K. The reaction is typically located after the water scrubber, where syngas is saturated with water at about 200−250°C. Furthermore, this catalyst can also convert carbonyl sulfide and other organic sulfur compounds into H_2S. The reactor operates adiabatically, reaching conversions of CO of around 97%. Apart from packed beds, lately membrane reactors are gaining attention. They are based on the particularly high diffusivity of hydrogen through metals;

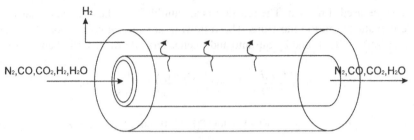

FIGURE 5.5

Membrane WGSR.

see Fig. 5.5. Thus, the reaction occurs in the inner pipe and the hydrogen crosses the palladium-based wall. The purity of the hydrogen produced reaches 99.99% (Osenwengie Uyi, 2007).

Let us illustrate the operation and yield of WGS reactors in a numerical example.

EXAMPLE 5.1

A WGS reactor works at 600K and the feed consists of 36% CO, 30% CO_2, 20% steam, and 14% hydrogen.

a. Compute the conversion of the reaction.
b. Plot the profile of the conversion with the temperature.

Hint: use the correlation for K_p given by the following equation:

$$K_p = 0.0265 e^{\left[\frac{33,010}{8.314T}\right]}$$

Solution

a. We write the equation of the constant (see Eq. 5.4) as follows:

$$K_p = \frac{(n_{CO_2} + X)(n + X)}{(n_{CO} - X)(n_{H_2O} - X)} = 0.0265 e^{\left[\frac{33,010}{8.314T}\right]}$$

Solving for X we choose the lowest solution: $X = 16.4$.
With respect to the limiting reactant, H_2O, the conversion is

$$X = \frac{16.4}{20} = 0.82.$$

b. For a range of temperatures, we compute the conversion following the same procedure as before (Fig. 5E1.1).

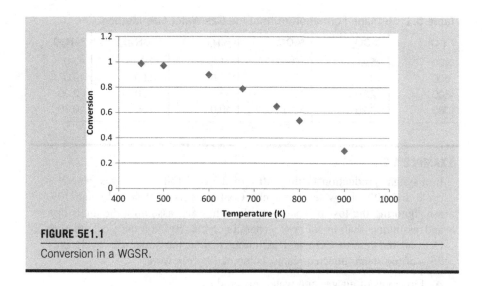

FIGURE 5E1.1

Conversion in a WGSR.

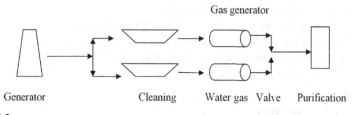

FIGURE 5.6

Coal-based syngas production.

Now that both reactions, the gas generator, and the water gas path have been evaluated, the production of syngas consists of using both to process the coal and adjust the composition of the syngas.

$$\text{Air gas } 3.76\,N_2 + O_2 + 2C \rightarrow 2CO + 3.76N_2 \quad \text{Exothermic}$$
$$\text{Water gas } H_2O + C \rightarrow CO + H_2 \quad\quad\quad\quad \text{Endothermic}$$

Fig. 5.6 shows the scheme for the process dividing the use of coal between the two processes. In Example 5.2 we evaluate the performance of such a process.

Unfortunately, at pressures above atmosphere and low temperatures, methane is produced following this reaction:

$$C_{(s)} + 2H_{2(g)} \leftrightarrow CH_{4(g)}$$

When the temperature rises, the equilibrium reverses. The equilibrium constants for the reaction for a couple of temperature values are $\text{Log}_{10}\,K_p\,(25°C) = 9$ and $\text{Log}_{10}\,K_p\,(100°C) = -2$. Table 5.1 shows the effect of temperature on methane production.

Table 5.1 Methane Formation in the Flue Gas/Water Gas Method

T (°C)	$\%CO_2$	$\%CO$	$\%H_2$	$\%CH_4$	$\%H_2O$
400	25.2	0.50	11.3	20.0	43.0
600	19.5	15.5	37.5	8.50	19.0
800	2.50	46.0	48.0	1.50	2.00
1000	0.20	50.0	49.0	0.50	0.30

EXAMPLE 5.2

In a syngas production facility, 40% of the coal follows the gas generator cycle and 60% the water gas cycle. Coal consists of 90% coke and 10% ash. Ignoring the loss of coke with the ashes and assuming the air is dry, and assuming that in the gas generator cycle 50% of the coal goes to CO and the rest to CO_2, and assuming that in the water gas cycle only 75% of the steam injected reacts, compute the following:

A. Flowrates of air gas and water gas produced.
B. Energy available in the two gas streams.
C. Fraction of the energy in the coal remaining at each gas flow.
D. Amount of air and water needed to process 10 t of coal.
E. Repeat the example assuming humid air at $T_s = 25°C$, $\varphi = 0.65$, and $P_T = 720$ mmHg.

Solution

Each cycle is analyzed separately.

Air Cycle (Gas Generator)
In the first case, the air is dry. Therefore, the only reaction that takes place is that given by the combustion of C. We assume that half of the C forms CO and the other half produces CO_2. The stoichiometry is as follows:

$$2C + \frac{3}{2}O_2 + \frac{3}{2}3.76N_2 \rightarrow CO + CO_2 + \frac{3}{2}3.76N_2$$

$$2C + 1.5O_2 + 5.64N_2 \rightarrow CO + CO_2 + 5.64N_2$$

We use 1 kg of coal as the basis. Thus, 0.9 kg of C and 0.1 kg of ash are processed; 40% is processed through the air cycle. Therefore, the carbon burnt is equal to 0.4×0.9 kg $= 0.36$ kg of C. Based on the stoichiometry of the reaction, the mass balances to the species are as follows:

$$C_{react} = 0.4 \cdot 0.90 \text{ kg} \frac{1 \text{ kmol}}{12 \text{ kg}} = 0.03 \text{ kmol C}$$

$$O_{2,needed} = \frac{1.5}{2}C_{react} = 0.0225 \text{ kmol } O_2$$

$$N_2 = 3.76\, O_{2,needed} = 0.0846 \text{ kmol } N_2$$

$$CO_{prod} = CO_{2,prod} = \frac{1}{2}C_{react} = 0.015 \text{ kmol}$$

Table 5E2.1 Gas Generator Composition

	Air Gas			
Species	kmol	kg	m^3 (c.n.)	%
N$_2$	0.0846	2.3688	1.894	73.8
CO$_2$	0.0150	0.66	0.336	13.1
CO	0.0150	0.42	0.336	13.1
Total	0.1146	3.4488	2.565	100.00

Table 5E2.2 Water Gas Composition

	Water Gas			
Species	kmol	kg	m^3 (c.n.)	%
H$_2$	0.045	0.090	1.007	42.85
CO	0.045	1.260	1.007	42.85
H$_2$O	0.015	0.270	0.336	14.29
Total	0.105	1.620	2.457	100.00

The composition of the gas produced is given in Table 5E2.1.

In the *water cycle* case, the reaction that takes place is the following, where we assume that no WGSR occurs (Table 5E2.2):

$$C + H_2O \rightarrow CO + H_2$$

$$C_{react} = 0.60 \cdot 0.9 \text{ kg} \frac{1 \text{ kmol}}{12 \text{ kg}} = 0.045 \text{ kmol C}$$

$$H_2O_{needed} = \frac{H_2O_{theoretical}}{0.75} = \frac{C_{react}}{0.75} = 0.060 \text{ kmol } H_2O$$

$$H_2O_{noreact} = 0.25 \ H_2O_{needed} = 0.015 \text{ kmol } H_2O$$

$$CO = C_{react} = 0.045 \text{ kmol CO}$$

$$H_2 = C_{react} = 0.045 \text{ kmol } H_2$$

The ratio between both is defined by this equation:

$$\text{Ratio} = \frac{\text{Generator gas}}{\text{Water gas}} = \frac{4.238 \text{ kg}}{1.62 \text{ kg}} = 2.61; \quad \frac{H_2}{CO} = \frac{0.045}{0.045 + 0.015} = 0.75$$

Energy in Gas Generator

The energy available in the gas is that resulting from burning the CO:

$$CO + \frac{1}{2}O_2 \rightarrow CO_2$$

$$\Delta H_r = \Delta h_{f,CO_2} - \Delta h_{f,CO} = -94.052 + 26.416 = -67.636 \ \frac{kcal}{mol \ CO}$$

The energy per unit volume is given as:

$$1.0 \ m^3_{cN}CO \rightarrow \frac{1 \ kmol \ CO}{22.4 \ m^3_{cN}CO} \ \frac{10^3 mol \ CO}{1 \ kmol \ CO} = 44.643 \ \frac{mol \ CO}{m^3_{cN}CO}$$

$$Pot.Cal_{Air \ gas} = 44.643 \ \frac{mol \ CO}{m^3_{cN} \ CO} \ 67.636 \ \frac{kcal}{mol \ CO} \ \frac{0.336 \ m^3_{cN}CO}{2.565 \ m^3_{cN} \ Generator \ gas}$$

$$= 395.2 \ \frac{kcal}{m^3_{cN} \ Generator \ gas}$$

For the *energy in water gas*, we can burn hydrogen and CO:

$$CO + \frac{1}{2}O_2 \rightarrow CO_2$$

$$\Delta H_r = \Delta h_{f,CO_2} - \Delta h_{f,CO} = -94.052 + 26.416 = -67.636 \ \frac{kcal}{mol \ CO}$$

Per unit volume it is as follows:

$$1.0 \ m^3_{cN}CO \rightarrow \frac{1 \ kmol \ CO}{22.4 \ m^3_{cN}CO} \ \frac{10^3 mol \ CO}{1 \ kmol \ CO} = 44.643 \ \frac{mol \ CO}{m^3_{cN}CO}$$

$$Pot.Cal_{Water \ gas,CO} = 44.643 \ \frac{mol \ CO}{m^3_{cN}CO} \ 67.636 \ \frac{kcal}{mol \ CO} \ \frac{1.007 \ m^3_{cN}CO}{2.350 \ m^3_{cN} Water \ gas}$$

$$= 1294 \ \frac{kcal}{m^3_{cN} Water \ gas}$$

For the hydrogen, the combustion energy is the enthalpy of formation of steam:

$$H_2 + \frac{1}{2}O_2 \rightarrow H_2O$$

$$\Delta H_r = \Delta h_{f,H_2O} = -57.7979 \ \frac{kcal}{mol \ H_2}$$

Given the per unit volume we have the following:

$$1.0 \; m^3_{cN}H_2 \; \frac{1 \; kmol \; H_2}{22.4 \; m^3_{cN}H_2} \; \frac{10^3 mol \; H_2}{1 \; kmol \; H_2} = 44.643 \; \frac{mol \; H_2}{m^3_{cN} \; H_2}$$

$$Pot.Cal_{Water \; gas,H_2O} = 44.643 \; \frac{mol \; H_2}{m^3_{cN}H_2} \; 57.7979 \; \frac{kcal}{mol \; H_2} \; \frac{1.007 \; m^3_{cN}H_2}{2.350 \; m^3_{cN} \; Water \; gas}$$

$$= 1104 \; \frac{kcal}{m^3_{cN} Water \; gas}$$

$$Pot.Cal_{Water \; gas} = Pot.Cal_{Water \; gas,CO} + Pot.Cal_{Water \; gas,H_2} = 2489 \; \frac{kcal}{m^3_{cN} Water \; gas}$$

Fraction of Energy in Original Raw Materials

Our raw material was coal that when completely burned could produce the following amount of energy:

$$C + O_2 \rightarrow CO_2$$
$$\Delta H_r = \Delta h_{f,CO_2} = -94.053 \; \frac{kcal}{mol \; C}$$

Using the same basis:

$$1000 \; g \; Coal \Rightarrow 0.9 \; \frac{g \; C}{g \; Coal} \; 1000 \; g \; Coal \; \frac{1 \; mol \; C}{12 \; g \; C} \; 94.052 \; \frac{kcal}{mol \; C} = 7054 \; kcal$$

$$\eta = \frac{395.2 \; \frac{kcal}{m^3_{cN} Generator \; gas} \; 2.565 \; m^3_{cN} Generator \; gas}{7054 \; kcal} \cdot 100 = 14.37\% \; In \; Gas \; generator$$

$$\eta = \frac{2398 \; \frac{kcal}{m^3_{cN} Water \; gas} \; 2.350 \; m^3_{cN} \; Water \; gas}{7054 \; kcal} \cdot 100 = 79.90\% \; In \; Water \; Gas$$

Coal Processing

To process 10 t of coal, simply multiply all the previous results by 10^4:

$$H_2O_{needed} = 0.06 \; kmol \; H_2O/1 \; kg \; C$$

$$H_2O_{needed,10 \; t} = 0.06 \frac{kmol \; H_2O}{kg \; Coque} \; \frac{18 \; kg}{kmol}\bigg|_{Agua} \; 10 \; t = 10.8 \; t$$

Table 5E2.3 Dry Air Used

Component	Dry Air per kg of Coal		
	kmol	kg	m^3_{cN}
O_2	0.0225	0.72	0.503
N_2	0.1128	3.15	2.525
Total	0.353	3.87	3.028

The air needed (Table 5E2.3) is this:

$$Air_{needed,10\,t} = 3.028\,\frac{m^3_{cN}Air}{kg\,Coal}\,10,000\,kg = 30,280\,m^3_{cN}Air$$

In case the air is humid, the reactions taking place are not only those corresponding to burning C to CO and CO_2, but also those of the water gas cycle:

$$2C + \frac{3}{2}(O_2 + 3.76N_2) \rightarrow CO + CO_2 + \frac{3}{2}(3.76N_2)$$

$$C + H_2O \rightarrow CO + H_2$$

For the initial basis of calculation of 1 kg of coal (90% of C), 40% is processed using this cycle; thus 30 mol of C are processed. We assume that the moisture of the air is completely consumed. The moisture in the air is computed as follows:

$$y = 0.62\,\frac{\varphi P_v}{P_T - \varphi P_v} = 0.62\,\frac{0.65 \cdot 23.547}{720 - 0.65 \cdot 23.547} = 0.013\,kg_{vapor}/kg_{dryair}$$

$$y\frac{29}{18} = y_m$$

$$4.76\,\frac{mol_{dryair}}{mol\,O_2}y_m = 0.1033\,\frac{mol\,H_2O}{mol\,O_2}$$

The elementary mass balance proceeds as follows:

$$30C + a(O_2 + 3.76N_2 + 0.1033H_2O) \rightarrow bCO + dCO_2 + eH_2 + f(N_2)$$

Balance to C

$$30 = b + d$$

Balance to O

$$2a + 0.133a = b + 2d$$

Balance to H

$$2 \cdot 0.1033a = 2e$$

CO comes from two reactions

$$b = d + e$$

Solving the system we have the following:

$a = 21.233$ mol
$b = 16.096$ mol
$d = 13.903$ mol
$e = 2.193$ mol
$f = 3.76 \cdot a = 79.83$ mol

The air required for this case—assuming ideal gases ($PV = nRT$)—is 2.304 Nm^3. The gas produced has a volume of 2.508 Nm^3. Thus, the energy contained in the gas is again due to the presence of hydrogen and CO:

$$CO + \frac{1}{2}O_2 \rightarrow CO_2$$

$$\Delta H_r = \Delta H_{f,CO_2} - \Delta H_{f,CO} = 94.052 - 26.416 = 67.636 \ \frac{kcal}{mol \ CO}$$

$$H_2 + \frac{1}{2}O_2 \rightarrow H_2O$$

$$\Delta H_r = \Delta H_{f,H_2O} = 57.7979 \ \frac{kcal}{mol \ H_2}$$

$$\Delta H_{1 \ m^3} = 44.6 \ mol \left(67.6 \ \frac{0.360 \ m^3 CO}{2.508 \ m^3 \ gas} + 57.8 \ \frac{0.049 \ m^3 \ H_2}{2.508 \ m^3 \ gas} \right) = 479 \ kcal$$

The part of the problem corresponding to the water cycle does not change. Therefore the total power for the generator gas is 1215 kcal, representing 17.2% of the energy in the coal.

5.2.1.2 Coal distillation

Coal distillation is the chemical process that decomposes a carbonous material by heating it up to high temperatures in the absence of air. It is basically a pyrolysis that allows the production of a number of species, from coke to coal gas, gas carbon, Buckmisterfullerene, and hydrogen—when operating from 800°C to 1000°C. As the coal is heated, several processes take place:

- Up to 100°C gases such as O_2, N_2, and even CH_4 and H_2O, are desorbed.
- From 100°C to 300°C some gases, such as H_2S, CO_2, CO, and small hydrocarbons, are released due to the breakage of radicals, sulfur ligands, and acid groups.
- The first liquid fractions appear at around 310°C. Gases are also produced due to the breakage of the liquids.

- From 400°C to 450°C the coal starts to melt, becoming plastic, so that the volume of the mass contracts. Paraffinic chains melt above 500°C, and gases and liquid hydrocarbons are also produced.
- From 550°C melting finishes and the coal is no longer a plastic. The fraction of gases and liquids produced is lower, and we produce coke.
- From 500°C to 600°C component A depolymerizes, generating C12 particles; hydrogen is also released.
- From 700°C to 1000°C there is only cracking. The main component of the gases is hydrogen that, as it is released, generates porosity in the coal mass, providing mechanical resistance. Liquid production is low.
- Above 1000°C coke transforms into graphite.

Typically, from 100 kg of coal we obtain 72 kg of coke, 22 kg of gas (hydrogen, ethylene, ethane, methane, CO, CO_2, and nitrogen), and 6 kg of tar containing mainly benzene, toluene, xylene, phenol, naphthalene, anthracene, and liquor.

5.2.1.3 Gasification

Gasification is the partial oxidation of a carbon substrate into a gas with a low heating value; it proceeds via a series of reactions that occur in the presence of a gasification agent such as air, oxygen, and/or steam. To produce hydrogen, gas reforming after gasification is required. There are a number of designs for the gasifiers. They are first classified as either *fixed or fluidized beds*. Fig. 5.7 shows four designs: (a) a fixed bed with *countercurrent* solids and gas circulation; (b) a *downdraft*-type fixed bed where the carbonous material and the gases exit the unit from the bottom; (c) a *bubbling*-type fluidized bed where only one unit is involved; and (d) a *circulating*-type fluidized bed that requires solid gas separation to recycle the solids. The product gas composition is highly dependent on the

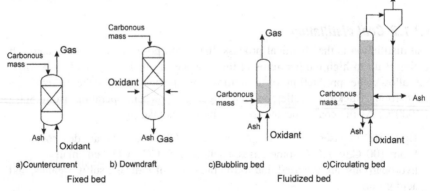

FIGURE 5.7

Gasifier designs.

configuration, the raw materials, and the operating conditions. Brigdwater (1995) presented some tables that gathered the yields of different designs. The operating temperatures range from 700°C to 1000°C and the pressure from slightly above atmospheric to 30 atm. Typically, the largest production capacity is obtained using fluidized beds.

Lately, gasification has been used to process biomass to obtain syngas. This has been used for the production of different biofuels such as ethanol, Fischer—Tropsch (FT) liquids, or hydrogen (Martín and Grossmann, 2013). The two basic designs are either the one based on the FERCO/Battelle design or the one by the Gas Technology Institute (GTI). Fig. 5.8 shows the scheme for the GTI gasifier. It is a bubbling-type gasifier that operates above atmospheric pressure (5—20 atm) and 800—900°C. It consists of one single unit, and in order to avoid gas dilution, oxygen (instead of air) and steam are used as fluidification agents. The gas produced has a low concentration of hydrocarbons, but the fraction of CO_2 is large. It is called Renugas. Fig. 5.9 shows the scheme of the FERCO/Battelle gasifier. The system consists of two units: the gasifier and the combustor. In the gasifier, steam is mixed with the carbonous material. The energy for the gasification is provided by hot sand. The solids, sand and char, are separated in the cyclone, and both are sent to a combustor where air is used to burn the char and reheat the sand. A cyclone separates the flue gas from the solids (the sand), which are recycled to the gasifier. The syngas produced in this system has a higher concentration of hydrocarbons but typically lower concentration of CO_2. Furthermore, since the combustion of the char takes place in a second unit, air can be used without diluting the syngas.

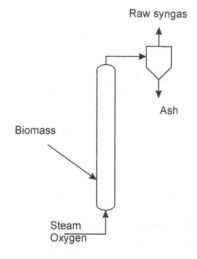

FIGURE 5.8

Renugas gasifier design.

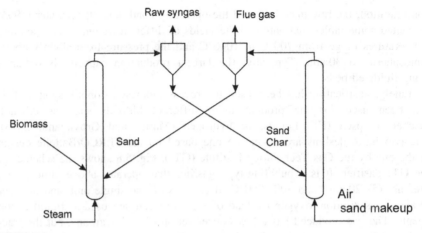

FIGURE 5.9

FERCO/Battelle gasifier system.

The main reactions can be divided into several types:

— Oxygen-based reactions:

$$C + O_2 \rightarrow CO_2$$
$$C + 0.5O_2 \rightarrow CO$$
$$CO + 0.5O_2 \rightarrow CO_2$$

— Steam-based reactions:

$$C + H_2O \rightarrow CO + H_2$$
$$C + 2H_2O \rightarrow CO_2 + 2H_2$$
$$CO + H_2O \rightarrow CO_2 + H_2$$

— Methane formation:

$$C + 2H_2 \rightarrow CH_4$$
$$2C + 2H_2O \rightarrow CH_4 + CO_2$$
$$CO + 3H_2 \rightarrow CH_4 + H_2O$$
$$CO_2 + 4H_2 \rightarrow CH_4 + 2H_2O$$

— The Boudouard reaction:

$$C + CO_2 \rightarrow 2CO$$

— Other reactions (eg, hydrocarbon production):

$$H_2 + 0.5O_2 \rightarrow H_2O$$
$$CH_4 + 2O_2 \rightarrow CO_2 + 2H_2O$$
$$C_2H_4 + 3O_2 \rightarrow 2CO_2 + 2H_2O$$
$$C_nH_{2n} + H_2 \rightarrow C_nH_{2n+2}$$
$$C_nH_m + nH_2O \rightarrow nCO + (n + m/2)H_2$$

If gasification is the first stage to hydrogen production, after gas purification, a WGS stage is needed.

5.2.1.4 Thermal decomposition/partial oxidation

1. Hydrocarbon + Heat → H_2 + Lamp Black

   This reaction is of interest for the production of lamp black more than for hydrogen; the hydrogen is not pure enough.

2. Hydrocarbon + Heat → H_2 + Unsaturated Hydrocarbon

   For instance, the production of ethylene from methane:

$$2CH_4 \leftrightarrow C_2H_2 + 2H_2$$

$$C_2H_2 \leftrightarrow 2C + H_2$$

3. Hydrocarbon + O_2 + Cooling → H_2 + CO   (partial oxidation)

Operating under limited availability of oxygen, this reaction not only produces energy, but also syngas. The yield to hydrogen is lower than when using steam. The oxygen needed can be fed as pure oxygen if the syngas is due to produce fuels, or air if nitrogen is required (eg, in the production of ammonia). The presence of nitrogen typically results in an increased size of all the units downstream, and it is not recommended.

5.2.1.5 Steam reforming of hydrocarbons

In the section on gasification, reforming was mentioned as a means to enhance the proportion of hydrogen by decomposing the hydrocarbons into CO and hydrogen. If the feedstock is already made of hydrocarbons, reforming is the first stage towards syngas:

$$\text{Hydrocarbon} + \text{Steam} \rightarrow H_2 + CO$$

$$C_2H_2 + 2H_2O \rightarrow 2CO + 3H_2$$

$$CH_4 + H_2O \rightarrow CO + 3H_2$$

The most representative case is methane steam reforming. The process involves a coupled equilibria, that of the decomposition of the methane and that of the WGS. The constants are presented below Davies and Lihou (1971):

$$CH_4 + H_2O \leftrightarrow CO + 3H_2 \quad K_p = 10^{\left[\frac{-11,650}{T} + 13.076\right]} \quad T[=]K \qquad (5.7)$$

$$CO_{(g)} + H_2O_{(g)} \leftrightarrow CO_{2(g)} + H_{2(g)} \quad K_p = 10^{\left[\frac{1910}{T} - 1.784\right]} \quad T[=]K \qquad (5.8)$$

Steam reforming is an endothermic process that requires a large amount of energy. Furthermore, the products show a larger number of moles. Therefore, pressure drives the equilibrium to the reactants, not to the product. The WGSR is exothermic and increases the proportion of hydrogen. Here we present the kinetics for this particular packed bed reactor, where y represents the molar fraction:

$$r_1 = \frac{\rho_b \cdot A_1 \cdot e^{[-Ea_1/RT]} \cdot P_{CH_4}}{(1 + K_a \cdot P_{H_2})}$$

(5.9)

$$r_2 = \rho_b \cdot A_2 \cdot e^{[-Ea_2/RT]} \left(y_{CH_4} y_{H_2} - \frac{y_{CO_4} y_{H_2}}{K_{eq_2}} \right)$$

$\rho_b = 1200 \ \text{kg/m}^3$
$A_1 = 5.517 \times 10^6 \ \text{mol/kg s atm}$
$Ea_1 = 1.849 \times 10^8 \ \text{J/mol}$
$R = 8.314 \ \text{J/mol K}$
$P = 30 \ \text{atm}$
$K_a = 4.053 \ \text{atm}^{-1}$
$A_2 = 4.95 \times 10^8 \ \text{mol/kg/s}$
$Ea_2 = 1.163 \ \text{J/mol}$
$K_{eq2} = \exp(-4.946 + 4897/T) \ T[=]K$

The *reforming of natural gas for the production of hydrogen* consists of four stages, namely, purification, reforming, WGS, and gas purification. Fig. 5.10 shows the scheme of the process. The natural gas must be purified to remove sulfur compounds that are poisonous to the catalysts downstream. Typically a hydrodesulfurization stage transforms the organic sulfur compounds into H_2S using as catalyst an alumina base impregnated with cobalt and molybdenum (usually called a CoMo catalyst) operating at 300–400°C and 30–130 atm in a fixed bed reactor. For example, ethanethiol (C_2H_5SH):

$$C_2H_5SH + H_2 \rightarrow H_2S + C_2H_6$$

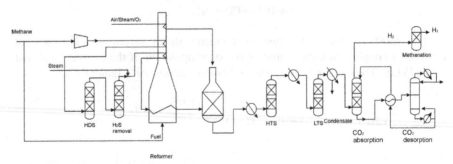

FIGURE 5.10

Production of hydrogen through steam reforming.

After that, a bed of ZnO, operating at 150–200°C, is used to remove H_2S:

$$ZnO + H_2S \rightarrow H_2O + ZnS$$

Alternatively, iron-based oxides (Fe_2O_3) operating at 25–50°C can also be used:

$$Fe_2O_3 + 3H_2S \rightarrow Fe_2S_3 + 3H_2O$$
$$2Fe_2S_3 + 3O_2 \rightarrow 2FeO_3 + 6S$$

At this point, the gas is fed to the reformer. Remember that hydrocarbon reforming is endothermic and a fraction of the feed is used to provide that energy. The primary reformer uses Ni/Al_2O_3 as a catalyst, and the secondary one, if used, is Ni supported on ceramics. Around 10% of the feed is typically used as fuel. In the reforming reaction, CO is produced, as seen above. Thus WGS is carried out in two stages: an HTS at 350°C using Fe_3O_4 as a catalyst, and an LTS at 200°C using CuO as catalyst, operating adiabatically, to achieve more than 97% conversion. Finally, the CO_2 generated is eliminated. Several technologies that will be presented in the following section can be used, such as absorption in ethanolamines or adsorption in fixed beds, pressure swing adsorption (PSA) systems. Finally, the traces of CO and CO_2 are eliminated by methanation. The reaction consumes a small amount of hydrogen using Ni/Al_2O_3 as catalyst:

$$CO + CO_2 + 7H_2 \rightarrow 2CH_4 + 3H_2O + Heat$$

The purity achieved is above 99.95 mol%. Fig. 5.11 shows a scheme of the reformer furnace. The process typically requires 1.3 Mg/d of steam at 2.6 MPa, and produces 1.9 Mg/d at 4.8 MPa.

In the *production of ammonia*, the reforming system consists of two reformers (see Fig. 5.11). A primary reformer is fed with desulfurized gas that has been preheated to 400–600°C. The furnace consists of a number of high-nickel chromium alloy tubes filled with a nickel-based catalyst. The heat to the process is provided by burning fuel, typically a part of the supply, requiring from 7 to 8 GJ per ton of ammonia. In the secondary reformer, a packed bed reactor, the nitrogen needed for preparation of the synthesis gas is added. Air is compressed and heated on the convection section of the primary reformer and fed to the secondary reformer. Next, HTS and LTS are carried out to produce more hydrogen by reducing the CO generated in the reforming stage. Throughout the chapter, the complete process will be described.

In Table 5.2 the comparisons between fired furnaces and gas-heated reformers are presented, justifying the use of the second ones.

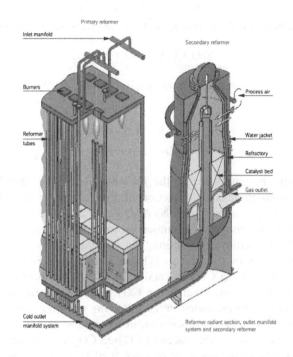

FIGURE 5.11

Reforming system.

With permission from Thyssenkrupp, Uhde, Ammonia ThyssenKrupp (www.uhde.com) http://www.thyssenkrupp-csa.com.br/fileadmin/documents/brochures/0a2d5391-b166-484d-847d-3cbfd941f06b.pdf, Uhde (2015).

Table 5.2 Advantages of Gas-Heated Reformers

Fired Furnace	Gas-Heated Reformers
Large volumes	Smaller volumes
Larger surface area and heat loss	Reduced surface area and heat loss
Complex instrumentation	Simplified instrumentation
High maintenance costs	Low maintenance costs
Large convection zone	No convection zone
Stack losses	No stack losses
High fixed capital costs	Low fixed capital costs
Reduced catalyst tube loss from high temperature and uneven heat distribution	Longer tube life due to uniform heat distribution
Increased downtime required for shutdown	Reduced downtime required for shutdown
Well-established process	Yet to gain wide acceptance

EXAMPLE 5.3

A novel process for the production of hydrogen from propane is being evaluated. In the reformer, two main reactions take place:

$$C_3H_{8(g)} + 3H_2O_{(g)} \rightarrow 3CO_{(g)} + 7H_{2(g)}$$

$$CO_{(g)} + H_2O_{(g)} \rightarrow CO_{2(g)} + H_{2(g)}$$

Propane steam reforming occurs in a reactor whose pipes are packed with nickel catalyst. The feed to the reactor consists of a mixture of steam–propane with a molar ratio of 6:1 at 125°C. The products of the reforming stage leave the reactor at 800°C. The excess of steam assures the complete consumption of propane.

To provide the energy for the endothermic reactions, flue gas consisting of N_2 and 22% of CO_2 is used. The gas is fed at a rate of 4.94 m^3/mol of C_3H_8 at 1400°C and 1 atm, and exits the reactor at 900°C. Assume adiabatic operation. This stream is further integrated to produce low-pressure steam at 125°C from water at 25°C. The flue gas leaves the heat exchanger at 250°C. Determine the molar composition of the gas product and the amount of steam produced.

Solution

Since we are not given the equilibrium constants, we formulate the problem based on elementary balances and an energy balance as follows:

$$Q_{gas} = \sum_{i=CO_2,N_2} n_i \int_{1673\ K}^{1173\ K} c_{p,i} dT$$

$$Q_{in} + Q_{react} = Q_{out}$$

$$Q_{gas} + \sum_{i=C_3H_8,H_2O} n_i \int_{298\ K}^{398\ K} c_{p,i} dT + \sum_i \Delta H_{r,\text{reactants}} =$$

$$\sum_i \Delta H_{r,\text{products}} + \sum_{i=CO_2,CO,H_2,H_2O} n_i \int_{298K}^{1073K} c_{p,i} dT$$

The elementary balances are as follows:

$$\text{Bal C:} 3n_{C_3H_8} = n_{CO} + n_{CO_2}$$

$$\text{Bal O:} n_{H_2O} = n_{CO} + 2n_{CO_2} + n_{H_2O,sal}$$

$$\text{Bal H:} 8n_{C_3H_8} + 2n_{H_2O} = 2n_{H_2} + 2n_{H_2O,sal}$$

Using 1 kmol of propane as the basis for the calculations, we solve the problem (Table 5E3.1).

Table 5E3.1 Results of Example 5.3

	A	B	C	D	E	F	G	H	I	J	K
16											
17			Inlet			Outlet					
18			kmol	kg	Sensible in	kmol	kg	Sensible Out	Reaction		Comp gas
19		C3H8	1	44	2019.08979	0	0	0	24861.244		
20		H2O	6	108	4886.81702	0.53258564	9.58654159	3705.05311	316005.615		4.096812645
21		CO				0.53258564	14.912398	3076.53663	-14094.4076		4.096812645
22		CO2				2.46741436	108.566232	20913.8727	-232438.696		18.98011043
23		H2			0	9.46741436	18.9348287	43516.1498	0		72.82626428
24		Temp	398			1073					
25	kmol gas	Q			6905.90681			71211.6123	94333.7549		
26	36.8007843	N2	28.7046118	803.72913	296519.481	28.7046118	803.72913	186769.663		45332.3503	
27	comp CO2	CO2	8.09617255	356.231592	127128.2	8.09617255	356.231592	78238.5573		20092.3604	
28	0.22	Temp	1673			1173				523	
29					423647.681			265008.221		65424.7107	
30					Q in + Q react		Qout				
31					165545.367	165545.367	-1.0215E-08				
32											
33		Balance al C	3			3					
34		Balance al O	6			6					
35		Balance al H	20			20					
36						kg	kmol				
37		Steam generated				84.7848386	4.71026881				

FIGURE 5E3.1

Solver model.

In EXCEL Solver, the model is written as in Fig. 5E3.1

5.2.1.6 Dry reforming and autoreforming

So far steam reforming and partial oxidation have been considered to produce syngas or hydrogen. There are two more alternatives. The first one is the combination of the previous two. *Autoreforming* uses the energy generated in the partial oxidation of a fraction of the feed to provide the energy required for the steam reforming reactions. In this way, the operation is adiabatic. The stoichiometry depends on the energy balances. For example:

$$3CH_4 + O_2 + H_2O \rightarrow 3CO + 7H_2$$

Dry reforming is a novel technology in which CO_2 is used to treat hydrocarbons to produce syngas. The reaction is endothermic and the yield to hydrogen is low, but it represents a process where CO_2 becomes a carbon source to be reintegrated in the process:

$$C_2H_2 + 2CO_2 \rightarrow 4CO + H_2$$

5.2.2 NITROGEN PRODUCTION

See Section 3.2, Air Separation.

5.3 STAGE II: GAS PURIFICATION

Typically, in the production of syngas, apart from CO and hydrogen, a number of species are also produced. We have already discussed the processing of hydrocarbons through reforming for their decomposition. However, the presence of CO_2 and sulfur compounds creates a poisonous environment for the reactions downstream (Reep et al., 2007). In this section, two cases are considered: the production of ammonia and the production of synthetic hydrocarbons and fuels.

5.3.1 AMMONIA PRODUCTION (SYNGAS $N_2 + H_2$)

The CO_2 in contact with ammonia and water produces carbonates. This was the desired reaction in the Solvay process, but not in this case since they deposit on the catalyst:

$$CO_2 + NH_3 + H_2O \rightarrow NH_4HCO_3 \rightarrow (NH_4)_2CO_3$$

The sulfur compounds, if not eliminated before, poison the catalyst since they target Fe.

The presence of CO_2, O_2, and H_2O also negatively affect the catalyst by deposition.

CO can react with the catalyst, producing copper formate.

5.3.2 SYNTHESIS OF FT TYPE OF FUELS (SYNGAS CO + H_2)

In this case the major concern is the removal of sour gases, CO_2, and H_2S, since the latter are poisonous to the catalyst and the former negatively affect the reaction.

5.3.2.1 *Sour gas removal*

Sour gas removal is a more general case of carbon capture. The technologies have been used at industrial scale over the last few decades, but only now are they attracting more attention Plasynski and Chen (1997), GPSA (2004).

5.3.2.1.1 Alkali solutions

These are used for the simultaneous removal of CO_2 and H_2S via absorption; Fig. 5.12 shows a scheme. The use of ethanolamines is based on a series of consecutive reactions between the sour gases and the alkali solution. The absorption occurs at low temperature (29°C) and moderate pressure (29 atm). To avoid dilution of the solution by the condensed water, it is separated upon cooling. There are a number of rules of thumb to determine the flow of amine as a function of the type. Typically, for monoethanolamine, a solution of 15−25% by weight with a correction factor of 0.25−0.4 based on a mol-to-mol-based reaction is used. The reaction is exothermic, and therefore the exit liquid heats up. The reactions for capturing the CO_2 are as follows:

$$2H_2O \leftrightarrow OH^- + H_3O^+$$

$$CO_{2\,(g)} \leftrightarrow CO_{2\,(aq)}$$

$$CO_{2(aq)} + 2H_2O \leftrightarrow HCO_3^- + H_3O^+$$

$$HCO_3^- + H_2O \leftrightarrow CO_3^{2-} + H_3O^+$$

$$CO_{2(aq)} + RNH_2 \leftrightarrow RNH_2^+COO^-$$

$$RNH_2^+COO^- + RNH_2 \leftrightarrow RNH_3^+ + RNHCOO^-$$

$$RNH_2^+COO^- + H_2O \leftrightarrow H_3O^+ + RNHCOO^-$$

$$RNH_2^+COO^- + OH^- \leftrightarrow H_2O + RNHCOO^-$$

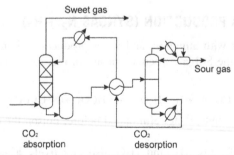

	Duty (BTU/h)	Area
Reboiler	72,000 GPM	11.30 GPM
Condenser	30,000 GPM	5.20 GPM
Amine feed to distillation	45,000 GPM	11.25 GPM

FIGURE 5.12

CO_2 capture using alkali solutions (GPSA, 2004).

where the absorption kinetics are given by the equations below (Kierzkowska-Pawlak and Chacuk, 2010):

$$r = k_2[\text{Amine}][CO_2] = k_{ob}[CO_2]$$

$$k_{ob} = k_2[\text{Amine}]_0 = 2.07 \cdot 10^9 \exp\left(\frac{-5912.7}{T(\text{K})}\right)[\text{Amine}]_0 \qquad (5.10)$$

$$k_2[=]\text{m}^3\text{kmol}^{-1}\text{s}^{-1}$$

As long as:

$$3 < Ha << E_i$$

$$3 < \frac{\sqrt{k_{ob}D_{CO_2}}}{k_L} << 1 + \frac{D_{\text{amine}}[\text{Amine}]}{\nu_{\text{amine}}D_{CO_2}[CO_2]_i}$$

$$[CO_2]_i = p_{CO_2}/H; \nu_{\text{amine}} = \text{stoichiometry}_{CO_2-\text{amine}}; D_i = \text{Diffusivities}(m/s); k_L = Sh \cdot D_{CO_2}/d_b$$

$$\log(H) = 5.3 - 0.035^\circ C_{\text{amine}} - 1140/T(\text{K})$$

$$(5.11)$$

The solution is regenerated by distillation. The column operates at 1.7 atm and is fed with the solution at 93°C. The distillate exits at 54°C and the bottoms at 125°C. The energy involved in the different heat exchangers can be computed based on rules of thumb as a function of the flow of amine in gallons per minute, see Figure 5.12 (GPSA, 2004)

EXAMPLE 5.4

A stream of 40 kmol/s at 34.5 atm and 298K of syngas with 10% CO_2 is processed to remove 99.5% of it before further stages. The equilibrium data for the system CO_2−amine are given in the table below. The *L/G* ratio is 1.5, the minimum. Determine the number of stages in a gas−liquid absorption tower for the removal of CO_2 from syngas assuming 25% efficiency per stage (Table 5E4.1).

Table 5E4.1 Equilibrium Data

mol CO_2/mol Amine	P_{CO2}(Pa)
0.025	65
0.048	197
0.145	1204
0.276	3728
0.327	4815
0.45	8253
0.9	26,325
10	1,480,620
15	2,918,315
20	4,722,992

Solution

We first draw the XY curve where X represents the moles of gas per mol of solution free of gas and Y is the moles of CO_2 in the gas per mol of stream free of CO_2:

$$X(\text{free of } x) = \frac{x}{1-x}$$

$$Y(\text{free of } y) = \frac{y}{1-y}$$

$$G' = G(1-y)$$

$$G'(Y_{n+1} - Y_1) = L'(X_n - X_o)$$

Since $y_{n+1} = 0.1$ and $y_1 = 0.005$, the corresponding Ys are as follows (Table 5E4.2):

$$Y_{n+1} = 0.11 \text{ and } Y_1 = 0.005$$

If we plot the equilibrium data for Y_{n+1}, we have $X_{max} = 16.5$, and thus:

$$\left.\frac{L'}{G'}\right|_{min} = \frac{Y_{n+1} - Y_1}{X_{max}} = 0.00641$$

$$G' = G(1-y) = 36 \text{ kmol/s}$$

$$L'/G' = 0.0096$$

$$L' = 4156 \rightarrow X_{max} = 11.02 \text{ mol/mol}$$

We also plot the operating line (triangles) and the minimum operating line (squares) (Fig. 5E4.1).

Therefore, we need three stages. With an efficiency of 0.25, 12 trays are needed for the operation.

Table 5E4.2 Equilibrium Data

mol CO_2/mol Amine	P_{CO2}(Pa)	mol CO_2/mol Inert
0.025	65.4479554	1.89705E-06
0.048	197	5.71018E-06
0.145	1204	3.48998E-05
0.276	3728	0.00010807
0.327	4815	0.000139585
0.45	8252.67269	0.000239265
0.9	26,324.9963	0.000763626
10	1,480,619.55	0.044840925
15	2,918,315.23	0.092405305
20	4,722,991.64	0.158612026

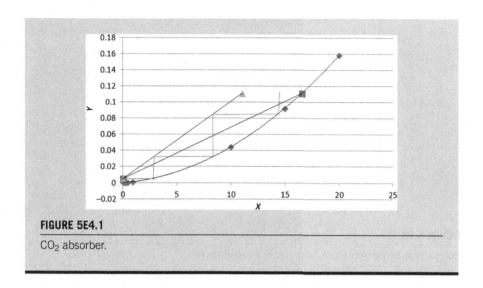

FIGURE 5E4.1

CO_2 absorber.

5.3.2.1.2 Physical absorption

For operating pressures above 50 atm, physical absorption is widely extended using different solvents that define the processes, such as Selexol (using polyethylene glycol dimethyl ether, UOP) or Rectisol (using methanol, Lurgi). In general, these processes operate at very low temperatures and high pressures. Thus, only if the gas is already at high pressure is it interesting to use them. The flowsheet is similar to the one presented for the use of alkali solutions, an absorber and a desorber column.

Currently, the *Selexol process* operates on a license by UOP LLC. The solvent, a mixture of dimethyl esters of polyethylene glycol, absorbs or dissolves the sour gases at 20−140 atm. The solvent is regenerated by reducing the pressure. The Selexol process can selectively separate CO_2 and H_2S so that the latter can be sent to the Claus process to obtain elementary sulfur and from it sulfuric acid; the CO_2 can be further used. Below 2 MPa the absorption capacity of the solvent is greatly reduced and alkali-based processes are more interesting.

The *Rectisol process (Lurgi)* uses methanol under cryogenic conditions (−40°C) and from 27.6 atm to 68.9 atm for the removal of H_2S and CO_2. It is also selective with the gases.

Fig. 5.13 shows the comparison between physical and chemical sorbents. Physical sorbents show a linear absorption capacity with the partial pressure of the gas, and for them to be competitive, the pressure must be higher than when using alkali solutions. However, the energy involved in the regeneration of the solvent is lower.

5.3.2.1.3 Pressure swing adsorption (PSA)

The operation of PSA systems for the removal of sour gases is similar to the one presented in Chapter 3, Air, where they were used to separate nitrogen

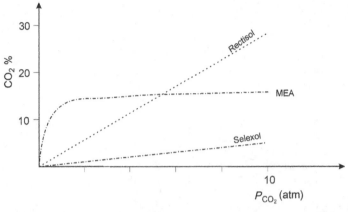

FIGURE 5.13

Comparison between physical and chemical absorption.

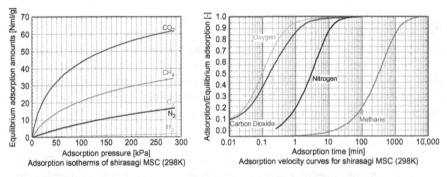

Adsorption isotherms of shirasagi MSC (298K)   Adsorption velocity curves for shirasagi MSC (298K)

FIGURE 5.14

Absorption isotherm and capacity of molecular sieves for CO_2 capture.

Reproduced with permission from Osaka Gas Chemicals. http://www.jechem.co.jp/shirasagi_e/tech/psa.html,

Osaka Gas Chemicals (2015).

and oxygen. The difference is the adsorbent bed. The operating conditions are low temperatures (25°C) and moderate pressures (4−6 atm). Typically, CO_2 is the one removed. In Fig. 5.14 the adsorption capacity for molecular sieve carbon is shown. Zeolites can also be used.

5.3.2.1.4 Membrane separation

The use of membranes is based on the different diffusion rate of the gases through the membranes; see Chapter 3, Air, for further discussion. The materials of the membranes are typically the same as those used as adsorbent beds so that the membranes are of the type known as gas absorption membranes. On the other side of the membrane, a carrier is used to withdraw the CO_2 molecules that diffuse through it. No hydrostatic pressure is required. The drawback is that the carrier must be regenerated, similar to when using alkali solutions.

5.3.2.1.5 Cryogenic separation

It is possible to liquefy and condense CO_2 by compression and cooling. However, the compression rates required and the small fraction of CO_2 in the gas result in high energy consumption. This method is only recommended for streams with concentrations over 60% in CO_2.

5.3.2.1.6 Mineral storage and capture

This process was described in Chapter 4, Water, when the equilibrium for the decomposition of the calcium carbonate was presented. Calcium carbonate as well as other chemicals can be used.

5.3.2.2 Removal of H_2S

This was briefly depicted for the reforming of natural gas. This stage takes place from 150°C to 400°C following the reaction below in an adsorbent bed or in a scrubber:

$$H_2S + ZnO \rightarrow ZnS + H_2O$$

Instead of using a bed of ZnO, it is possible to use iron compounds. They operate optimally at 25−50°C. Condensation of water on the catalyst should be avoided since it reduces the contact area.

$$Fe_2O_3 + 3H_2S \rightarrow Fe_2S_3 + 3H_2O$$

$$2Fe(OH)_3 + 3H_2S \rightarrow Fe_2S_3 + 6H_2O$$

In the *Claus process* the recovered H_2S can be further processed to recover the S in it and produce H_2SO_4. This process was discovered by Carl Friedrich Claus, who obtained a patent in 1883. The process has a thermal and a catalytic stage.

In the *thermal stage*, the H_2S is burned to produce SO_2 that reacts with the feed of H_2S to produce sulfur, as in the following reactions:

$$2H_2S + 3O_2 \rightarrow 2SO_2 + 2H_2O$$

$$2H_2S + SO_2 \rightarrow 3S + 2H_2O$$

$$10H_2S + 5O_2 \rightarrow 2H_2S + SO_2 + 7/2S_2 + 8H_2O$$

Two thirds of the H_2S are converted to elementary sulfur.

In the *catalytic stage*, the Al_2O_3 or titanium oxide IV is used as catalyst. The reaction progresses as follows:

$$2H_2S + SO_2 \rightarrow 3S + 2H_2O$$

The sulfur obtained can be in the form of S_6, S_7, S_8, or S_9. This stage can be subdivided into three phases: heating up, catalytic reaction, and cooling with

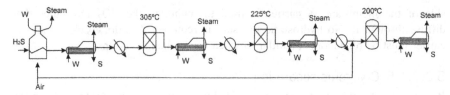

FIGURE 5.15

Scheme of the Claus process (W, water; S, sulfur).

condensation. Heating up avoids condensation of sulfur on the catalytic bed. The first catalytic stage operates at 315–330°C. For the next two stages, the bottom of the bed is at 240°C and 200°C, respectively, since the conversion is favored. Finally, the gas is cooled down to 130–150°C and the excess energy is used to produce steam. Fig. 5.15 shows the scheme of the process.

5.3.2.3 Removal of CO (ammonia synthesis)

The WGSR is used to convert CO into CO_2. A two-step method for sweet gases is employed. First the gas stream is passed over a Cr/Fe_3O_4 catalyst at 360°C, and later over a $Cu/ZnO/Cr$ catalyst at 210°C.

The remaining CO can be separated at low temperatures or by absorption in copper solutions. In the first case, after the removal of CO_2, methane and CO are cooled down to $-180°C$ at 40 atm, condensing both. The mixture is expanded to 2.5 atm and distilled to obtain CO from the top and methane from the bottom. The second method can use several copper aqueous solutions such as CuCl in HCl (at 300 atm, recovering the CO by expansion at 50°C), and solutions of CuCl and $AlCl_3$ in toluene operating at 25°C and 20 atm. The CO is recovered at 100°C and 1–4 atm.

5.3.2.4 Water removal (ammonia synthesis)

The gas mixture is cooled to 40°C, at which temperature the water condenses out and is removed.

5.3.2.5 Removal of carbon oxides: methanation (ammonia synthesis)

The remaining CO_2 and CO after carbon capture are converted to methane (methanation) using an Ni/Al_2O_3 catalyst at 325°C:

$$2CO + 3H_2 \rightarrow CH_4 + H_2O$$

$$CO_2 + 4H_2 \rightarrow CH_4 + 2H_2O$$

The water produced in the reactions is removed by condensation at 40°C.

5.4 STAGE III: SYNTHESIS

5.4.1 AMMONIA

5.4.1.1 Introduction

History: Ammonia was already known back in ancient times. There are up to three possible origins for the word ammonia. It is typically related to the discovery of sal ammoniac (ammonium chloride) near the Temple of Zeus Ammon, in the Siwar Oasis, Libya. The word means "sand" in Greek. It was reported that the salt was found beneath the sand. Pliny also reported the existence of another hammoniacum close to the oracle of Ammon. It was a plant secretion deposited in the sand as droplets. The last possible origin is also related to the Temple of Zeus Ammon since the priest, while burning camel dung for fuel, observed the formation of some white crystals. It was not until 1716 that ammonia was linked to rotten food. A few years later, in 1727, it was obtained from lime and ammonium chloride, as reported in the Solvay process:

$$CaO + H_2O \rightarrow Ca(OH)_2$$
$$Ca(OH)_2 + NH_4Cl \rightarrow CaCl_2 + NH_4OH$$
$$NH_4OH \leftrightarrow NH_3 + H_2O$$

Free ammonia was first prepared in 1774 by J. Priestly using ammonium carbonate, and in 1785, the chemical formula was discovered by C. Berthollet. In the 1840s ammonia was a byproduct in coke production. The current process for the production of ammonia is based on the research by Fritz Haber from 1880 to 1900 on the equilibrium of the reaction between nitrogen and hydrogen. In 1908 BASF and Haber came to an agreement to design an industrial plant. In the 5 years working with Karl Bosch, they developed the so-called Haber–Bosch process. The first plant dates back to 1913 in Ludwigshafen, with a capacity of 30 t/day, a process that displaced the use of coal and charcoal as sources for ammonia. In 1953, 143 facilities were already in operation (The HaberBosch Heritage. 1997, www.science.uva.nl).

Properties: Ammonia is a colorless gas with a characteristic odor whose density is 0.7714 g/cm^3. However, it can easily be liquefied at moderate pressure. In Table 5.3 we can see some values for the vapor pressure as a function of the temperature. The heat of vaporization is 5580 cal/mol and the critical temperature is 132.9°C. It is highly soluble in water, in particular at high pressure and low temperature. At 20°C and 1 atm, a solution of 28% can be obtained; it is sold commercially. It solidifies at −78°C in the form of colorless crystals.

$$NH_3 + H_2O \rightarrow NH_4OH$$
$$NH_4OH \rightarrow NH_4^+ + OH^-$$

Table 5.3 Vapor Pressure of Ammonia

T (°C)	−33.5	−20	0	20
P_v (mmHg)	760	1427	3221	6429

From a health and safety point of view, a feature worth mentioning is the fact that ammonia—air mixtures may explode with a hot point. Furthermore, in the gas phase it is toxic since it increases the blood pressure if breathed in. The eyes are sensitive to it, and it burns the skin.

Production: Ammonia is naturally produced from rotten, nitrogenated organic species. For decades it was obtained from the distillation of carbonous materials. It is currently produced from the reaction of nitrogen and hydrogen. This is the process that will be analyzed below.

Usage: Ammonia is the basic raw material for the nitrogen industry. Nitrogen fertilizers are the most widely used fertilizers, which constitute 80% of the ammonia market. The increase in consumption is 18% over the several last years. Urea, caprolactama, and explosives, as well as nitric acid, are other well-known products.

Storage: Ammonia is stored as a compressed liquid in sphere tanks made of chromium—nickel stainless steel; the use of Zn or Cu would produce soluble amines.

5.4.1.2 Stage III reaction (H₂ + N₂)

The reaction is shown below. We are going to evaluate its thermodynamics and kinetics as well as the design of the converter Appl (1998). Finally, we will describe different commercial processes for the production of ammonia.

$$\frac{1}{2}N_2 + \frac{3}{2}H_2 \leftrightarrow NH_3; \quad \Delta H_r = -48 \text{ kcal/mol}_{NH_3}$$

5.4.1.2.1 Thermodynamics of the reaction

Ammonia production is an exothermic reaction and thus low temperatures favor conversion. However, the kinetics sets the lower bound of the operating temperature. On the other hand, there is a decrease in the number of moles as the reaction progresses. Therefore, the higher the pressure, the larger the conversion. The ideal operation would be isothermal, since high conversions could be achieved. However, removing the energy generated as the reaction advances is a technical challenge and the design of the reactors look for an optimal operating line. The typical pressures and temperatures are 100—1000 atm and 400—600°C (Vancini, 1961). In Table 5.4 we see some values of the equilibrium constant as a function of both pressure and temperature. Alternatively, in the literature we can find correlations (Eqs. 5.12 and 5.13) to compute the equilibrium constant either for atmospheric pressure, or as a function of it; the parameters b and u can be seen in Table 5.5.

$$\log_{10}(K_p) = \frac{2250.322}{T} - 0.85430 - 1.51049 \log_{10}T - 2.58987 \times 10^{-4}T + 1.48961 \times 10^{-7}T^2$$

(5.12)

$$\log_{10}(K_p) = \frac{2074.8}{T} - 2.4943\log_{10}T - b \cdot T + 1.8564 \times 10^{-7}T^2 + u \quad T[=]K \quad (5.13)$$

Table 5.4 Values for the Equilibrium Constant

P (atm)	Temperature (°C)				
	325	350	400	450	500
1.0	0.0401	0.0266	0.0129	0.0060	0.0038
100			0.0137	0.0072	0.0040
300				0.0088	0.0049
600				0.0130	0.0065
1000				0.0233	0.0099

Table 5.5 Values for Parameters b and u

P (atm)	b	u
300	1.256×10^{-4}	2.206
600	1.0856×10^{-4}	3.059
1000	2.6833	4.473

The equilibrium constant is a pseudoconstant since it is computed with partial pressures instead of fugacities. Thus, for the stoichiometry presented, the equilibrium is as follows Reklaitis (1983):

$$K_p = \frac{P_{NH_3}}{P_{N_2}^{0.5} P_{H_2}^{1.5}} \quad K_f = \frac{f_{NH_3}}{f_{N_2}^{0.5} f_{H_2}^{1.5}} \tag{5.14}$$

$$f_{NH_3} = u_{NH_3} P_{NH_3} = u_{NH_3} P x_{NH_3} \tag{5.15}$$

$$K_f = \frac{f_{NH_3}}{f_{N_2}^{0.5} f_{H_2}^{1.5}} = \frac{\nu_{NH_3}}{\nu_{N_2}^{0.5} \nu_{H_2}^{1.5}} \frac{x_{NH_3}}{x_{N_2}^{0.5} x_{H_2}^{1.5}} = K_u K_p \Rightarrow K_p = \frac{K_f}{K_u} \tag{5.16}$$

EXAMPLE 5.5

Compute the ammonia fraction at equilibrium for an operating pressure equal to 1 atm. The reaction for the production of ammonia is given as follows:

$$0.5N_2 + 1.5H_2 \longleftrightarrow NH_3$$

The feed enters the reaction in stoichiometric proportions. Assume that no inerts accompany the feed and that the correlation for the equilibrium constant is given by the following equation:

$$\log(K_p) = \frac{2250.322}{T} - 0.85340 - 1.51049 \log(T) - 2.58987 \times 10^{-4} T$$

$$+ 1.48961 \times 10^{-7} T^2$$

Solution

For this example, we use the following nomenclature: T = temperature, $P_T = P$, and the molar fraction for the species involved is detoned as x, y and z for NH_3, N_2 and H_2 respectively. Therefore:

$$P_{NH_3} = xP_T; P_{N_2} = yP_T; P_{H_2} = zP_T;$$

$$K_p = \frac{P_{NH_3}}{P_{N_2}^{0.5} \cdot P_{H_2}^{1.5}} = \frac{x \cdot P_T}{(y \cdot P_T)^{0.5}(z \cdot P_T)^{1.5}} = \frac{1}{P_T} \frac{x}{y^{0.5} \cdot z^{1.5}}$$

$$K_p \cdot P_{total} = \frac{x}{y^{0.5} z^{1.5}}$$

$$z = 3 \cdot y$$
$$x + y + z = 1$$
$$x + y + 3 \cdot y = 1$$
$$x + 4 \cdot y = 1 \Rightarrow y = \frac{1-x}{4}$$
$$z = 3 \cdot y = \frac{3}{4}(1-x)$$

Thus:

$$K_p = \frac{x}{P_T\left(\frac{1-x}{4}\right)^{0.5}\left(\frac{3}{4}(1-x)\right)^{1.5}} = \frac{4^2}{3^{1.5}}\frac{x}{P_T(1-x)^2} = 3.0792\frac{x}{P_T(1-x)^2}$$

We can solve the equation for x as follows:

$$x = \frac{\left(2 + \frac{3.0792}{K_p \cdot P_T}\right) - \sqrt{\left(\left(2 + \frac{3.0792}{K_p \cdot P_T}\right)^2 - 4\right)}}{2}$$

We compute, for different temperatures, the fraction of ammonia given by the variable x (Table 5E5.1).

Assuming that the correlation is valid for different pressures, Fig. 5E5.1 presents the ammonia fraction of the product gas as a function of the pressure and temperature. As predicted by Le Chatelier's principle, the conversion increases with the pressure and for lower temperatures. As described above, the optimal operation in the reaction would be isothermal, since high conversion could be obtained in one stage. Adiabatic operation leads us to find the limit given by the equilibrium line. The reader can also use the values for K_p given in Table 5.4 to compute the ammonia fraction obtained.

Table 5E5.1 Equilibrium Constant Values

P (atm)	T (K)	K_p	x
1	473	0.59515055	0.14221511
1	523	0.17715127	0.051733
1	573	0.06429944	0.02005278
1	623	0.02717032	0.00867146
1	673	0.01293599	0.00416616
1	723	0.00677637	0.00219106

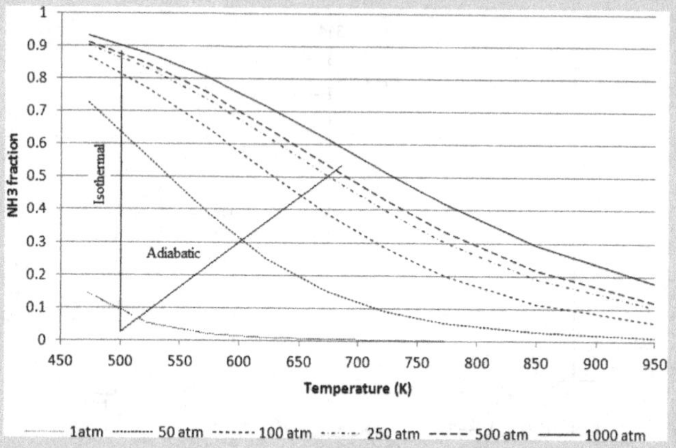

FIGURE 5E5.1

Effect of pressure and temperature on the fraction of ammonia produced.

5.4.1.2.2 Reaction kinetics

The kinetics are based on the process of adsorption of the molecules on the surface of the catalyst and subsequent desorption. The mechanisms can be divided into four stages; the scheme can be seen in Fig. 5.16:

1. Diffusion of the nitrogen and hydrogen from the bulk of the gas to the surface of the catalyst and the porosity of the grain.
2. Chemical adsorption of the gases on the catalyst.
3. Interaction between the atoms of nitrogen and hydrogen. The nitrogen accepts the electrons from the catalyst and the hydrogen atom gives them away to compensate for the charge. In a sequence, imide (NH), amide (NH_2), and ammonia (NH_3) are produced.
4. Desorption of ammonia and diffusion to the gas phase.

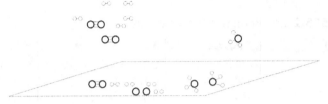

FIGURE 5.16

Catalyzed production of ammonia.

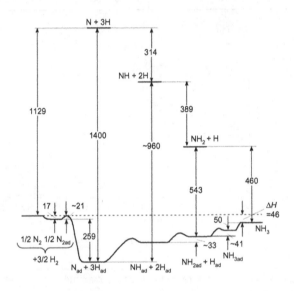

FIGURE 5.17

Potential diagram of species (kJ/mol).

With permission Appl, M., 1999. Ammonia Principles and Industrial Practice. Wiley-VCH. Weinheim.

The mechanism can be described as follows:

$$H_2 \leftrightarrow 2H_{ad}$$

$$N_2 \leftrightarrow N_{2,ad} \leftrightarrow 2N_s$$

$$N_s + H_{ad} \leftrightarrow NH_{ad}$$

$$NH_{ad} + H_{ad} \leftrightarrow NH_{2,ad}$$

$$NH_{2,ad} + H_{ad} \leftrightarrow NH_{3,ad}$$

$$NH_{3,ad} \leftrightarrow NH_3$$

where the relative stability of the species can be seen in Fig. 5.17.

The main challenges of the kinetics are the following: (1) there is a high dependence on the dissociation of the nitrogen molecules; (2) there is an effect of the diffusion on the process, which relies on the catalyst structure; (3) the limiting stage is either the physical or chemical adsorption and ammonia desorption; and (4) the relative distance to equilibrium has impact (ie, the greater the distance, the quicker rate).

Although the kinetics is complex, it is widely assumed that the slowest stage is ammonia desorption. Thus, calling r' the formation rate of ammonia and r'' the desorption rate, we have as the global rate of ammonia produced:

$$\text{Global rate} = r' - r''$$

Tiomkin and Pyzhev (Wallas, 1959) suggested the rate of production of ammonia as per Eq. (5.17):

$$r = \frac{dP_{NH_3}}{d\theta} = k_1 \frac{P_{N_2}^{0.5} P_{H_2}^{1.5}}{P_{NH_3}} - k_2 \frac{P_{NH_3}}{P_{H_2}^{1.5}} \tag{5.17}$$

In Fig. 5.18 we see that the rate increases with the temperature up to 500°C but there is a maximum due to the equilibrium, and with pressure, since higher pressure helps in the gas adsorption. The kinetic constants depend on the catalyst used. It is assumed that r' is limited by the chemical adsorption of nitrogen on the catalyst, while r'' is controlled by the desorption of the ammonia produced.

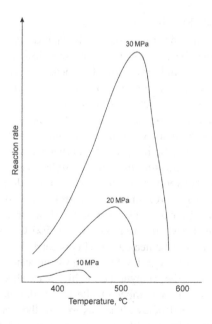

FIGURE 5.18

Relative reaction rate with P and T.

Based with permission of Appl, M., 1999. Ammonia Principles and Industrial Practice. Wiley-VCH. Weinheim.

The *catalysts* used for the production of ammonia, unlike what is expected, are not those appropriate for hydrogenation processes, but are those capable of producing nitrates. Typically, transition metals such as W, Os, Fe, Mo, and U can be used, but for economic reasons Fe is used in its crystalline α structure. The structure of Fe depends on the temperature:

770°C   Fe α
770°C−900°C   Fe β
900°C−1400°C   Fe γ
1400°C−1433°C   Fe δ

Due to the high operating temperature during the reaction, it would be possible to reach the more compact structure γ, which represents a problem in terms of species diffusion. To avoid that compaction, and to stabilize the catalyst, small amounts of Al_2O_3 are added to the bulk. Thus, the preparation of the catalyst is as follows:

1. Production of the catalytic mass by oxidation of melted Fe under a flow of O_2 to which small amounts of Al_2O_3 are added.
2. Activation of the catalytic mass by reduction with synthetic H_2.

These catalysts are sensitive to CO_2 and H_2S, and therefore both must be eliminated before synthesis. It is also possible to poison a catalyst in a previous reactor so that FeS and CH_4 are produced at low temperature in a controlled stage.

5.4.1.2.3 Converter design

Operation with hydrogen under high pressure is what defines reactor structure and design. Hydrogen under the reaction pressure conditions is capable of removing carbon from carbon steel and producing hydrocarbons such as methane. As a result, the material loses resistance, and over time can crack and even explode. Furthermore, iron at high temperature is permeable to hydrogen. This permeability increases with pressure. To address these challenges, the converter is made of iron with almost no carbon. Since this material cannot handle high pressure or even maintain the hydrogen inside, the internal structure is surrounded by a second structure made of chromium−nickel stainless steel. As the designs will show, a flow of gas between both tubes will drag along the hydrogen that has escaped the inner structure, similar to current designs for energy efficient buildings where a flow of air prevents cold or hot flows from entering.

The operating conditions are from 400°C to 500°C, as seen in Fig. 5.18. Therefore, the feedstock must be heated up. This is carried out in the bottom part of the reactor where a heat exchanger kind of design is allocated using hot products as hot stream. The reaction in the catalytic bed is highly exothermic. Thus, the energy generated must be removed. A number of alternative designs are available in the industry, as will be presented below. The gas exiting the reactor contains 8−12% ammonia. The product is recovered by condensation, and the unreacted gas is recycled. The various designs can be classified as either multibed or tubular converters (Couper et al., 2009, Appl, 1999).

Multibed designs. *Multibed reactors with direct cooling*, also known as quench converters, are devices that consist of a series of catalytic beds operating adiabatically. The various beds are separated by mixing areas where the hot converted gas is cooled down with cold fresh feed gas. Therefore, only a fraction of the feed is heated up to 400°C. The packed bed is designed so that the final temperature of the gases is around 500°C. This gas is cooled down to 400°C using cold, unconverted gas. One of the main drawbacks of the system is that the feed flow does not cross the catalytic beds, and therefore most of the ammonia is produced within a flow already containing ammonia. This means that the reaction rate is lower and a larger amount of catalyst is required.

In Fig. 5.19 it can be seen that the last bed is larger than the previous ones. In the reactor, the feed enters from the top, it is guided between the inner and outer tubes, heated in the bottom heat exchanger up to 400°C, and then rises through an internal pipe to be fed to the first bed. In Fig. 5.19 we see that the gas has been gradually heated up along the way. The conversion in the first bed increases the temperature and the amount of ammonia. Once the gas reaches 500°C, it is cooled down with unconverted gas. Furthermore, the unconverted gas dilutes the concentration of ammonia before entering the second bed; see the third picture in Fig. 5.19. We see that the profile of ammonia concentration tries to follow a path parallel to the equilibrium line. After the fourth bed, the gas is used as a hot stream for the initial heating of the feedstock. Heat integration is inherent in the design of these reactors, and these converters are used in high-capacity facilities. One of the most famous designs of the quench converter is the four-bed Kellogg converter, which has been installed more than 100 times (see Fig. 5.20). We can

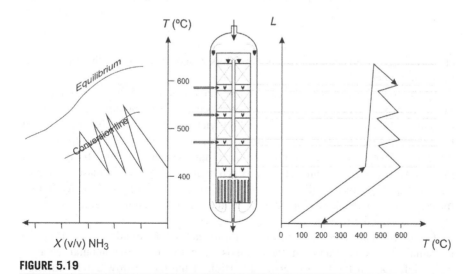

FIGURE 5.19

Multibed converter with quench cooling.

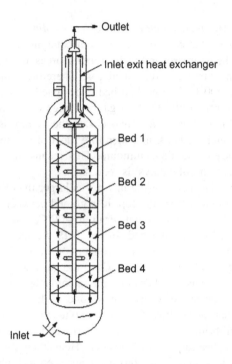

FIGURE 5.20

Kellogg four-bed vertical quench converter.

With permission from Czuppon, T.A., Knez, S.A., Rovner, J.M., 2000. Kellogg Company, M.W. Ammonia, Kirk-Othmer Encyclopedia of Chemical Technology. John Wiley and Sons. New York, Czuppon et al. (2000) http://dx.doi.org/10.1002/0471238961.0113131503262116.a01 Copyright © 2004 by John Wiley & Sons, Inc.

classify different designs of multibed reactors based on the flow direction across the catalytic beds.

Axial flow converters, such as the one shown in Fig. 5.20, cannot increase their capacity by increasing their diameter because of economic and technical reasons, and therefore they can achieve high yield only by increasing the depth of the catalytic bed. However, this solution increases the pressure drop. In order to mitigate it, larger catalytic particles are used, but they have lower activity. *Radial flow converters* do not demonstrate this problem, so they can increase their production capacity using smaller particles with low-pressure drop and moderate diameters. The drawback of this design is the need for better sealing to avoid leakages across the materials.

Casale presented the concept of *axial and radial flow* to address the problem of radial flow converters. In these kinds of converters, the annular catalyst bed is left open at the top to permit a portion of the gas to flow axially through the catalyst. The remainder of the gas flows radially through the bulk of the

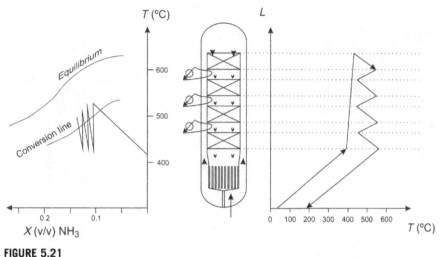

FIGURE 5.21

Multibed converter with indirect cooling.

catalyst bed. *Crossflow* presents an alternative design to allow for a low-pressure drop, even with small catalytic particles. It was used by Kellogg in their horizontal quench converter. The catalysis beds are arranged side-by-side in a removable cartridge so that it is easy to load and unload catalyst. The gas flows from top to bottom.

Multibed reactors with indirect cooling are known by the classical Montecatini process. The design uses pipes with hot pressure water in a closed circuit to cool down the gases. Fig. 5.21 shows the operation. The gas is fed from the bottom and the feedstock is heated up in the lower heat exchanger and led to the first bed through the interspace between the inner and outer tubes. In the first bed, the gas reacts and is allowed to heat up until its temperature reaches 500°C, at which point the catalytic bed ends and the gas is cooled using hot water (Appl, 1999). Since there is no contact between the cooling agent and the gas (first picture of Fig. 5.21), after each bed the composition in ammonia remains constant. After four beds, the hot converted gas is used as a hot stream to heat up the feedstock. Heat integration is therefore already part of the reactor design.

Apart from this basic design, there are some modified converters that also use indirect cooling such as the Topsoe Series 200, a radial flow reactor that has internal heat exchangers (Fig. 5.22A), the Topsoe Series 300 (Fig. 5.22B), the Casale design (Fig. 5.22C), and the Kellogg alternative horizontal design for indirect cooling (Fig. 5.22D).

Multibed reactors such as the ones depicted in Figs. 5.19 and 5.21 can be analyzed following Example 7.7 by means of an adiabatic energy balance to each of the beds and the mass balance provided by the equilibrium.

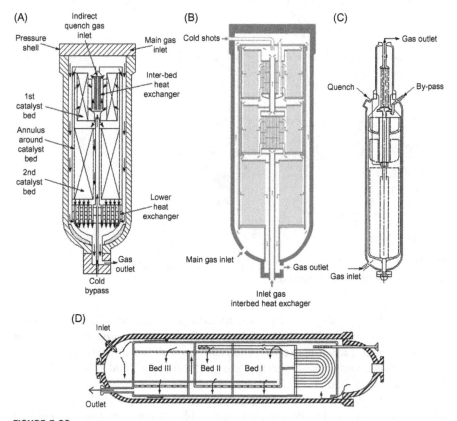

FIGURE 5.22

Industrial designs of indirect cooled converters.

> With permission from Czuppon, T.A., Knez, S.A., Rovner, J.M., 2000. Kellogg Company, M.W. Ammonia,
> Kirk-Othmer Encyclopedia of Chemical Technology. John Wiley and Sons. New York, Czuppon et al. (2000)
> http://dx.doi.org/10.1002/0471238961.0113131503262116.a01 Copyright © 2004 by John Wiley &
> Sons, Inc.; with permission from Haldor Topsoe http://www.topsoe.com/sites/default/files/
> latest_developments_in_ammonia_production_technology.pdf.

Tube converters. Tube converters provide an alternative design for the removal
of the energy generated in the reaction. In this case the catalyst is packed into
pipes that are the ones responsible for heating up the feed. The designs available
are appropriate for low-capacity facilities. Furthermore, temperature control is
slow and it is difficult to mitigate oscillations in the temperature. Two designs are
presented, Fig. 5.23 for countercurrent flow, and Fig. 5.24 for concurrent flow.
In the first design, the main feed stream enters from the bottom of the converter.
It is heated up in the lower heat exchanger using the hot product gases,
and ascends through pipes as it is further used to cool down the reaction gases.

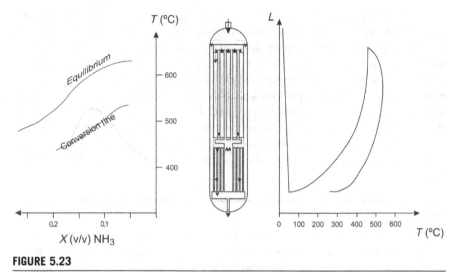

FIGURE 5.23

Countercurrent tube-cooled converter (TVA converter).

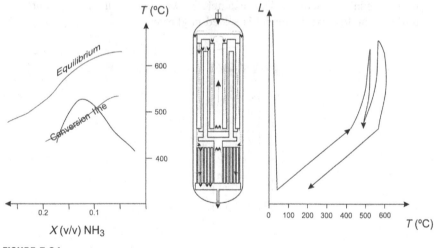

FIGURE 5.24

Cocurrent-flow, tube-cooled converter.

Then it is fed to the catalytic pipes where it reacts. The heat integration, as seen in the central picture of Fig. 5.23, is carried out with the same gas as that which is going to react later. Additional feed is used to cool down the vessel walls. The conversion accelerates across the pipes and there is a peak in temperature along the catalyst. In Example 5.6, a simplified version of this reactor is modeled.

Fig. 5.24 shows an alternative design where the cooling gas and the reacting mixture flow parallel and in the same direction. The feed enters the reactor from the top. It is also used to control the leakages of hydrogen by forcing its flow between the inner and outlet structures of the reactor. Next, it is heated up to reaction temperature in the lower heat exchanger, and by means of a system of pipes, fed parallel to the pipes containing the catalyst so that both gases descend and exchange heat. The hot feed is then guided to enter the catalyst. In the second picture of Fig. 5.24 is the profile of temperature across the reactor, which presents two peaks due to the change in direction of the feed. In the third picture of the same figure, the content of ammonia is presented as it matches a subequilibrium line.

EXAMPLE 5.6

The original problem was described by Murase et al. (1970) and models the operation of an ammonia reactor with the configuration shown in Fig. 5E6.1. It is assumed that there is no axial mass or heat transfer, that the temperature at the catalytic region is the same as at the catalytic particle, the heat capacities of the gases are constant, and the activity of the catalyst is 1. We also assume that the pressure drop can be computed using Ergun's Equations. This example is well-known in the literature, however, in this text the model is solved in gProms.

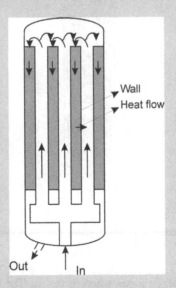

FIGURE 5E6.1

Reactor geometry.

Energy balance to the feed:

$$\frac{dT_f}{dx} = \frac{US_1}{Wc_{p,f}}(T_g - T_f)$$

U: Overall heat transfer coefficient (500 kcal/m² h·°C)
S_1: Surface area of cooling tubes per unit length of reaction, 10 m
T_g: Temperature of reacting gas (°C)
T_f: Temperature of the feed (°C)
$c_{p,f}$: Heat capacity of the feed gas 0.707 kcal/kg K
X: Distance along the axis
W: Mass flow rated 26,400 kg/h

Reactant gas energy balance:

$$\frac{dT_g}{dx} = -\frac{US_1}{Wc_{p,f}}(T_g - T_f) + \frac{(-\Delta H)S_2}{Wc_{p,g}} \cdot f[\text{ra}]$$

ΔH: Heat of reaction per mol of N_2 ($-26{,}000$ kcal/kmol N_2)
S_2: Cross-sectional area of catalyst zone 0.78 m²
ra: Reaction rate (kmol N_2, h m²)
$c_{p,g}$: Heat capacity of the reacting gas 0.719 kcal/kg·K

Nitrogen mass balance:

$$ra = \frac{dN_{N_2}}{dx} = -f\left[K_1 \frac{P_{N_2} \cdot P_{H_2}^{1.5}}{P_{NH_3}} - K_2 \frac{P_{NH_3}}{P_{H_2}}\right]$$

$$K_1 = 1.78954 \times 10^4 \exp\left(\frac{-20{,}800}{RT_g}\right)$$

$$K_2 = 2.5714 \times 10^{16} \exp\left(\frac{-47{,}400}{RT_g}\right)$$

f: Catalyst activity
P_i: Partial pressure of the species
K_i: rate constants,
where

$$P_{N_2} = P_o(\text{atm}) \cdot \left[\frac{N_{N_2}}{N_{\text{total}}}\right]$$

$$P_{H_2} = 3 \cdot P_{N_2}$$

$$P_{NH_3} = P_o(\text{atm}) \cdot \left[\frac{N_{NH_3}}{N_{\text{total}}}\right]$$

$$N_{NH_3} = N_{NH_3,o} + 2(N_{N_2,o} - N_{N_2})$$

$$N_{\text{total}} = N_{N_2} + 3 \cdot N_{N_2} + N_{NH_3,o} + N_{\text{inertes}} + 2(N_{N_2,o} - N_{N_2})$$

and the pressure drop is given by Ergun's equation:

$$\frac{dP}{dz} = -\frac{G(1-\phi)}{\rho g_c D_p \phi^3}\left[\frac{150(1-\phi)\mu}{D_p} + 1.752G\right]$$

$$\frac{dP}{dz} = -\frac{G(1-\phi)}{\rho_o g_c D_p \phi^3}\left[\frac{150(1-\phi)\mu}{D_p} + 1.752G\right]\frac{P_o}{P}\frac{T}{T_o}\frac{F_T}{F_{T_o}}$$

$$G = \frac{\sum_i F_{i\,o}M_i}{A_c} = 1307.6\,\frac{\text{lb}}{\text{ft}^2\text{h}}$$

$$gc = 4.17 \times 10^{-8}[=]\text{psi}$$

$$\mu \approx cte$$

G: Mass flow rate per unit of area (lb/ft^2 h)
ϕ: Porosity
D_p: Particle diameter (ft)
μ: Gas viscosity (lb/ft·h)
ρ: Gas density (lb/ft^3)

$T_f = 694K$, $T_g = 694K$, $P = 286$ atm, $N_{N2} = 701.2$ kmol/h m^2
Feed: 21.75% N_2; 62.25% H_2; 5% NH_3; 4% CH_4; 4% Ar.
We model the reactor in gProms. For a further description of the modeling of reactors using gProms, we refer the reader to Martín (2014). We define a model block as:

```
PARAMETER
S1 as real
S2 as real
eff as real
cooling as real
U as real
Pto as real
ph as real
rhoo as real
Dp as real
visc as real
gc as real
Tgo as real
Cpf as real
Cpg as real
deltaHr as real
R as real

VARIABLE
K1 as notype
K2 as notype
```

```
NN2o as concentration
Ni as concentration
NNH3o as concentration
Nto as concentration
Ntotal as concentration
NN2 as concentration
Tf as temperature
Tg as temperature
Presion as Pressure
PN2 as Pressure
PH2 as Pressure
PNH3 as Pressure
ra as notype
G as notype

SET
S1: = 10;
S2: = 0.78;
eff: = 1;
cooling: = 26400;
U: = 500;
Pto: = 286;
ph: = 0.45;
rhoo: = 0.054;
Dp: = 0.015;
visc: = 0.090;
gc: = 4.17e8;
Tgo: = 694;
Cpf : = 0.707;
Cpg: = 0.719;
deltaHr: = - 26000;
R: = 1.987;

EQUATION
K1 = 1.78954e4*exp(-20800/(R*Tg));
K2 = 2.5714e16*exp(-47400/(R*Tg));
NN2o = 0.2125*26400/10.5;
Ni = 0.08*26400/10.5;
NNH3o = 0.05*26400/10.5;
Nto = 4*NN2o + Ni + NNH3o;
Ntotal = Ni + NNH3o + 2*NN2o + 2*NN2;
PN2 = NN2*Presion/Ntotal;
PH2 = 3*PN2;
PNH3 = Presion*((NNH3o + 2*(NN2o-NN2))/(Ntotal));
ra = eff*(K1*PN2*PH2^1.5/PNH3-K2*PNH3/(PH2^1.5));
G = 0.454*(Ni*16 + NN2o*28 + 3*NN2o*2)/((S2/0.3048^2));
```

```
$NN2 = -ra;
$Tf = (U*S1*(Tg-Tf)/(cooling*Cpf));
$Tg = -(U*S1*(Tg-Tf)/(cooling*Cpg)) + (ra)*(-deltaHr)*S2*eff/
  (Cpg*cooling);
$Presion = -G*(1-ph)*Pto*Tg*(150*(1-ph)*visc/Dp + 1.75*G)*(Nto/
  (Ni + NNH3o + 2*NN2o + 2*NN2))/(14.7*rhoo*gc*Dp*ph^3*Tgo*Presion);

PROCESS
Unit R101 as Reactor

INITIAL
  WITHIN R101 DO
    NN2 = 0.2125*26400/10.5;
    Tg = 694;
    Tf = 694;
    Presion = 286;
END # Within

SOLUTIONPARAMETERS
  ReportingInterval := 0.1;

SCHEDULE
  CONTINUE FOR 5
```

The temperature profiles are prsented in (Fig. 5E6.2).

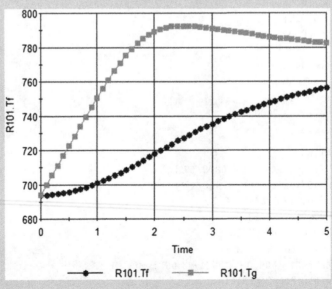

FIGURE 5E6.2

Profile of the temperature along the pipes.

5.4.1.2.4 Production processes

The various processes differ from one to another based on the raw material used, and therefore, the procedure to obtain the syngas and the reactor design.

Kellogg process (medium pressure). Fig. 5.25 shows a flowsheet for the Kellogg process. In this figure the different sections can easily be recognized. Natural gas is the typical feedstock. After sulfur removal, we have primary and secondary reformers followed by high-temperature and low-temperature sweet shift. Once the syngas is obtained, CO_2 is removed. Selexol or MDEA processes are among the selections for the Kellogg design. The traces of CO and CO_2 are removed in a methanator. The synthesis occurs in their proprietary converter designs using ruthenium catalyst (Kaap), which has recently substituted the Fe-based ones (after 90 years). The reaction operates at 300°C and 100−130 atm. Next, ammonia is separated. There are some rules that determine the type of separation, condensation, or absorption, depending on the operating conditions. Condensation is recommended above 100 atm in spite of the high fix costs. Absorption typically follows Raoults' Law (see Chapter 2: Chemical processes).

Haber−Bosch process (medium pressure). This is one of the most extended processes for the production of ammonia. There are versions that use natural gas, but coal was commonly the raw material. Fig. 5.26 shows the scheme of the process. The syngas is produced using steam, air, and coke. There are a few preparation stages including the scrubber, WGSR, and CO_2 capture. The converter operates at 200−350 atm and 550°C. The ammonia concentration in the outlet stream from the converter is from 13% to 15%, and the ratio between the recycle and the feed is 6−7 to 1. The converter is 12 m tall, with an inner diameter of 0.8 m. The wall is 0.16 m thick and uses 6 t of double-promoted iron catalyst (see Fig. 5.27). The purge is around 5%. The ammonia is absorbed in water that is distilled to recover the ammonia (Lloyd, 2011).

Claude's process (high pressure). George Claude's process is characterized by a high operating pressure at the reactor (900−1000 atm), and temperatures in the range of 500−650°C using a promoted iron catalyst. No recirculation is present due to the high conversion achieved—higher than 40%. Since there is no recycle and no purge, the unreacted hydrogen is lost. The catalyst is iron oxidized in magnesia containing 5−10% calcium oxide. Magnesium oxide also dissolves in the mixture. The reactor includes a guard bed to absorb poisons and convert residual CO into methanol. As a result of the pressure, water can be used to condense the produced ammonia out of the product stream. The drawbacks of the process are the compression costs and shorter converter life. Thirty percent of US plants used the Claude process. The reactor is 5 m tall, with an inner diameter of 0.5 m and a wall thickness of 0.3 m. Fig. 5.28 shows a scheme of the original reactor (Lloyd, 2011).

Uhde's process. The first stages of this process are common to others (see Fig. 5.29). The raw material is typically natural gas that is processed to remove the sulfur compounds. Next, it is mixed with steam to produce the syngas over nickel catalysts. The first reformer operates at 40 atm and 800−850°C to save compression energy later on. This reformer is top-fired,

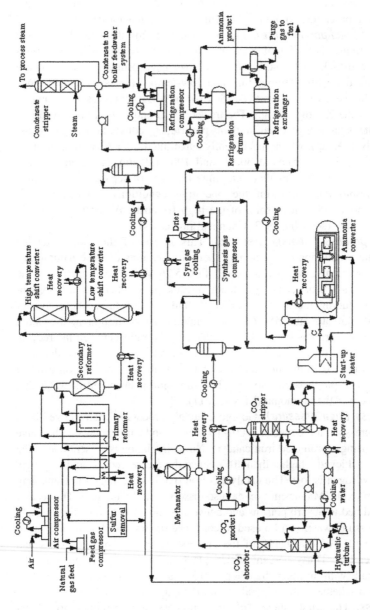

FIGURE 5.25

Kellogg process for the production of ammonia. a) Feed gas compressor; b) Desulfurization; c) Heat integration; d) Primary reformer; e) Air compressor; f) Secondary reformer; g) Heat recovery; h) High-temperature shift converter; i) Low-temperature shift converter; j) Condensate stripper; k) CO2 absorber; l) CO2 flash drum; m) Recycle compressor; n) Semi-lean Pump; o) Stripper (other options: Benfield or BASF aMDEA); p) Stripper air blower; q) CO2 lean pump; r) Methanator feed preheater; s) Methanator; t) Synthesis gas compressor; u) Dryer; v) Purge gas H2 recovery; w) Ammonia converter; x) Start-up heater; y) Refrigeration exchanger; z) Refrigeration compressor.

Reproduced with permission from Czuppon, T.A., Knez, S.A., Rovner, J.M., 2000. Kellogg Company, M.W. Ammonia, Kirk-Othmer Encyclopedia of Chemical Technology. John Wiley and Sons. New York, Czuppon et al. (2000) http://dx.doi.org/10.1002/0471238961.0113150326116.a01 Copyright © 2004 by John Wiley & Sons, Inc.

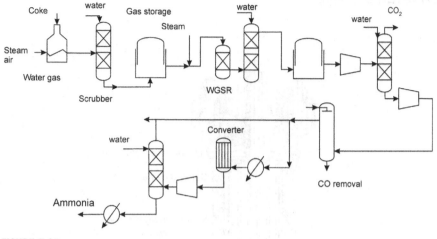

FIGURE 5.26

Haber—Bosch process.

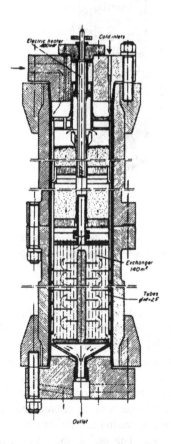

FIGURE 5.27

Haber converter.

Reproduced with permission from Wallas, S.M., 1990. Chemical Process Equipment. Selection and Design. Butterworth-Heinemann Series in Chemical Engineering. Washington, Wallas (1990). Butterworth-Heinemman Copyright 1990.

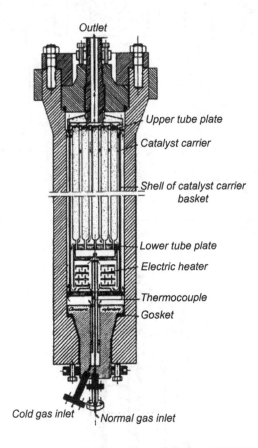

Outlet

Upper tube plate

Catalyst carrier

Shell of catalyst carrier basket

Lower tube plate

Electric heater

Thermocouple

Gosket

Cold gas inlet   Normal gas inlet

FIGURE 5.28

Claude reactor.

Reproduced with permission from Wallas, S.M., 1990. Chemical Process Equipment. Selection and Design.
Butterworth-Heinemann Series in Chemical Engineering. Washington, Wallas (1990). Butterworth-
Heinemman Copyright 1990.

and its tubes are made of a stainless steel alloy for better reliability. In the second reformer, air is added. After gas cooling, the CO is transformed into CO_2 using the sweet shift combining HTS and LTS. The process uses ethanolamines for the removal of the CO_2, while the traces of CO and CO_2 are eliminated via methanation. This process is characterized by the use of two converters with three catalytic beds operating under radial flow conditions; see Fig. 5.30 for a scheme of the reactor. The first reactor consists of an intercooled three bed reactor operating at 110 atm. The second one, operating at 210 atm, is responsible for two thirds of the total ammonia production. The use of this system allows a limited pressure drop. The energy generated is used to produce high-pressure steam. The ammonia is recovered by condensation. The purge is processed to recover the hydrogen, and the rest of the gas is used as flue gas.

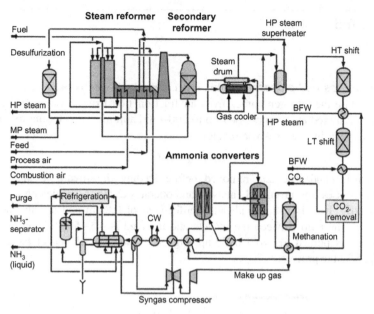

FIGURE 5.29

Uhde's ammonia process.

With permission from Thyssenkrupp, Uhde, Ammonia ThyssenKrupp (www.uhde.com) http://www.thyssenkrupp-csa.com.br/fileadmin/documents/brochures/0a2d5391-b166-484d-847d-3cbfd941f06b.pdf, Uhde (2015).

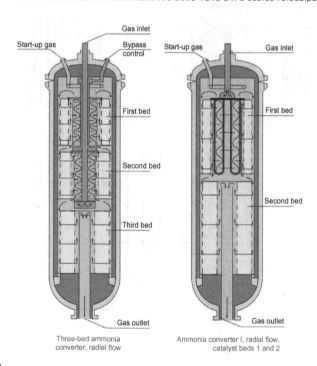

Three-bed ammonia
converter, radial flow

Ammonia converter I, radial flow,
catalyst beds 1 and 2

FIGURE 5.30

Uhde reactor: http://www.thyssenkrupp-industrial-solutions.com/fileadmin/documents/brochures/TKIS_Ammonia.pdf.

With permission from Thyssenkrupp, Uhde, Ammonia ThyssenKrupp (www.uhde.com) http://www.thyssenkrupp-csa.com.br/fileadmin/documents/brochures/0a2d5391-b166-484d-847d-3cbfd941f06b.pdf, Uhde (2015).

Fauser process (old process). In this process the hydrogen is produced via electrolysis while the nitrogen comes from air fractionation Ernest et al. (1925). Both streams are mixed and compressed up to 200–300 atm after processing to remove oil and oxygen in a deoxo-type reactor.

$$H_2 + O_2 \rightarrow H_2O$$

Water is condensed and removed before feeding the stream to the reactor. The reactor uses iron-promoted catalysts and operates at 500°C, achieving a conversion of 20%. It is 14 m tall with an interdiameter of 0.85 m and a wall thickness of 0.16 m. The ratio between the recycle and the feed is 4.5–5 to 1, and a part of the recycle is purged to avoid build-up. Fig. 5.31 shows a scheme of the reactor, and Fig. 5.32 the process.

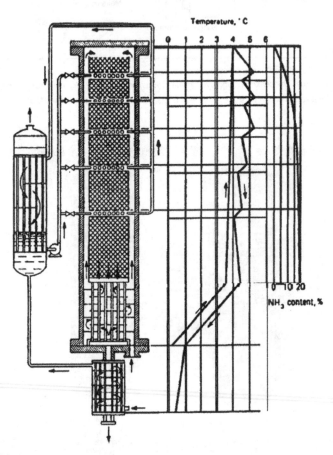

FIGURE 5.31

Fauser reactor.

Reproduced with permission from Wallas, S.M., 1990. Chemical Process Equipment. Selection and Design.
Butterworth-Heinemann Series in Chemical Engineering. Washington, Wallas (1990). Butterworth-
Heinemman Copyright 1990.

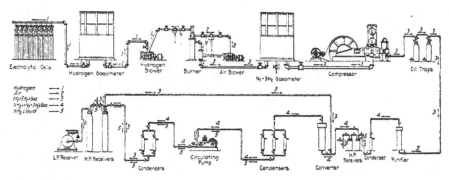

FIGURE 5.32

Fauser process. Ernest et al. (1925).

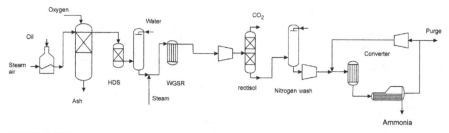

FIGURE 5.33

Texaco process.

Texaco process. The Texaco process is characterized by the technology used in the production of the syngas, the partial oxidation of hydrocarbons. It is called the Texaco syngas generation process (TSGP). Next, WGSR is carried out. Subsequently, the CO_2 is captured using the Rectisol process. Fig. 5.33 shows the stages for syngas production in the Texaco process.

Gasification. It is also possible to gasify coal or other carbonous mass for syngas production and later use it to produce ammonia. Gasification was responsible for 90% of the share of ammonia production by the 1940s.

Casale process. The main characteristic of this process, developed by Luigi Casale, is the reactor; a scheme can be seen in Fig. 5.22. It operates at 600 atm and 500°C, and the control of temperature in the reaction is achieved by recycling the gas so that 2−3% ammonia in the feed of the converter is allowed. As a result, the rate of ammonia production is reduced, and with it, the energy generation. The high operating pressures allow the use of cooling water to condense and separate the ammonia. The reactor is typically 10 m tall, with an inner diameter of 0.6 m and a wall thickness of 0.25 m. The ammonia concentration at the reactor exit is around 20% and the ratio between the recycle and the feed is 4−5 to 1 (Vancini, 1961, Lloyd, 2011).

Gas generator/water gas process. This process is characterized by the production of the syngas using the combination of the generator gas and water gas cycles that were presented in the beginning of the chapter. After the gas

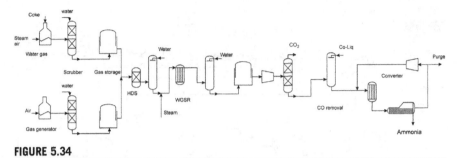

FIGURE 5.34

Gas generator/water gas reaction.

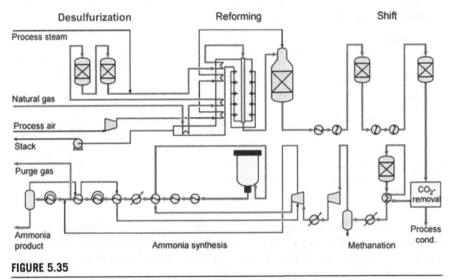

FIGURE 5.35

Haldor–Topsoe process for ammonia production.

With permission from Haldor Topsoe http://haldor.dk/Business_areas/Ammonia.

is generated, sulfur is removed using hydrodesulfurization and later H₂S removal stages. Next, WGS is carried out to remove the CO and increase the production of hydrogen. Subsequently, CO₂ is captured and the traces of CO are removed using copper. Finally, ammonia is synthesized. Fig. 5.34 shows the flowsheet.

Haldor–Topsoe process. The main feature of this process is that the raw material can be either natural gas or liquid hydrocarbons. After hydrodesulfurization and H₂S removal, primary and secondary reforming are performed. Air is injected in the second reformer. Next, HTS and LTS are carried out, and after CO₂ removal using Selexol or MDEA, methanation removes the traces of CO and CO₂. The main feature of the process is also the use of proprietary radial flow reactors operating at 140 atm. Ammonia can be condensed using water, and 20% of the unconverted gas is lost. Fig. 5.35 shows the flowsheet.

EXAMPLE 5.7

In an ammonia production facility, the converter is fed with a stoichiometric mixture of syngas $N_2 + 3H_2$. The conversion at the reactor is 25% to ammonia. Assume that all the ammonia is separated in a condensation stage. The unconverted gases are recycled to the reactor. The initial feedstock contains 0.2 moles of Ar per 100 mol of mixture. At the entrance of the reaction, a maximum of 5 moles of ammonia per 100 mol of syngas is allowed. Compute the purge.

Solution

The synthesis loop is shown in the figure below. We formulate the mass balance from point 1 onwards, assuming that R is the syngas recycle (Fig. 5E7.1).

For the *balance to the converter*, the reaction taking place is as follows:

$$N_2 + 3H_2 \rightarrow 2NH_3$$

Note that we can consider the syngas a mixture, and that per 4 moles of reactants, 2 moles of ammonia are produced. Using this consideration, only three species are evaluated. Let "Recy" be the recycled syngas (Tables 5E7.1 and 5E7.2):

The *balance to the gas–liquid separator* is covered in Table 5E7.2.

For the *balance to the divider (the purge)* we assume that a fraction of the recycle, α, is purged (Table 5E7.3).

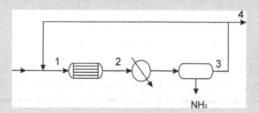

FIGURE 5E7.1

Synthesis loop.

Table 5E7.1 Balance to Converter

	In	Reacts	Out
$N_2 + 3H_2$	100 + Recy	$-(100 + \text{Recy}) \cdot 0.25$	$(100 + \text{Recy}) \cdot 0.75$
NH_3		$(2/4)(100 + \text{Recy}) \cdot 0.25$	$(100 + \text{Recy}) \cdot 0.125$
Ar	$(100 + \text{Recy}) \cdot 0.05$		$(100 + \text{Recy}) \cdot 0.05$

Table 5E7.2 Balance to Separator

	In	Ammonia Removal	Out
$N_2 + 3H_2$	$(100 + Recy) \cdot 0.75$		$(100 + Recy) \cdot 0.75$
NH_3	$(100 + Recy) \cdot 0.125$	$(100 + Recy) \cdot 0.125$	
Ar	$(100 + Recy) \cdot 0.05$		$(100 + Recy) \cdot 0.05$

Table 5E7.3 Balance to Divider

	In	Purge	Recycle
$N_2 + 3H_2$	$(100 + Recy) \cdot 0.75$	$(100 + Recy) \cdot 0.75 \cdot \alpha$	$(100 + Recy) \cdot 0.75 \, (1 - \alpha)$
NH_3			
Ar	$(100 + Recy) \cdot 0.05$	$(100 + Recy) \cdot 0.05 \cdot \alpha$	$(100 + Recy) \cdot 0.05 \cdot (1 - \alpha)$

Table 5E7.4 Results of the Mass Balance

	1	2	3	4
$N_2 + 3H_2$	388	291	291	3
NH_3	0	48.5	0	0
Ar	19.4	19.4	19.4	0.2

Under steady-state conditions, the Ar fed to the system must be eliminated through the purge. Furthermore, the recycle, assumed to be R, is also a function of the purge. Thus, we have a system of two equations and two variables:

$$0.2 = (100 + Recy) \cdot 0.05\alpha$$

$$Recy = (1 - \alpha) \cdot (100 + Recy) \cdot 0.75$$

Solving the system we have $R = 288$ and $\alpha = 0.0103$. The compositions of the streams at the four positions along the flowsheet are given in Table 5E7.4.

EXAMPLE 5.8

The gas exiting the reactor at 400°C and 300 atm contains ammonia in a concentration that corresponds to 30% of the equilibrium value under those conditions. Assume that the pressure does not change across the facility. Assuming humid air calculations with ammonia as the moisture:

a. Compute the composition of the gases at the reactor exit.
b. Determine the gas composition after a cooling stage at 5°C.
c. Compute the gas composition after a second cooling at −25°C

Hint: The equilibrium constant at 400°C and 300 atm is 0.0153.

Solution

As presented in the section describing the equilibrium, the constant is as follows:

$$K_p = \frac{x}{P_T \left(\frac{1-x}{4}\right)^{0.5} \left(\frac{3}{4}(1-x)\right)^{1.5}} = \frac{4^2}{3^{1.5}} \frac{x}{P_T(1-x)^2} = 3.0792 \frac{x}{P_T(1-x)^2} = 0.0153$$

$$x = 0.45$$

Since only 30% of the conversion at equilibrium is achieved,

$$x_{real} = 0.135$$

$$BC = 100 \text{ kmol}$$

$$\underset{75-3y}{3H_2} + \underset{25-y}{N_2} \leftrightarrow \underset{2y}{2NH_3}$$

$$\frac{2y}{100-2y} = 0.135$$

$$y = 5.947$$

$$H_2 = 57.159$$

$$N_2 = 19.053$$

$$NH_3 = 11.895$$

the molar gas composition is as follows:

$$H_2 = 64.87\%$$

$$N_2 = 21.63\%$$

$$NH_3 = 13.50\%$$

Ammonia vapor pressure at the two temperatures, from Antoine correlation,

$$P_v = \exp\left(16.9471 - \frac{2132.5}{T(K) - 32.98}\right)$$

is given in Table 5E8.1:

Table 5E8.1 Vapor Pressures of Ammonia at the Example Pressures

P (atm)	1.5	5
T (°C)	−25	5

After the condensation, the gas is saturated, and thus:

$$P_{NH_3} = P_T \cdot y_{NH_3}$$
$$5 = 300 \cdot y_{NH_3}$$
$$y_{NH_3} = 0.0167$$

Thus, the uncondensed moles are computed as follows:

$$0.0167 = \frac{n_{NH_3}}{19.053 + 57.159 + n_{NH_3}}$$
$$n_{NH_3} = 1.2944$$

Therefore, 10.6 moles have condensed. After the second condensation, we proceed similarly to compute the ammonia in the gas:

$$P_{NH_3} = P_T \cdot y_{NH_3}$$
$$1.5 = 300 \cdot y_{NH_3}$$
$$y_{NH_3} = 0.005$$
$$0.005 = \frac{n_{NH_3}}{19.053 + 57.159 + n_{NH_3}}$$
$$n_{NH_3} = 0.383$$

Therefore, 0.91 kmol have been condensed. In total, 11.51 kmol have been recovered out of the 11.8945, around 97% recovery.

EXAMPLE 5.9

The stream from the reactor consists of 281 kmol of syngas, 9.12 kmol of Ar, and 62 kmol of ammonia. Ammonia is cooled and condenses. Assuming a condensation pressure of 50 atm, compute the temperature of the cooling for recovering 75% of the ammonia produced. Use flash calculations to compute the separation.

Solution

The problem is formulated as a gas–liquid equilibrium using the following nomenclature:

L: Liquid exit (kmol)
V: Vapor exit (kmol)
K_i: Equilibrium constant for noncondensables
K_j: Equilibrium constant for condensables
x_i: Liquid molar fraction
y_i: Vapor molar fraction

A global balance and a component balance are used to compute the molar fraction of the gas and liquid phases:

$$F = V + L$$
$$Fz_i = Vy_i + Lx_i$$
$$y_i = K_i \cdot x_i$$

Combining the equations:

$$y_i = \frac{z_i}{\dfrac{V}{F} + \left(1 - \dfrac{V}{F}\right) \cdot \left(\dfrac{1}{K_i}\right)}$$

$$x_i = \frac{z_j}{(K_j - 1) \cdot \left(\dfrac{V}{F}\right) + 1}$$

$$K_i >> 1 ===> V \cdot y_i \approx F \cdot z_i$$

$$===\rightarrow$$

$$K_j << 1 === > L \cdot x_i \approx F \cdot z_j$$
$$V \approx \sum f_i; \; L \approx \sum f_j$$

$$y_i = \frac{f_i}{V} = \frac{f_i}{\sum f_i} \quad \text{and} \quad x_i = \frac{y_i}{K_i} = \frac{f_i}{K_i \cdot \sum f_i}$$

The vapor compositions are given by the following equations:

$$l_i = L \cdot x_i = \frac{f_i \cdot \sum f_j}{K_i \cdot \sum f_i}$$

$$v_i = f_i - l_i = f_i \left(1 - \frac{\sum f_j}{K_i \sum f_i}\right)$$

The liquid compositions are computed as follows:

$$v_j = V \cdot y_j = V \cdot K_j \cdot x_j = \frac{K_j \cdot f_j \cdot \sum f_i}{\sum f_j}$$

$$l_j = f_j - v_j = f_j \cdot \left(1 - \frac{K_j \cdot \sum f_i}{\sum f_j}\right)$$

where K_i or K_j is equal to P_v/P_{total}. Thus, we fix l_j to be 75% of f_j for ammonia and solve the system of equations. Alternatively, we can just solve $v_j = 0.25 \, f_j$ to obtain K_j, and from that the temperature. Table 5E9.1 presents the results. We see that the Ks are within the error of the approximation.

Table 5E9.1 Results of the Gas–Liquid Separation

	T (K)	261.484564							
	P (bar)	50							
	Entrada FlasI	Pv	Ki	fi (no cond)	fj (cond)	vj (cond)	lj (cond)	vi	li
H2	211.11	446980.716	11.7626504	211.11				207.275315	3.83468474
N2	70.37	310037.494	8.15888142	70.37				68.5271797	1.84282027
Ar	9.12	266419.736	7.01104569	9.12				8.8420683	0.2779317
NH3	62.09	2029.77955	0.05341525		62.09	15.5224721	46.5675279		
					0.75000045		0.75	4.4987E-07	

EXAMPLE 5.10

Compute the liquid flow required to recover 98% of the ammonia content in the product gases of the converters using an absorption column. The total gas flow is 10 kmol/s with 20% ammonia by volume. The stream is at 10 atm and 25°C. The gas flow is 50% of the minimum required. Calculate the number of stages.

Solution

Assume that Raoult's law holds. Therefore:

$$xP_v = yP_T \leftarrow \rightarrow y = Hx$$

The vapor pressure of ammonia is computed using the Antoine correlation as follows:

$$P_v = \exp\left(16.9471 - \frac{2132.5}{T(K) - 32.98}\right)$$

We formulate a mass balance to the ammonia using as control volume that of the scrubber:

$$G_{out} \, y_{out} + L_{out} \cdot x_{out} = L_{in} \cdot x_{in} + G_{in} \cdot y_{in}$$

$$10 \cdot 0.2 = 2 \cdot 0.02 + L_{out} \cdot x_{out}$$

The concentration of ammonia in the liquid is in equilibrium with that of the gas in the inlet, $y = 0.2$.

$$xP_v = 0.2 \cdot 10 \cdot 760$$

$$x = \frac{0.2 \cdot 7600}{\exp\left(16.9471 - \frac{2132.5}{298 - 32.98}\right)} = 0.207$$

$L_{in} = 7.72$ kmol/s. Thus, the flow used is 11.58 kmol/s. y_i is the composition of the exiting gas consisting of 8 kmol/s of air and 0.04 kmol of ammonia, $y_i = 0.005$.

$$N_{min} = \frac{\log\left[\left(\frac{y_{n+1} - Hx_0}{y_1 - Hx_0}\right)\left(\frac{A-1}{A}\right) + \frac{1}{A}\right]}{\log A} = \frac{\log\left[\frac{0.2}{0.005}\left(\frac{A-1}{A}\right) + \frac{1}{A}\right]}{\log(1.20)}$$

$$= 11.1 = 12(\text{say})$$

$$A = \frac{L}{V \cdot H} = \frac{L}{V \cdot \frac{P_v(T)}{P_T}} = \frac{11.58}{10 \cdot \frac{7336}{7600}} = 1.20$$

5.4.2 FISCHER–TROPSCH TECHNOLOGY FOR FUEL AND HYDROCARBON PRODUCTION

5.4.2.1 Introduction

FT synthesis dates back to 1902 when Sabatier and Senderens found that the CO could be hydrogenated over catalysts of Co, Fe, and Ni to produce methane. In 1920 Franz Fischer, founding director of the Kaiser Wilhelm Institute of Coal Research in Mülheim an der Ruhr, and his head of department, Dr. Hans Tropsch, presented the formation of hydrocarbons and solid paraffins on Co–Fe catalysts at 250–300°C. FT technology was further developed and became commercial in Germany during the Second World War when their access to crude was denied. Lately, apart from the hydrogenation of CO, the large production of CO_2 has made it widely available as a carbon source. Therefore, research has focused on the possibility of using it as a source of methane, methanol, or fuels Eckhard et al. (1998), Weissermel and Arpe (1978).

5.4.2.2 Methanol production

In this section the mechanism of the formation of methanol is depicted. Next, the equilibrium for the production is evaluated, followed by the reactor kinetics, the reactor designs, and a description of the production process.

5.4.2.2.1 Mechanisms

The scheme for the reaction is as follows, where the limiting reactions are marked as "rds". The mechanism is given by the hydrogenation of CO and CO_2 together with the WGS reaction (Vanden Bussche and Froment, 1996):

$$H_2(g) + 2s \leftrightarrow 2Hs \quad (K_{H_2})$$
$$CO_2(g) + s \leftrightarrow Os + CO(g) \quad (k_1, K_1) \text{ rds}$$
$$CO_2(g) + Os + s \leftrightarrow CO_32s \quad (K_2)$$
$$CO_32s + Hs \leftrightarrow HCO_32s + s \quad (K_3)$$
$$HCO_32s + s \leftrightarrow HCO_22s + Os \quad (K_4)$$
$$HCO_22s + Hs \leftrightarrow H_2CO_22s + s \quad (k_{5a}) \text{ rds}$$
$$H_2CO_22s \leftrightarrow H_2COs + Os \quad (K_{5b})$$
$$H_2COs + Hs \leftrightarrow H_3COs + s \quad (K_6)$$
$$H_3COs + Hs \leftrightarrow CH_3OH(g) + 2s \quad (K_7)$$
$$Os + Hs \leftrightarrow OHs + s \quad (K_8)$$
$$OHs + Hs \leftrightarrow H_2Os + s \quad (K_9)$$
$$H_2Os \leftrightarrow H_2O(g) + s \quad (K_{H_2O})$$

5.4.2.2.2 Equilibrium

In the reactor, a series of reactions takes place. Only two of the three are linearly independent, and thus the equilibrium model for the reaction can be formulated using the elementary balances to C, O, and H, and the following two equilibrium constants. Note that the WGSR reaction is present. The methanol synthesis is affected by pressure. High pressure drives the reaction to products.

$$CO + 2H_2 \leftrightarrow CH_3OH$$
$$CO_2 + H_2 \leftrightarrow CO + H_2O$$
$$CO_2 + 3H_2 \leftrightarrow CH_3OH + H_2O$$

The equilibrium constants are given as follows:

$$K_{p1} = \frac{P_{CH_3OH}}{P_{CO}P_{H_2}^2} = 9.74 \times 10^{-5}$$

$$\exp\left[21.225 + \frac{9143.6}{T} - 7.492 \ln T + 4.076 \times 10^{-3}T - 7.161 \times 10^{-8}T^2\right]$$

$$(5.18)$$

$$K_{p3} = \frac{P_{CO}P_{H_2O}}{P_{CO_2}P_{H_2}}$$

$$= \exp\left[13.148 - \frac{5639.5}{T} - 1.077 \ln T - 5.44 \times 10^{-4}T + 1.125 \times 10^{-7}T^2 + \frac{49,170}{T^2}\right]$$

$$(5.19)$$

The elementary mass balance is given by Eq. (5.20):

$$H: \left. 2 \cdot n_{H_2} + 2 \cdot n_{H_2O} \right|_{in} - \left. \left(2 \cdot n_{H_2} + 2 \cdot n_{H_2O} + 4 \cdot n_{CH_3OH} \right) \right|_{out} = 0;$$

$$C: \left. n_{CO} + n_{CO_2} \right|_{in} - \left. \left(n_{CO} + n_{CO_2} + n_{CH_3OH} \right) \right|_{out} = 0; \qquad (5.20)$$

$$O: \left. n_{CO} + 2 \cdot n_{CO_2} + n_{H_2O} \right|_{in} - \left. \left(n_{CO} + 2 \cdot n_{CO_2} + n_{H_2O} + n_{CH_3OH} \right) \right|_{out} = 0;$$

There are two main operating variables that the feed to the reactor must meet for the optimal production of methanol:

- The ratio of hydrogen to CO should be $1.75 \leq \dfrac{H_2}{CO} \leq 3$.
- The role of CO_2 in the reaction mechanism is not well-known or understood. However, it is considered that the concentration of CO_2 should be $2-8\%$, and the ratio of the syngas components involving CO_2 should be the following:

$$1.5 \leq \frac{H_2 - CO_2}{CO + CO_2} \leq 2.5 \qquad (5.21)$$

5.4.2.2.3 Kinetics of methanol production

There are a number of models in the literature for the production of methanol from syngas. We consider the work by Vanden Bussche and Froment (1996) as reference. The rates for the production of methanol and the reverse WGS are given by:

$$r_{MeOH} = k'_{5a} \cdot K'_2 \cdot K_3 \cdot K_4 \cdot P_{CO_2} \cdot P_{H_2} \left(1 - \frac{1}{K_1^*} \frac{P_{H_2O} \cdot P_{CH_3OH}}{P_{H_2}^3 \cdot P_{CO_2}} \right) \beta^3$$

$$r_{RWGS} = k'_1 P_{CO_2} \left(1 - K_3^* \frac{P_{H_2O} P_{CO}}{P_{CO_2} P_{H_2}} \right) \beta \qquad (5.22)$$

where

$$\beta = \frac{C_s}{C_t} = \frac{-b + \sqrt{b^2 + 4a}}{2a}$$

$$a = K'_2 K_3 K_4 \sqrt{K_{H_2}} P_{CO_2} \sqrt{P_{H_2}} + \frac{K'_2 K_{H_2O}}{K_{H_2} K_8 K_9} \frac{P_{H_2O} P_{CO_2}}{P_{H_2}} \qquad (5.23)$$

$$b = 1 + \frac{K_{H_2O}}{K_{H_2} K_8 K_9} \frac{P_{H_2O}}{P_{H_2}} + \sqrt{K_{H_2} P_{H_2}} + K_{H_2O} P_{H_2O}$$

$$K'_2 = K_2 c_t$$

$$k'_{5a} = k_{5a} c_t^2 \qquad (5.24)$$

$$k'_1 = k_1 \cdot c_t$$

The equilibrium constants are given by Eq. (5.25),

$$\log_{10} k_1^* = \frac{3066}{T} - 10.592$$

$$\log_{10} \frac{1}{K_3^*} = \frac{-2073}{T} + 2.029$$

(5.25)

and the kinetic constants are of the form given by Eq. (5.26):

$$k(i) = A(i) \exp\left(-\frac{B(i)}{R}\left(\frac{1}{T_{av}} - \frac{1}{T}\right)\right)$$

(5.26)

$$T_{av} = 501.57K$$

Therefore, the rates are as follows:

$$r_{MeOH} = \frac{k'_{5a} \cdot K'_2 \cdot K_3 \cdot K_4 \cdot K_{H_2} \cdot P_{CO_2} \cdot P_{H_2}\left(1 - \frac{1}{K_1^*}\frac{P_{H_2O} \cdot P_{CH_3OH}}{P_{H_2}^3 \cdot P_{CO_2}}\right)}{\left(1 + \left(\frac{K_{H_2O}}{K_{H_2}K_8K_9}\right)\left(\frac{P_{H_2O}}{P_{H_2}}\right) + \sqrt{K_{H_2}P_{H_2}} + K_{H_2O}P_{H_2O}\right)^3}$$

(5.27)

$$r_{RWGS} = \frac{k'_1 P_{CO_2}\left(1 - K_3^*\frac{P_{H_2O}P_{CO}}{P_{CO_2}P_{H_2}}\right)}{1 + \left(\frac{K_{H_2O}}{K_{H_2}K_8K_9}\right)\left(\frac{P_{H_2O}}{P_{H_2}}\right) + \sqrt{K_{H_2}P_{H_2}} + K_{H_2O}P_{H_2O}}$$

The parameters required are in the following table:

$\sqrt{K_{H_2}}$	A	0.499
	B	17,197
K_{H_2O}	A	6.62×10^{-11}
	B	124,119
$\frac{K_{H_2O}}{K_{H_2}K_8K_9}$	A	3453.38
	B	–
$k'_{5a} \cdot K'_2 \cdot K_3 \cdot K_4 \cdot K_{H_2}$	A	1.07
	B	36,696
k'_1	A	1.22×10^{10}
	B	−94,765

EXAMPLE 5.11

Assume a feed composition of 0.2CO, 5H$_2$, and 0.4CO$_2$, and no inerts. The reactor operates at 493K and 50 atm. Model the reactor.

Solution

```
Reactor model

PARAMETER
A_rKH2 as Real
A_KH2O as Real
A_quot as Real
A_k5K2K3K4KH2 as Real
A_k1 as Real
B_rKH2 as Real
B_KH2O as Real
B_quot as Real
B_k5K2K3K4KH2 as Real
B_k1 as Real
R as Real

VARIABLE
Temp as Temperature
Press as Pressure
rKH2 as notype
KH2O as notype
quot as notype
k5K2K3K4KH2 as notype
k1 as notype
#% Algebraic equations - equilibrium constants
K as notype
K3 as notype
# Algebraic equations - Partial pressures as ideal gas mixture
P_CO as Pressure
P_H2O as Pressure
P_MeOH as Pressure
P_H2 as Pressure
P_CO2 as Pressure
# Algebraic equations - Reaction rates
r_MeOH as notype
r_RWGS as notype
molCO as molar
molH2O as molar
molMeOH as molar
molH2 as molar
molCO2 as molar
moltotal as molar
```

```
SET
A_rKH2:=0.499;
A_KH2O:=6.62e-11;
A_quot:=3453.38;
A_k5K2K3K4KH2:=1.07;
A_k1:=1.22e10;
B_rKH2:=17197;
B_KH2O:=124119;
B_quot:=0;
B_k5K2K3K4KH2:=36696;
B_k1:=-94765;
R:=8.314;

EQUATION
rKH2 = A_rKH2*exp(B_rKH2/(R*Temp));
KH2O = A_KH2O*exp(B_KH2O/(R*Temp));
quot = A_quot*exp(B_quot/(R*Temp));
k5K2K3K4KH2 = A_k5K2K3K4KH2*exp(B_k5K2K3K4KH2/(R*Temp));
k1 = A_k1*exp(B_k1/(R*Temp));
#% Algebraic equations - equilibrium constants
K = 10^(3066/(Temp)-10.592);
K3 = 1/(10^(-2073/(Temp)+2.029));
# Algebraic equations - Partial pressures as ideal gas mixture
moltotal = molCO + molH2O + molMeOH + molH2 + molCO2;
P_CO = molCO/moltotal*Press;
P_H2O = molH2O/moltotal*Press;
P_MeOH = molMeOH/moltotal*Press;
P_H2 = molH2/moltotal*Press;
P_CO2 = molCO2/moltotal*Press;
# Algebraic equations - Reaction rates
r_MeOH = (k5K2K3K4KH2*P_CO2*P_H2*(1-(1/K)*(P_H2O*P_MeOH/
  (P_H2^3*P_CO2))))/((1+quot*P_H2O/P_H2 + rKH2*(P_H2)^(1/2) +
  KH2O*P_H2O)^3);
r_RWGS = (k1*P_CO2*(1-K3*(P_H2O*P_CO/(P_CO*P_H2))))/
  (1+quot*P_H2O/P_H2 + rKH2*(P_H2)^(1/2) + KH2O*P_H2O);
#Differential continuity equations
$molCO = r_RWGS;
$molH2O = r_RWGS + r_MeOH;
$molMeOH = r_MeOH;
$molH2 = -r_RWGS-3*r_MeOH;
$molCO2 = -r_MeOH-r_RWGS;

Process
Unit R102 as Methanol
assign
```

```
Within R102 do
Temp: = 493;
Press: = 50;
end

INITIAL
   WITHIN R102 DO
   molCO = 0.2;
   molH2O = 0;
   molMeOH = 0;
   molH2 = 5;
   molCO2 = 0.4;
END # Within

SOLUTIONPARAMETERS
   ReportingInterval : = 0.05 ;

SCHEDULE
   CONTINUE FOR 1
```

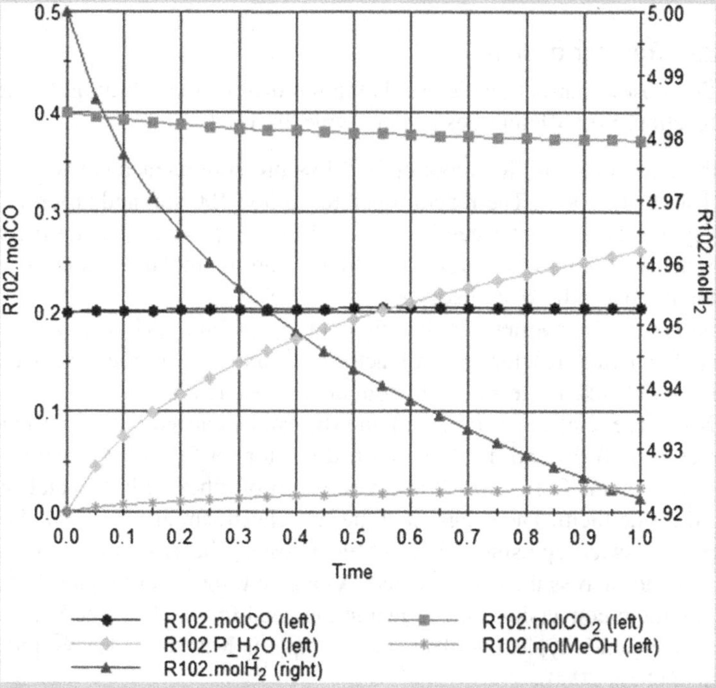

FIGURE 5E11.1

Concentration profile in the reactor.

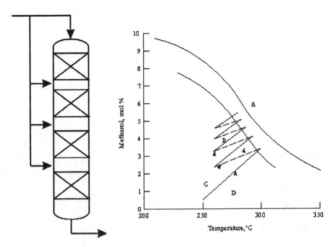

FIGURE 5.36

Direct cooled multibed reactors. (A) Equilibrium line; (B) maximum rate line; (C) quench line; and (D) intrabed line.

5.4.2.2.4 Reactor design

One of the most important challenges is the removal of the energy generated during the reaction. Six different designs are typically used:

a. Multibed reactors with intercooling. ICI low pressure quench converters (English et al., 2000). The reactor operates at 50–100 atm and around 270°C. Cold syngas is fed in between beds to cool the hot product. Since syngas is fed at different stages, the product is diluted and the molar fraction of methanol reduces between beds, as seen in Fig. 5.36.

b. A Kellogg-type converter is indirectly cooled; see Fig. 5.37. The feed enters a series of spherical reactors. Each reactor is separated from the next one by means of heat exchangers used to cool down the stream.

  (Note: The analysis of types (a) and (b) can be carried out as presented in Chapter 7, Sulfuric acid, for the multibed reactor for SO_3 production.)

c. Tubular reactors. These consist of a vessel full of tubes with the catalyst packed inside them. The syngas and the recycle stream are fed to the bottom where it is heated up using the hot product gases. The syngas turns at the top and descends across the catalytic bed. As a result, the reaction path follows the maximum conversion line, reducing the need for catalyst; see Fig. 5.38. The conversion is typically 14%. One example is Mitsubishi Heavy Industries (English et al., 2000).

d. Rising steam converters, Lurgi technology, operate in nearly isothermal conditions. Catalyst can be placed in the tubes region or in the shell region. These are the most efficient from the thermodynamic point of view since the catalyst volume is smaller. The good temperature control reduces the formation of byproducts; see Fig. 5.39.

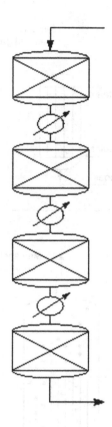

FIGURE 5.37

Indirect cooled converter.

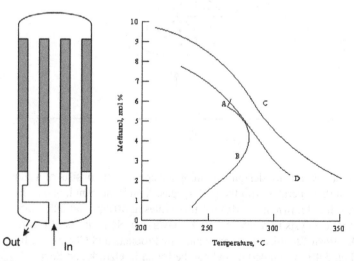

FIGURE 5.38

Tubular reactor for methanol synthesis.

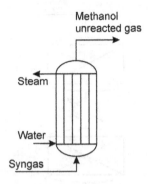

FIGURE 5.39

Rising steam converter.

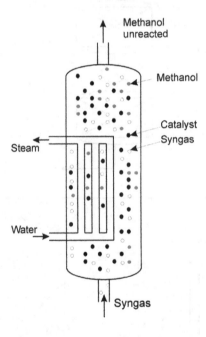

FIGURE 5.40

Slurry reactor.

e. Lately a new reactor design is gaining acceptance. It consists of a slurry bubble column that operates with the gas, syngas, liquid, methanol, and solid, catalyst, phases. It is known as LPMEOH. It operates at 50 atm and 225–265°C using Cu/ZnO catalysts to reach 15–40% conversion. Some users of this technology are Eastman Chemical Company and Air Products and Chemicals, Inc.; see Fig. 5.40. A detailed model can be found in Ozturk and Shah (1984).

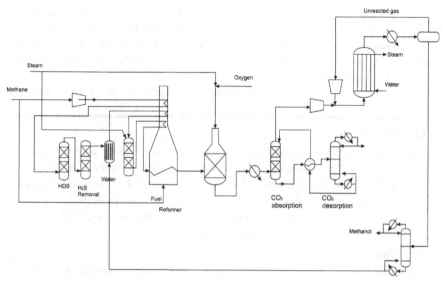

FIGURE 5.41

Methanol production flowsheet.

f. The last alternative consists of using a Gas−Solid−Solid Trickle Flow Reactor (GSSTFR) with an adsorbent material (SiO/Al_2O_3) to remove the methanol in situ, driving the equilibrium to products (Spath and Dayton, 2003; Verma, 2014).

5.4.2.2.5 Production process

There are three alternatives based on the operating pressure at the reactor: high pressure (250−300 atm), medium pressure (100−250 atm), and low pressure (50−100 atm). In all cases the process consists of three stages; see Fig. 5.41 for a typical configuration.

Syngas production: Fig. 5.41 shows a scheme of the production of methanol from natural gas reforming consisting of desulfurization of the methane and removal of H_2S, saturation of the methane with water, prereforming, and primary and secondary reforming. Part of the methane is burned to provide the energy to the endothermic reaction, and also to heat up the natural gas. The proper H_2-to-CO ratio is around 2. Other technologies for syngas production can be used as presented in the first section of this chapter.

Methanol synthesis: Step reaction that is 50% of the equilibrium at most, operating at 200−300°C and under pressure. Methanol and water are condensed and the unconverted gases recycled:

$$CO + 2H_2 \leftrightarrow CH_3OH$$

$$CO_2 + H_2 \leftrightarrow CO + H_2O$$

High-pressure operation: Operating at 25 MPa, zinc oxide–chromium oxide catalysts are used in five bed reactors. Dimethyl ether and water are also produced.

Low-pressure operation: Since 1923 these have been the typical plants. Operating around 50 atm and using copper-based catalysts, this operation uses reactors such as the one in Fig. 5.36.

Distillation of a water–methanol mixture: The liquid mixture is separated.

EXAMPLE 5.12

The production of methanol yields a stream with 45% molar methanol and the rest water. The requested purity of methanol is 97%, and the residue is assumed to have 5% methanol. The column processes a saturated liquid mixture with an *L/D* ratio of 1.5. Calculate the flow of distillate and residue, and the number of trays, assuming that each tray has an efficiency of 70%. Assume that the mixture behaves as ideal.

Solution.

Assuming that Raoult's Law holds, and using Antoine's correlation for the vapor pressures, we draw the XY diagram for the mixture:

$$\ln(P_v) = \left(A - \frac{B}{(T\,(°C) + C)} \right)$$

where

$$x = \frac{P_{Total} - P_{Vap,H_2O}}{P_{Vap,CH_3OH} - P_{Vap,H_2O}}$$

$$\rightarrow \text{We draw the diagram}$$

$$y = \frac{P_{Vap,CH_3OH}}{P_{Total}} x$$

Assuming that the liquid and vapor flows remain almost constant along the columns in each region, we have the rectifying and stripping operating lines by performing a mass balance to the condenser and the reboiler as follows:

$$y_n = \frac{L}{V} x_{n-1} + \frac{D}{V} x_D$$

$$y_m = \frac{L'}{V'} x_{m-1} + \frac{W}{V'} x_W$$

Since we have the information on the reflux ratio, we compute the slope and the intercept. We perform a mass balance to the condenser as follows:

$$\frac{L}{V} = \frac{1}{1 + 1/(L/D)}$$

$$\frac{D}{V} = \frac{1}{1 + L/D}$$

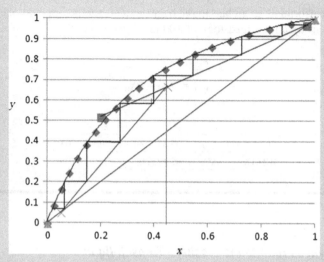

FIGURE 5E12.1

Number of theoretical trays.

Thus,

L/D	1.5
L/V	0.6
D/V	0.4

$x_D = 0.97$, according to the information provided. Thus we plot the operating line at the rectifying section. The operating lines intercept at the q line.

$$\Phi_L = \frac{L' - L}{F}$$

$$\Phi_V = \frac{V - V'}{F}$$

$$q = \frac{\Phi_V - 1}{\Phi_V}x_{n.} + \frac{1}{\Phi_V}x_f$$

Since the feed is saturated liquid, the q line is vertical, and with the rectifying section operating line we can also plot the stripping section operating line; see Fig. 5E12.1 The intercept is given for $x = 0.45$. Solving y_n from the operating line, $y_n = 0.67$ (Fig. 5E12.1).

We need eight theoretical trays. The actual number is given as follows:

$$n_{real} = \frac{n_{theoretical} - 1}{\eta} + 1 = \frac{7}{0.7} + 1 = 11 \text{ trays}$$

By performing a global mass balance and another one to the most volatile, using a total feedstock of 100 kmol, we have:

$$F = D + R$$

$$Fx_f = DX_D + RX_R$$

	F	D	R
Total	100	43.4782598	56.5217402
CH_3OH	45	42.173912	2.82608701

5.4.2.3 Syngas reaction to Fischer–Tropsch liquids

The desired products determine the operating conditions in terms of temperatures and pressures (from 10 to 40 bar), and the catalyst type employing either cobalt- or iron-based catalysts. For the production of gasoline and small hydrocarbons, currently iron-based catalysts are used, operating at high temperatures (300–350°C)—ie, high-temperature Fischer–Tropsch (HTFT). If diesel or heavier products are to be obtained, either cobalt-based or iron-based catalysis can be used. The reactors will operate at lower temperature (200–240°C), ie, low-temperature Fischer–Tropsch (LTFT). The iron catalyst provides high selectivity for C_{10}–C_{18}, which means a higher yield of diesel. Cobalt catalysts typically do not allow WGSR, and require higher H_2-to-CO ratios. Moreover, the reactions with iron catalyst are usually conducted at 30 bar. Furthermore, FT synthesis requires careful control of the H_2:CO ratio to satisfy the stoichiometry of the synthesis reactions as well as to avoid deposition of carbon on the catalysts (coking). An optimal H_2:CO ratio from 1:1 to 2:1 for the production of diesel and gasoline is recommended (Dry, 2002, Swanson et al., 2010).

5.4.2.3.1 Mechanisms

The main reactions can be seen below. The first reaction corresponds to methanation, the second is the WGS reaction, and the third reaction is the Boudouard reaction. The methanation reaction and the Boudouard reaction are undesirable:

$$CO + 3H_2 \leftrightarrow CH_4 + H_2O; \quad \Delta H_{298} = 247 \text{ kJ/mol}$$
$$CO + H_2O \leftrightarrow CO_2 + H_2; \quad \Delta H_{298} = -41 \text{ kJ/mol}$$
$$2CO \leftrightarrow C + CO_2; \quad \Delta H_{298} = -172 \text{ kJ/mol}$$

The reaction to produce hydrocarbons given below is the desired one, and is the most dominant reaction when applying cobalt-based FT catalyst. When using

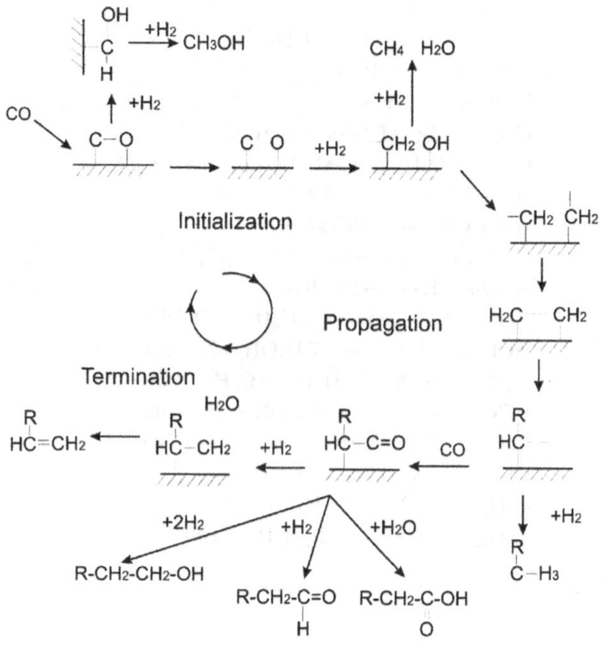

FIGURE 5.42

FT liquid production mechanism.

Reproduced with permission from Martín, M., Grossmann, I.E., 2011. Process optimization of FT-diesel production from lignocellulosic switchgrass. Ind. Eng. Chem. Res. 50, 13485–13499, Martín and Grossmann (2011).

iron-based catalyst, the WGS reaction also readily occurs, enabling the operation at a lower temperature.

$$nCO + \left(n + \frac{m}{2}\right)H_2 \rightarrow C_nH_m + nH_2O$$

$$CO + 2H_2 \rightarrow -CH_2- + H_2O; \qquad \Delta H_{FT} = -165 \text{ kJ/mol}$$

The details for the elementary chemical steps on the surface of the catalysts are unclear, but it is widely accepted that a stepwise chain growth process is involved. Fig. 5.42 shows the polymerization-based mechanism. Large hydrocarbon chains are believed to be produced through the dissociation of CO. At each stage of the growth process, the surface species (at first, surface carbon) have the option of desorbing to produce alkenes, or to be hydrogenated to desorb as an alkane, or to continue the chain growth process by adding another CH_2. Fischer–Tropsch liquids can be refined to various amounts of renewable (green) gasoline, diesel fuel, and aviation fuel depending upon process conditions.

Botao et al. (2007) presented the following kinetic mechanism where "rds" represents limiting steps.

$$CO + s \leftarrow \rightarrow CO\text{-}s$$
$$CO\text{-}s + H\text{-}s \leftarrow \rightarrow C\text{-}s + OH\text{-}s$$
$$C\text{-}s + H_2 \leftarrow \rightarrow CH_2\text{-}s$$
$$H_2 + 2s \leftarrow \rightarrow 2H\text{-}s$$
$$CH_2\text{-}s + H\text{-}s \rightarrow CH_3\text{-}s + s \quad (rds)$$
$$CH_2\text{-}s + C_nH_{2n+1}\text{-}s \rightarrow C_nH_{2n+1}CH_2\text{-}s + s \quad (rds)$$
$$HO\text{-}s + H\text{-}s \leftarrow \rightarrow H_2O + 2s$$
$$H\text{-}s + CO\text{-}s \leftarrow \rightarrow HCO\text{-}s + s$$
$$C_nH_{2n+1}\text{-}s + CO\text{-}s \leftarrow \rightarrow C_nH_{2n+1}CO\text{-}s + s$$
$$HCO\text{-}s + H_2 \leftarrow \rightarrow HCHOH\text{-}s$$
$$C_nH_{2n+1}CO\text{-}s + H_2 \leftarrow \rightarrow C_nH_{2n+1}CHOH\text{-}s$$
$$HCHOH\text{-}s + H\text{-}s \leftarrow \rightarrow CH_3OH + 2s \quad (rds)$$
$$C_nH_{2n+1}CHOH\text{-}s + H\text{-}s \leftarrow \rightarrow C_nH_{2n+1}CH_2OH\text{-}s$$
$$HCO\text{-}s + OH\text{-}s \leftarrow \rightarrow HCOOH + 2s \quad (rds)$$
$$C_nH_{2n+1}CO\text{-}s + OH\text{-}s \leftarrow \rightarrow C_nH_{2n+1}COOH + 2s$$
$$CH_3\text{-}s + H\text{-}s \rightarrow CH_4 + 2s$$
$$C_nH_{2n+1}\text{-}s + H\text{-}s \rightarrow C_nH_{2n+2} + 2s$$
$$C_nH_{2n+1}\text{-}s \leftarrow \rightarrow C_{n+1}H_{2n} + H\text{-}s \quad (rds)$$

5.4.2.3.2 Reactor types

Four main types of reactors are used within FT facilitates. From left to right in Fig. 5.43 we have circulating bed, typically used from 1950 to 1987 for HTFT, with a production capacity of 2−6.5 kbdp. To the right there is the multitubular reactor, used from 1950 to 1985 for LTFT, with production capacities of 0.5−0.7 kbdp. The second generation of reactors from the 1980s−90s includes

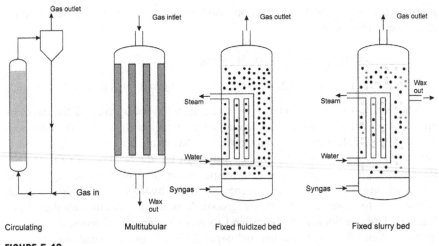

Circulating   Multitubular   Fixed fluidized bed   Fixed slurry bed

FIGURE 5.43

FT reactor designs.

Table 5.6 Alternative Processes for FT Production

Syngas Production	Reactor	Catalyst	Company
Partial oxidation	Slurry	Co-based	Energy Int
Catalytic partial oxidation	Slurry	Co-based	Exxon
Partial oxidation, steam reforming, autothermal reforming	Slurry	Fe-based	Rentech
Partial oxidation, steam reforming, coal gasification	Slurry	Co, LTFT	Sasol
	Slurry	Fe, LTFT	
	Fluidized	Fe, HTFT	
Partial oxidation	Fixed	Co	Shell
Autothermal reforming using air	Fixed	Co	Syntroleum

the fixed fluidized bed (Synthol reactor) from 1989, with a production capacity of 11−20 kbdp, used for HTFT; and the fixed slurry bed, introduced in 1993 for LTFT, with production capacities of 2.5−17 kbdp.

5.4.2.3.3 Syngas production and synthesis

The alternative technologies for the production of FT liquids came from the combination of the various technologies available for each step, from the processing of the various raw materials, gas cleaning, catalysts employed, and the type of reactor. In Table 5.6, a few of the main companies producing FT liquids are compared with regards to the decision on the technologies that constitute their characteristic flowsheets.

The entire flowsheet for the production of FT liquids consists of syngas production, purification, and composition adjustment. After that, the FT reactor is used to produce the hydrocarbon mixture. There are two alternatives as mentioned above: LTFT and HTFT. The product distribution is a direct function of them and the ratio of the H_2-to-CO feed to the reactor.

5.4.2.3.4 Product distribution

To determine the composition of the products as a function of the operating conditions at the reactor, the Anderson−Schulz−Flory (ASF) distribution has typically been used. It assumes that the FT reactor operates as a polymerization reactor. The fraction of mass of the hydrocarbons i = number of C, and w_i depends on the probability of chain growth, α.

$$w_i = \alpha^{i-1}(1-\alpha)^2 \cdot i \tag{5.28}$$

According to the studies by Song et al. (2004), α is a function of the temperature and the ratio CO-to-H_2, as follows:

$$\alpha = \left(0.2332 \cdot \left(\frac{y_{CO}}{y_{CO}+y_{H_2}}\right) + 0.633\right) \cdot (1 - 0.0039 \cdot ((T\,(^\circ C) + 273) - 533)) \tag{5.29}$$

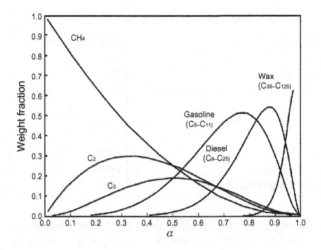

FIGURE 5.44

Product distribution as a function of chain length.

Reproduced with permission from Martín, M., Grossmann, I.E., 2011. Process optimization of FT-diesel production from lignocellulosic switchgrass. Ind. Eng. Chem. Res. 50, 13485–13499, Martín and Grossmann (2011).

Fig. 5.44 shows the effect of α on the composition of the products. Diesel is produced as the main component for values of α around 0.9 using LTFT. Gasoline requires slightly lower values of α, 0.7–0.8, and HTFT.

5.4.2.3.5 Product upgrading and refinery processes

The product stream from an FT reactor is separated and treated to obtain fuels. This stream can be very similar to crude oil, and the same upgrading methods are common to a regular refinery.

HTFT: The exit of the FT reactor is separated, obtaining the decanted oil (DO) fraction, the stabilized light oil (SLO) fraction, the condensate, and the aqueous phase. The condensate allows production of ethylene, propylene, and monomers using cold separation. From the SLO fraction, an atmospheric distillation unit allows the production of chemicals and gasoline. The heavy products from the atmospheric distillation are fed to the vacuum column together with the DO fraction, which allows production of light vacuum gas oil and heavy vacuum gas oil. From the bottoms we get wax. The most widely used method for upgrading the wax produced in FT reactors is hydrocracking, which allows the production of liquefied petroleum gases (LPGs): gasoline and diesel. This process combines the hydrogenation and cracking of the heavy oil. Therefore, the catalyst is bifunctional. Cracking is obtained using acidic support—ie, zeolites or amorphous silica alumina—while hydrogenation is imparted by metals such as Pd, Pt, Mo, Wo or Co, and Ni. The advantage is

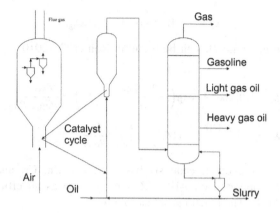

FIGURE 5.45

Fluid catalytic cracking unit.

that it is possible to process a wide range of feedstocks (with high distillate selectivity) to naphtha and diesel—typically around 60% diesel and 30% naphtha—operating at 370°C and 7 MPa. It is an exothermic process, but it is expensive. Fluid catalytic cracking (FCC) is another option to break down large hydrocarbons. The process uses a sand-like material as catalyst, and heat. The catalyst is circulated from the reactor and the regenerator. The reactor typically operates at about 715°C and 3.4 bar. On the catalyst, carbon is deposited, and the regenerator, operating at 1.7 bar, burns the coke, heating up the catalyst. This energy is used to vaporize the feedstock and to carry out the endothermic cracking reactions. The product stream, consisting of a gas phase containing light hydrocarbons, gasolines, light and heavy oils, and slurry oil, is sent to fractionation. Fig. 5.45 shows a scheme of the operation of such an upgrading process.

LTFT: The wax and condensate obtained from the FT reactor are cracked to obtain LPGs (gasoline and diesel) using hydrocracking or FCC.

5.4.2.4 Use of CO$_2$ to chemicals

Lately, various catalysts have been developed for the hydrogenation of CO$_2$ to fuels. Typically, those that allow reverse WGSR are the ones suitable for this operation. Thus, sufficient concentration of hydrogen is required (Riedel et al., 1999). Below we present the main equilibria involved in the production of the same chemicals as above, but from CO$_2$. The mechanism consists of a large number of steps (up to 49) involving more than 30 species (Grabow and Mavrikakis, 2011).

Methane has been produced using metal catalysts such as Ni, Ru, Fe, Co, Pd, or Rh supported on Al$_2$O$_3$ (Janke et al., 2014) following these reactions:

$$CO + 3H_2 \leftrightarrow CH_4 + H_2O \tag{5.28}$$

$$CO_{2(g)} + H_{2(g)} \leftrightarrow CO_{(g)} + H_2O_{(g)} \tag{5.29}$$

The equilibrium constants can be found in Roh et al. (2010), where T is in K and P in kPa.

$$K_{p1} = 10,266.76 \cdot \exp\left(-\frac{26,830}{T} + 30.11\right) = \frac{P_{CO} \cdot P_{H_2}^3}{P_{CH_4} \cdot P_{H_2O}}$$

$$K_{p2} = \exp\left(\frac{4400}{T} - 4.063\right) = \frac{P_{CO_2} \cdot P_{H_2}}{P_{CO} \cdot P_{H_2O}} \tag{5.30}$$

For *methanol* production, catalysis based on Cu, Zn, Cr, and Pd is used, such as Cu/ZnO supported on Al_2O_3. Zr can be used as an efficient promoter (Jadhav et al., 2014). The reactions are as follows:

$$CO_2 + 3H_2 \leftrightarrow CH_3OH + H_2O$$

$$CO_2 + H_2 \leftrightarrow CO + H_2O$$

The values of the equilibrium constants for reactions are computed using Eqs. (5.31) and (5.32) (Chinchen et al., 1988).

$$\frac{[P_{CH_3OH}][P_{H_2O}]}{[P_{CO_2}][P_{H_2}]^3} = K_{a2}e^{\left[22.225 + \frac{9143.6}{T} - 7.492 \ln(T) + 4.076 \times 10^{-3} \cdot T - 7.161 \times 10^{-8} \cdot T^2\right]}$$

$$\tag{5.31}$$

$$\frac{[P_{CO}][P_{H_2O}]}{[P_{CO_2}][P_{H_2}]} = K_{a2}$$

$$= \exp\left[13.148 - \frac{5639.5}{T} - 1.077 \ln T - 5.44 \times 10^{-4}T + 1.125 \times 10^{-7}T^2 + \frac{49,170}{T^2}\right] \tag{5.32}$$

For *FT fuels*, the reactions are as follows:

$$CO_2 + 3H_2 \rightarrow -CH_2- + 2H_2O + 125 \text{ kJ/mol}$$

$$CO_2 + H_2 \leftrightarrow CO + H_2O$$

So far the yield is nowhere comparable to the one from syngas (Riedel et al., 1999, Rodemerck et al., 2013).

5.4.2.5 Methanol to gasoline (MTG)

This is a technology based on the dehydration of methanol to dimethylether (DME) and its subsequent use to produce olefins (Exxonmobil, 2014). Methanol is vaporized and fed to a DME reactor, which is a fluidized bed reactor that uses ZSM-5 catalyst. The feed enters at 300–320°C and 14.5 bar. It is highly exothermic, 1740 kJ/kg of methanol. Therefore, heat removal is an issue. The effluent of this reactor, at 400–420°C, is mixed with the recycle gas from the methanol to gasoline (MTG) reactors and fed to the MTG reactors. This recycle helps control the temperature:

$$2CH_3OH \leftrightarrow CH_3OCH_3 + H_2O$$

$$(n/2) \cdot CH_3OCH_3 \rightarrow C_nH_{2n} - nH_2O \rightarrow n[CH_2] + nH_2O$$

The conversion of the process is close to 100% and yields 1% gas, 5% LPG, 38% gasoline, and 56% water. For the organic phase we have a composition of 0.7% CH_4, 0.4% C_2H_6, 0.2% propylene, 4.3% propane, 1.1% butylenes, 10.9% butane, 82.3% naphtha, and 0.1% oxygenates.

The water produced is removed from the organic phase and sent to a train of columns to recover the C_2 fraction (de-ethanizer column), LPG (stabilizer or de-butanizer column), and light and heavy gasoline (splitter). While most of the de-butanizer bottoms (stream containing C5 +) are sent to the gasoline splitter, a small portion of the de-butanizer is recycled to the de-ethanizer column to act as a lean oil solvent. This stream helps in the separation of the gases. After this stage, the heavy gasoline containing C9 + hydrocarbons is treated. This stream contains durene. It is an undesirable product that is responsible for carburetor "icing" due to its high melting point. The treatment is conducted with a Heavy Gasoline Treater (HGT). This stage removes durene via hydrogenation. The liquid obtained is stabilized in a stripper column and combined with the light gasoline from the de-butanizer. (Jones and Zhu, 2009).

Such a plant produces 3.2 bbl of gasoline per day and a ton of methanol, 0.42 bbl of LPG/day t, and 0.097 bbl/day t of fuel gas.

5.5 PROBLEMS

P5.1. Coal is processed using air and steam to produce a syngas with an H_2-to-N_2 ratio of 3. The coal has 10% ash and 90% C. 1% of the carbon is lost with the ash. In the water gas cycle, 80% of the steam reacts. In the gas generator cycle, 65% of the carbon produces CO_2 and the rest CO. In the water gas cycle, the carbon generates CO. The air used is at 25°C and 760 mmHg with 55% relative humidity.

Determine:

a. Relative amount of air and steam.
b. Heating power of the gas produced per ton of coal.
c. Amount of air required per ton of coal.

P5.2. Syngas with an H_2-to-CO ratio of 2 is to be produced from C_2H_2. Steam reforming of the hydrocarbon, fed at 900K, is carried out using 50% excess of water with respect to the stoichiometric one. In the process, 100% conversion of the hydrocarbon is obtained. A fraction of the gas product is sent to a WGSR, maintaining the temperature from the previous stage, to increase the yield to hydrogen. Assume isothermal operation of the WGSR. See Fig. P5.2 for the scheme of the flowsheet. Determine the temperature and composition of the gases leaving the reformer, which requires 75,000 kcal/kmol of hydrocarbon and the fraction of the product gas that is fed to the WGSR.

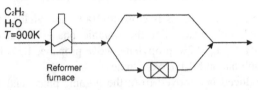

FIGURE P5.2

Scheme for the production of syngas.

P5.3. The feedstock for the production of ammonia contains 0.5 mol of Ar per 100 mol of N_2–H_2 mixture. After mixing it with the recycle, the stream is fed to the converter. The syngas has stoichiometric proportions of nitrogen and hydrogen ($N_2 + 3H_2$). The gas also contains 5 moles of Ar per 100 moles of syngas. The reactor is operated at 200 atm and 740K isothermally where the equilibrium shown below takes place:

$$0.5N_2 + 1.5H_2 \leftarrow \rightarrow NH_3$$

The equilibrium constant (K_p) can be computed as a function of the temperature using the following equation:

$$\log (K_p) = \frac{2250.322}{T} - 0.85340 - 1.51049 \log(T) - 2.58987 \times 10^{-4}T + 1.48961 \cdot 10^{-7}T^2$$

$$T[=]K, \quad K_p[=]atm^{-1}$$

The ammonia produced can be completely recovered by condensation. The unreacted gas is recycled. A fraction of the recycle stream is purged. Determine the flow of each stream (1, 2, 3, and 4) in Fig. P5.3 per 100 moles of syngas fed to the system.

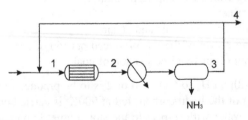

FIGURE P5.3

Scheme for the ammonia synthesis loop.

P5.4. In a coal-based synthesis gas facility, 45% of the fuel is consumed in the generator gas cycle while the rest follows the water gas cycle. The coal consists of 90% C and 10% ash. Assume that the losses of C with the ashes are negligible.

The generator gas cycle uses atmospheric air ($T_s = 23°C$, $\varphi = 0.62$, and $P_T = 720$ mmHg), and the reaction takes place at 973K and 1 atm. Assume that no water gas equilibrium with steam is reached, just reaction. The equilibrium constant as a function of the temperature can be computed using the following equation:

$$\text{Boudouard reaction: } \log(K_p) = 9.1106 - \frac{8841}{T} \quad T[=]K$$

The water gas cycle uses steam so that only 40% of it reacts. The equilibrium for the WGS among the resulting gases establishes at 800K. The equilibrium constant can be determined using the following equation:

$$\text{Water gas reaction } \log K_p = \left(\frac{2073}{T\,(K)} - 2.029 \right) T[=]K$$

$$P_v = \exp\left(18.3036 - 3844/(227 - T(°C)) \right)$$

Determine:
- Relative amounts of generator gas and water gas produced.
- Heating value of both gases.
- Fraction of the initial heating power in C remaining in both gases.

P5.5. The ammonia converter is fed with a stoichiometric mixture of nitrogen and hydrogen. The mixture carries 5 moles of Ar per 100 moles of synthesis gas. The reactor operates at 230 atm and 800K isothermally.

$$0.5N_2 + 1.5H_2 \leftarrow \rightarrow NH_3$$

The value for the equilibrium constant as a function of the temperature can be computed using the following equation:

$$\log(K_p) = \frac{2250.322}{T\,(K)} - 0.85340 - 1.51049 \log(T(K)) - 2.58987$$
$$\times 10^{-4} T(K) + 1.48961 \times 10^{-7} T(K)^2$$

The ammonia is condensed in a heat exchanger at the reaction pressure. Determine the final temperature to recover 96% of the ammonia produced. Assume humid air ideal behavior in the ammonia condensation. The ammonia vapor pressure is seen below (Fig. P5.5):

$$\log_{10}(P_v) = A - (B/(T + C))$$
P_v = vapor pressure (bar)
T = temperature (K)

Temperature (K)	A	B	C
164.0 – 239.6	3.18757	506.713	−80.78
239.6 – 371.5	4.86886	1113.928	−10.409

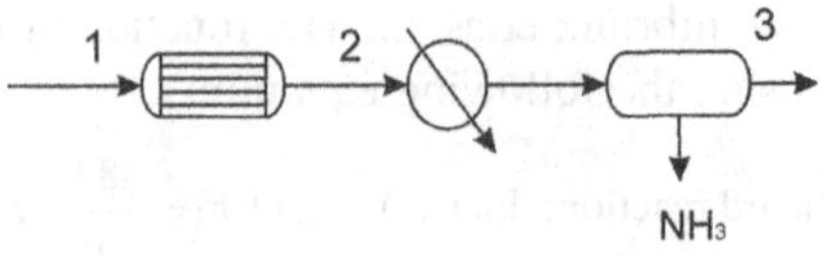

NH₃

FIGURE P5.5

Scheme for the ammonia synthesis and separation.

P5.6. In a thermal plant, coal is burned with dry and stoichiometric air. After sulfur removal, the CO_2 produced in the combustion is expected to be further used. The CO_2 stream is fed to a coal bed to recover part of its synthetic potential. The aim is to achieve 99% CO_2 conversion. Determine the pressure and temperature of operation, as well as the coal consumption during this operation, not accounting for the thermal plant operation, per kmol of dry air used to burn the original coal. The equilibrium constant for the operation of the coal bed is given by the following equation:

$$\log(K_p) = 9.1106 - \frac{8841}{T\,[=]K}$$

P5.7. In a coal-based synthetic gas production facility, 40% of the fuel follows the generator gas cycle and the rest, 60%, is used in the water gas cycle. The coal consists of 90% carbon and 10% ash. Assume negligible carbon losses in the ash. In the generator gas cycle, half of the carbon produces CO and the other half produces CO_2. The oxygen is completely consumed. Dry air is used. The water gas cycle operates at 1000K and 1.09 atm. Assume that the carbon is completely consumed and the steam is fed in 50% excess.

　　Determine:
- Composition of the gas generator and water gas.
- Operating temperature of the gas generator cycle.
- Heating values for the two gases per kg of carbon fed.
- The H_2-to-CO ratio of the mixed gases.

P5.8. Syngas with the appropriate H_2-to-CO ratio for ethanol production is produced. An electrolytic cell operating at 80°C is used for the

production of hydrogen. The current yield of the cell is 55%.
The hydrogen is at 80°C. On the other hand, a flowrate of 36 kmol/h
of flue gas from a thermal plant is available. The composition, once
cleaned, consists of 66% molar CO_2, and the rest is CO. Determine the
consumption of water in the electrolysis (kg/h) and the energy required
to be applied to the electrolytic cell (kW).

$$\text{Data: } K_p = \frac{(CO_2)(H_2)}{(CO)(H_2O)} = e^{\left[\frac{2073}{T(K)} - 2.029\right]}$$

P5.9. A purified CO_2 stream from a thermal plant is to be reused for the
production of hydrogen from CH_4. The methane and the CO_2 are fed at
25°C, while the gases exit at 600°C. Compute the amount of methane
per mol of hydrogen produced considering an adiabatic reforming
process and assuming 100% methane conversion. Determine the
fraction of methane used to provide the required energy for the
reaction.

P5.10. Methane is autoreformed adiabatically with oxygen and steam. The
methane and the oxygen are fed at 25°C. The steam is fed at 233°C. The
exit gas contains CO, CO_2, and hydrogen. The temperature is 600°C and
the ratio of H_2O vap/O_2 fed is equal to 2. Determine the amount of
water and oxygen required to perform the autorefoming of methane and
the composition of the product gas.

P5.11. Methanol is produced using hydrogen and CO. The feed gas comes from
the partial oxidation of hydrocarbons. The molar composition is given in
dry basis in Table P5.11.
1. This gas is to be processed to obtain a hydrogen-to-CO ratio of 2. A
   fraction of the feed is bypassed. The rest is subjected to WGS adding
   steam. The stream contains 1.6% moles of CO. Determine the
   fraction of the gas being bypassed.
2. Next, CO_2 is removed to obtain the syngas. Compute the moles of
   CO_2 removed from the raw syngas per 100 moles of initial gas.
3. Finally, in a third stage, the syngas produces methanol. Assume
   90% conversion; the methanol is recovered. Determine the
   composition of the unreacted gases and the volume to produce
   100 kmol of methanol.

Table P5.11 Molar Composition of Syngas

H_2	CO	CO_2	CH_4	N_2
45.8%	46.2%	5.0%	0.6%	2.4%

P5.12. FT diesel is produced using hydrogen and CO. The feed gas comes from the partial oxidation of hydrocarbons. The molar composition is given in dry basis in Table P5.12.

1. This gas is to be processed to obtain a hydrogen-to-CO ratio of 2. A fraction of the feed is bypassed. The rest is subjected to WGS adding steam. The stream contains 1.6% moles of CO_2. Determine the fraction of the gas being bypassed.
2. Next, CO_2 and H_2S are removed, as well as the water. Compute the moles of CO_2 removed from the raw syngas per 100 moles of initial gas.
3. Finally, in a third stage, the syngas produces FT diesel. Determine the operating temperature to maximize the diesel fraction, assuming that the product distribution can be predicted using the Anderson–Schulz–Flory model. Compute the moles of liquid fuels produced assuming that only the -CH_2- synthesis reaction occurs with 70% conversion.

Table P5.12 Molar Composition of Syngas

H_2	CO	CO_2	CH_4	H_2S
45.8%	46.2%	5.0%	0.6%	2.4%

P5.13. Syngas in stoichiometric proportions ($1:3$ $N_2:H_2$) is fed to the system. See Fig. P5.3 for the scheme of the synthesis loop. It also contains Ar, 0.2 moles per 100 moles of mixture ($N_2 + 3H_2$). Assuming the following, determine the purge of the system:

1. The reactor conversion is 25%.
2. Only 75% of the ammonia produced is recovered.
3. The converter can only handle Ar at 2.43 moles per 100 moles of syngas mixture.

P5.14. The expected increase in hydrogen demand results in the development of propane steam reforming. In the furnace, two main reactions take place, the steam reforming and the WGS:

$$C_3H_{8(g)} + 3H_2O_{(g)} \rightarrow 3CO_{(g)} + 7H_{2(g)}$$

$$CO_{(g)} + H_2O_{(g)} \rightarrow CO_{2(g)} + H_{2(g)}$$

The reforming reaction takes place over Ni catalyst in the tubes of the furnace. The feed consists of a mixture of steam and propane with a molar ratio of 5:1 at 125°C. The products leave the furnace at 800°C. Assume complete propane conversion.

The reaction is endothermic, and thus energy must be provided. A hot flue gas with a composition of 22% CO_2 and the rest nitrogen is fed at 1400°C and 1 atm, and exits at 900°C. The flue gas is further used to produce steam by cooling it down to 250°C. Water is fed at 25°C and the stream is produced at 125°C and 1 atm. Determine the molar composition of the gas product and the flue gas required to produce the steam required by the reforming stage.

P5.15. Methanol is produced using syngas, a mixture ($CO + 2\ H_2$) containing CO_2 as an inert impurity. Originally there were 0.25 mol per 100 mol of syngas. The reactor operates at 25 bar and 232°C. Assume that 90% of the methanol can be recovered condensing it. No water is produced since Co-based catalysts are used. The reactor can handle up to 5 moles of CO_2 per 100 moles of syngas. Fig. P5.15 shows the scheme of the synthesis loop. Determine:

1. The reactor conversion assuming that the equilibrium constant of the process taking place in the reactor can be computed using the following equation:

$$K_p = \frac{P_{CH_3OH}}{P_{CO}P_{H_2}^2} = 10^{(3921/T(K)-7.971\cdot\log_{10}(T(K))+0.002499\cdot(T(K))-0.0000002953\cdot(T(K))^2+10.2)}$$

2. The purge fraction.

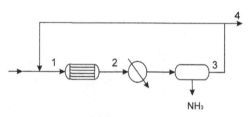

FIGURE P5.15

Scheme of the methanol synthesis loop.

P5.16. Stream number 2 in the flowsheet above is separated using a flash, the system between 2 and 4 in the figure. Compute the composition of the gas and liquid phases if the stream is cooled down to 27°C, maintaining the reactor pressure.

P5.17. Stream number 2 in the flowsheet above is separated using a flash, the system between 2 and 4 in the figure. Compute the cooling temperature to recover 98% of the methanol produced. Assume that the reactor pressure is maintained in the flash.

P5.18. A methanation reaction takes place in a multibed, Co-based catalytic reactor where the WGS does not take place. The reactants are fed with a hydrogen-to-CO ratio of 6, and the conversion per pass is 75% of the equilibrium conversion. Determine the temperature of the feed to the first bed to reach 25% conversion. Data for the equilibrium and the energy balance:

$$CO + 3H_2 \Leftrightarrow CH_4 + H_2O$$

$$K_p(kPa) = \frac{P_{CH_4} \cdot P_{H_2O}}{P_{CO} \cdot P_{H_2}^3} = \left(10,266.76 \exp^{\left[\frac{26,830}{T(K)}+30.11\right]}\right)^{-1}$$

$$\Delta H_r = \Delta H(CH_4) + \Delta H(H_2O) - \Delta H(CO) - 3 \cdot \Delta H(H_2)$$
$$\bar{c}_p = 29 \frac{kJ}{kmol\ C}$$

Hint: See Example 7.7.

P5.19. Ammonia synthesis is performed in a multibed catalytic reactor; see Fig. P5.19. Each bed operates adiabatically and the reactor employs indirect cooling. Compute the conversion reached after two catalytic

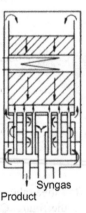

Syngas
Product

FIGURE P5.19

Scheme of the reactor.

beds if only 50% of the equilibrium conversion is reached. The feed to the first bed is 400°C, and after each bed the feed is cooled to 400°C. Total pressure is 300 atm. Data for the equilibrium and the energy balance:

$$0.5 \cdot N_2 + 1.5H_2 \Leftrightarrow NH_3$$

$$K_p = \frac{P_{NH_3}}{P_{N_2}^{0.5} \cdot P_{H_2}^{1.5}}$$

$$\log_{10}(K_p) = \frac{2250.322}{T(K)} - 0.85430 - 1.51049 \log_{10} T(K)$$

$$- 2.58987 \times 10^{-4} T(K) + 1.48961 \cdot 10^{-7} T(K)$$

$$\Delta H_r = \Delta H(NH_3) = -45,720 \text{ kJ/kmol}$$

$$\bar{c}_p = 28.8 \frac{kJ}{kmol \, C}$$

Hint: See Example 7.7.

P5.20. A methanation reaction takes place in a multibed Co-based catalytic reactor where the WGS does not take place. The reactants are fed at 500°C with a hydrogen-to-CO ratio of 7. After each bed, the products are cooled to the initial temperature. The pressure of operation is 500 kPa. Compute the number of beds so that the final conversion is 75%.

Data for the equilibrium and the energy balance:

$$CO + 3H_2 \Leftrightarrow CH_4 + H_2O$$

$$K_p(kPa) = \frac{P_{CH_4} \cdot P_{H_2O}}{P_{CO} \cdot P_{H_2}^3} = \left(10,266.76 \exp \left[\frac{26,830}{T} + 30.11 \right] \right)^{-1}$$

$$\Delta H_r = \Delta H(CH_4) + \Delta H(H_2O) - \Delta H(CO) - 3 \cdot \Delta H(H_2)$$

$$\bar{c}_p = 29 \frac{kJ}{kmol \, C}$$

Hint: See Example 7.7.

P5.21. Ammonia synthesis is performed in a two-bed catalytic reactor; see Fig. P5.21. Each bed operates adiabatically and the reactor employs

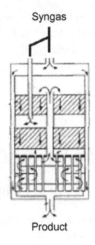

Syngas

Product

FIGURE P5.21

Scheme of the reactor.

direct cooling. The syngas is at 20°C. The heated syngas is fed to the first bed at 400°C. Total pressure is 300 atm. Determine the fraction of the syngas that is directly fed to the second bed so that the initial temperature of all beds is 400°C and there is total conversion.

Data for the equilibrium and the energy balance:

$$0.5 \cdot N_2 + 1.5H_2 \Leftrightarrow NH_3$$

$$K_p = \frac{P_{NH_3}}{P_{N_2}^{0.5} \cdot P_{H_2}^{1.5}}$$

$$\log_{10}(K_p) = \frac{2250.322}{T(K)} - 0.85430 - 1.51049 \log_{10} T(K)$$

$$- 2.58987 \times 10^{-4} T(K) + 1.48961 \times 10^{-7} T(K)^2$$

$$\Delta H_r = \Delta H(NH_3) = -45,720 \text{ kJ/kmol}$$

$$\bar{c}_p = 28.8 \frac{\text{kJ}}{\text{kmol C}}$$

Hint: See Example 7.7.

P5.22. An ammonia production plant operates an absorption column with 15 trays. It processes a gas stream of 15 kmol/s with 25% ammonia in countercurrent flow with water. The flow of water is 20 kmol/s. The column operates at 10 atm and 25°C to recover 98% of the ammonia. Calculate the tray efficiency and the excess of liquid flow with respect to the minimum. Assume that Raoult's Law holds.

P5.23. An ammonia production facility uses an absorption tower with 10 trays with an efficiency of 80%. It processes a gas stream of 15 kmol/s, 25% ammonia with a water stream in countercurrent at 10 atm and 25°C to recover 98% of the ammonia. Determine the water flow rate and its excess with respect to the minimum. Assume that Raoult's Law holds.

P5.24. Captured CO_2 with nitrogen is to be used as a carbon source for the production of methanol (as in the figure). The reaction over Co catalysts is as follows:

$$CO_{2(g)} + 3H_{2(g)} \leftarrow \rightarrow CH_3OH_{(g)} + H_2O_{(g)}$$

$$\frac{[P_{CH_3OH}][P_{H_2O}]}{[P_{CO_2}][P_{H_2}]^3} = e^{\left[22.225 + \frac{9143.6}{T(K)} - 7.492 \ln(T(K))4.076 \times 10^{-3} \cdot T(K) - 7.161 \times 10^{-8} \cdot T(K)^2\right]}$$

The feed to the system is in stoichiometric proportions containing 0.5% nitrogen. The reactor operates with syngas in stoichiometric proportions at 25 atm. It can handle up to 3 moles of nitrogen per 100 mol of syngas ($CO_2 + 3H_2$). A conversion of 20% per pass is expected. Assuming that 90% of the methanol and all the water produced are recovered, determine the operating temperature at the reactor and the purge fraction. Fig. P5.24 shows the synthesis loop.

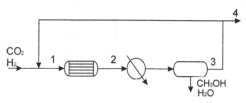

FIGURE P5.24

Synthesis loop for methanol production.

P5.25. In a methanol production plant the distillation column has 9 trays with an efficiency of 75% certified by the vendor. We process a mixture of 45% methanol and the rest water. The distillate has a composition of 97% methanol, and the residue 3% methanol. Calculate the reflux ratio (*L/D*) if the feed is saturated liquid and the mixture behavior is assumed to be ideal:

$$\ln(P_v) = \left(A - \frac{B}{(T\,(°C) + C)}\right)$$

P5.26. In a methanol production plant the distillation column has 12 trays, but we do not trust the efficiency provided by the vendor. We process a mixture of 45% methanol and the rest water. The distillate has a composition of 97% methanol, and the residue 3% methanol. The column is operated with a reflux ratio (L/D) of 1.75 and the feed is saturated liquid. Determine the efficiency of the trays assuming ideal behavior of the mixture.

P5.27. A stream of 40 kmol/s (34.5 bar and 298K) of syngas with 10% CO_2 is processed in a 12-tray absorption column to remove 99.5% of it before further stages. The equilibrium data for the system CO_2–amine is given in Table P5.27. The L/G ratio is 1.5, the minimum one. Determine the efficiency of the trays.

Table P5.27 Equilibrium Data

mol CO_2/mol Amine	PCO_2(Pa)
0.025	65
0.048	197
0.145	1204
0.276	3728
0.327	4815
0.45	8253
0.9	26,325
10	1,480,620
15	2,918,315
20	4,722,992

P5.28. Repeat Example 5.9 using any commercial process simulator and compare the results; T reactor (300°C).

P5.29. Simulate the methanol and ammonia synthesis loops using process simulators. Evaluate the effect of using CO_2 as a carbon source instead of CO.

P5.30. A novel process for the production of hydrogen from propane is being evaluated. In the fired furnace reformer, two main reactions take place:

$$C_3H_{8(g)} + 3H_2O_{(g)} \rightarrow 3CO_{(g)} + 7H_{2(g)}$$
$$CO_{(g)} + H_2O_{(g)} \rightarrow CO_{2(g)} + H_{2(g)}$$

Propane steam reforming occurs in a reactor whose pipes are packed with nickel catalyst. The feed to the reactor consists of a mixture, steam–propane with a molar ratio of 6:1 at 125°C. The products of the reforming stage leave the reactor at 800°C. The excess of steam assures the complete consumption of propane. The syngas has a molar ratio of CO-to-CO_2 of 5. Determine the fraction of the initial propane needed to operate the furnace adiabatically, and the syngas molar composition.

P5.31. Select between the use of a flash or an absorption column to recover 80% of the ammonia in a stream consisting of 211, 70.4, 9.12, and 62.1 kmol/s of H_2, N_2, Ar, and NH_3, respectively. The initial temperature and pressure are 400°C and 100 bar. For the absorption column, it operates at 100°C and 100 bar with a liquid flow twice the minimum and an efficiency of 57% for the stages.

Cost of cooling 1€/kW (c_p(kJ/kg K): H_2 = 14; N_2 = 1; Ar = 0.52; NH_3 = 2.2; λ = 1369 kJ/kg). Recovering ammonia from liquid, 1500€/(kg/s). Cost of vessels, 25,000€/stage. The flash is 1 stage.

P5.32. Using a process simulator (ie, CHEMCAD), compute the conversion of a multibed reactor with direct cooling; see Fig. 5.19 for reference. The feed to the system is 1 kmol/s of N_2, 3 kmol/s of H_2, 0.2 kmol/s of Ar at 30 MPa and 373K. The reactor consists of three beds, and the first is fed at 673K. The feed to each of the following beds should be around 673K. Determine the feed rates of unconverted syngas to each of the beds assuming that they are at 293K, 30 MPa, and the same composition as the initial feed.

REFERENCES

Appl, M., 1998. Ammonia. ULLMANN'S "Encyclopedia of Industrial Chemistry". Wiley-VCH.

Appl, M., 1999. Ammonia Principles and Industrial Practice. Wiley-VCH, Weinheim.

Botao, T., Chang, J., Wan, H., Lu, J., Zheng, S., Liu, Y., et al., 2007. A corrected comprehensive kinetic model of Fischer–Tropsch synthesis. Chin. J. Catal. 28 (8), 687–695.

Brigdwater, A.V., 1995. The technical and economic feasibility of biomass gasification for power generation. Fuel 14 (5), 631–653.

Chinchen, G.C., Denny, R.J., Jennings, J.R., Spencer, M.S., Waugh, K.C., 1988. Synthesis of methanol: part 1. catalysts aid killetics. Appl. Catal. 36, 1.

Couper, J.R., Penney, W.R., Fair, J.R., 2009. Chemical Process Equipment: Selection and Design. Butterworth-Heinemann.

Czuppon, T.A., Knez, S.A., Rovner, J.M., Kellogg Company, M.W., 2000. Ammonia. Kirk-Othmer Encyclopedia of Chemical Technology. John Wiley and Sons, New York.

Davies, J., Lihou, D., 1971. Optimal design of methane steam reformer. Chem. Proc. Eng. 52, 71e80.

Dry, M.E., 2002. The Fischer–Tropsch process: 1950–2000. Catal. Today 71, 227–241.

Eckhard, F., Grossmann, G., Burkard Kersebohm, D., Weissy, G., Witte, C., 1998. Methanol. ULLMANN'S "Encyclopedia of Industrial Chemistry". Wiley-VCH.

English, A., Rovner, J., Brown, J., Davies, S., 2000. Methanol. Kirk-Othmer Encyclopedia of Chemical Technology. John Wiley and Sons, New York.

Ernest, F.A., Reed, F.C., Edwards, W.L., 1925. A direct synthetic ammonia plant. Ind. Eng. Chem. 17 (8), 775−788.

Exxonmobil, 2014. http://cdn.exxonmobil.com/~/media/global/files/catalyst-and-licensing/2014-1551-mtg-gtl.pdf.

GPSA, 2004. Gas Processors Suppliers Association Engineering Data book, twelfth ed. Tulsa, Oklahoma.

Graaf, G.H., Sijtsema, P.J.J.M., Stamhuis, E.J., Joosten, G.E.H., 1986. Chemical equilibria in methanol synthesis. Chem. Eng. Sci. 41, 2883−2890.

Grabow, L.C., Mavrikakis, M., 2011. Mechanism of methanol synthesis of Cu through CO_2 and CO hydrogenation. ACS Catal. 2011 (1), 365−384.

Jadhav, S.G., Vaidya, P.D., Bhanage, B.M., Joshi, J.B., 2014. Catalytic carbon dioxide hydrogenation to methanol: a review of recent studies. Chem. Eng. Res. Des. 92 (11), 2557−2567.

Janke, C., Duyar, M.S., Hoskins, M., Farrauto, R., 2014. Catalytic and adsorption studies for the hydrogenation of CO_2 to methane. Appl. Catal. B: Environ. 152−153, 184−191.

Jones, S.B., Zhu, Y., 2009. Techno-Economic Analysis for the Conversion of Lignocellulosic Biomass to Gasoline via the Methanol-to Gasoline (MTG) Process. US Dept. of Energy. PNNL-18481.

Kierzkowska-Pawlak, H., Chacuk, A., 2010. Kinetics of carbon dioxide absorption into aqueous MDEA solutions. Ecol. Chem. Eng. 17 (4), 463−475.

Lloyd, L., 2011. Handbook of Industrial Catalysts. Springer, Bath.

Martín, M., 2014. Introduction to Software for Chemical Engineers. CRC Press, Boca Raton, FL.

Martín, M., Grossmann, I.E., 2011. Process optimization of FT-diesel production from lignocellulosic switchgrass. Ind. Eng. Chem. Res. 50, 13485−13499.

Martín, M., Grossmann, I.E., 2013. On the systematic synthesis of sustainable biorefineries. Ind. Eng. Chem. Res. 52 (9), 3044−3064.

Murase, A., Roberts, H.L., Converse, A.O., 1970. Optimal thermal design of an autothermal ammonia synthesis reactor. Ind. Eng. Chem. Des. Dev. 9, 503−513.

Ortuño, A.V., 1999. Introducción a la Química Industrial. Reverté, Barcelona.

Osaka Gas Chemicals, 2015. http://www.jechem.co.jp/shirasagi_e/tech/psa.html.

Osenwengie Uyi, I., 2007. H_2 Production in Palladium and Palladium-Copper Membrane Reactors at 1173K in the Presence of H_2S. Doctoral Dissertation, University of Pittsburgh.

Ozturk, S.S., Shah, Y.T., 1984. I-C methanol synthesis process. DOE report 60054 http://www.fischer-tropsch.org/DOE/DOE_reports/60054/doe_pc_60054-t9/doe_pc_60054-t9-C.pdf.

Plasynski, S.I., Chen, Z.Y., 1997. Review of CO_2 capture technologies and some improvement opportunities https://web.anl.gov/PCS/acsfuel/preprint%20archive/Files/45_4_WASHINGTON%20DC_08-00_0644.pdf.

Reep, M., Cornelissem, R.L., Clevers, S., 2007. Methanol catalyst poisons. A literature study. RportIND07002 06.25 Chrisgas. Versie 1.1.

Reklaitis, G.V., 1983. Introduction to Material and Energy Balances. Wiley, New York.

Reinmert, R., Marchner, F., Renner, H.-J., Boll, W., Supp, E., Brejc, M., et al., 1998. Gas production 2. processes. ULLMANN'S "Encyclopedia of industrial chemistry". Wiley-VCH.

Riedel, T., Claeys, M., Schulz, H., Schaub, G., Nam, S.S., Jun, K.W., et al., 1999. Comparative study of FTS with H_2/CO and H_2/CO_2 syngas using Fe and Co catalysts. Appl. Catal. A 186, 201–213.

Rodemerck, U., Holena, M., Wagner, E., Smejkal, Q., Barkschat, A., Baerns, M., 2013. Catalyst development for CO_2 hydrogenation to fuels. Chem. Cat. Chem. 5 (7), 1948–1955.

Roh, H.-S., Lee, D.K., Koo, K.Y., Jung, U.H., Yoon, W.L., 2010. Natural gas steam reforming for hydrogen production over metal monolith catalyst with efficient heat-transfer. Int. J. Hydrogen 35 (3), 1613–1619.

Song, H.-S., Ramkrishna, D., Trinh, S., Wright, H., 2004. Operating Strategies for Fischer–Tropsch reactors: a model-directed study. Korean J. Chem. 2004 (21), 308–317.

Spath, P.L., Dayton, D.C., 2003. Preliminary screening—technical and economic assessment of synthesis gas to fuels and chemicals with emphasis on the potential for biomass-derived syngas NREL/TP − 510 34929.

Susanto, H., Beenackers, A., 1996. A moving-bed gasifier with internal recycle of pyrolysis gas. Fuel 75 (11), 1339–1347.

Swanson, R.M., Platon, A., Satrio, J.A., Brown, R.C., 2010. Techno-economic analysis of biomass-to-liquids production based on gasification. Fuel 89 (Suppl. 1), S11–S19.

The Haber–Bosch Heritage: The Ammonia Production Technology. Max Appl 50th Anniversary of the IFA Technical Conference September 25, 1997.

Uhde, 2015. Ammonia ThyssenKrupp (www.uhde.com) http://www.thyssenkrupp-csa.com.br/fileadmin/documents/brochures/0a2d5391-b166-484d-847d-3cbfd941f06b.pdf.

Vancini, C.A., 1961. La sintesi dell'ammoniaca. Hoepli, Milano.

Vanden Bussche, K.M., Froment, G.F., 1996. A steady-state kinetic model for methanol synthesis and the water gas shift reaction on a commercial $Cu/ZnO/Al_2O_3$ catalyst. J. Catal. 161, 1–10.

Verma, A.K., 2014. Process Modelling and Simulation in Chemical, Biochemical and Environmental Engineering. CRC Press, Boca Raton, LA.

Wallas, S.M., 1959. Reactor Kinetics for Chemical Engineers. McGraw-Hill, New York, NY.

Wallas, S.M., 1990. Chemical Process Equipment. Selection and Design. Butterworth-Heinemann Series in Chemical Engineering, Washington.

Weissermel, K., Arpe, H.J., 1978. Industrielle Organische Chemie. 2. Auflage Bedeutende Vor- und Zwischenprodukte. Verlag Chemie, Weinheim.

<http://www.topsoe.com/sites/default/files/leckel.pdf>.

<http://www.infomine.com>.

<http://www.worldcoal.org/bin/pdf/original_pdf_file/coal_matters_2_-_global_availability_of_coal(16_05_2012).pdf>.

Nitric acid

6

6.1 INTRODUCTION

History: Nitric acid was already known by alchemists in the Middle Ages. However, it was not until the 12th century that its preparation was described in the "De Inventioni Veritatis." Nitric acid was produced by distillation of a mixture consisting of 1 g of vitriolum cyprium ($CuSO_4 \cdot 5H_2O$), 1.5 g of potassium nitrate KNO_3, and 0.25 g of potassium alum ($KAl(SO_4)_2 \cdot 12H_2O$). It was again described by Albert the Great in the 13th century and by Ramon Lull, who prepared it by heating niter and clay and called it "eau forte" (aqua fortis). A few centuries later, in 1776, Lavoisier showed that it contained oxygen. Cavendish, in 1785, was able to produce it by an electric discharge in humid air, proving that it contained nitrogen and oxygen. Finally, the complete composition was determined by Gay-Lussac and Bethollet in 1816. The actual use only expanded when sulfuric acid became commercial.

Production: The industrial process developed by Glauber in 1698 was used for centuries. It consisted of the distillation of a mixture of potassium nitrate (KNO_3) and sulfuric acid (H_2SO_4). Potassium nitrate was substituted by $NaNO_3$ (Chile saltpeter) during the great war. Although it had been produced from NO obtained in air combustion, the current method based on the catalytic oxidation of ammonia was patented in 1902 by Wilhem Ostwald. Nitric acid must be kept from light and heat since NO_2 gases are produced, giving it a yellow color. The major world producers are Germany, France, the United Kingdom, Belgium, Canada, and Spain (Ortuño, 1999).

Uses: In 2013 fertilizers used 80% of the total nitric acid produced. Among them, 96% was ammonium nitrate. Nonfertilizer use represents around 20%. The main products are nitrobenzene (3.6%), dinitrotoluene (2.8%), adipic acid (2.7%), and nitrochlorobenzenes (1.8%). The total world production is around 55 MMt/yr. Fig. 6.1 presents the major products obtained from nitric acid (Clarke and Mazzafro, 1993).

Industrial Chemical Process Analysis and Design. DOI: http://dx.doi.org/10.1016/B978-0-08-101093-8.00006-9

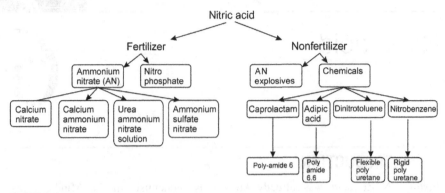

FIGURE 6.1

Major products from nitric acid.

6.2 PRODUCTION PROCESSES

6.2.1 FROM NITRATES

In this production process, sodium nitrate and sulfuric acid are mixed in a smelter. The sodium sulfate obtained remains in the vessel while the nitric acid is distilled.

$$2NaNO_3 + H_2SO_4 \rightarrow NaSO_4 + 2HNO_3$$

Since sulfuric acid is used in excess, up to 60–70% of sodium bisulfate is also produced. The advantage is that it is easier to handle.

$$NaNO_3 + H_2SO_4 \rightarrow NaHSO_4 + HNO_3$$

Furthermore, sodium pyrosulfate is also produced. This chemical is undesirable because it is a foaming agent that requires large vessels to contain it.

$$2NaHSO_4 \rightarrow Na_2S_2O_7 + H_2O$$

In the distillation, impurities such as HCl and nitrosyl chloride are removed first.

6.2.2 AIR AS RAW MATERIAL

The advantage that this method exploits is its cheap and widely available raw material: air. However, it exhibits a major drawback too—the large amount of energy required, which is provided by an electric arc. Therefore, although the method is technically interesting, it has not been developed industrially due to its high cost. The Nernst equation can provide an estimate of the production of NO as a function of the working temperature (see Table 6.1). The Birkeland and Eyde electric arc was installed in Norway in 1905 and was operated for 35 years. The energy consumption added up to 14,000–17,000 kWh per ton of nitric acid produced.

Table 6.1 NO Produced Using an Electric Arc

T (°C)	1227	1427	1527	1827	2027	2227	2427	2627	2927
%NO	0.10	0.23	0.46	0.79	1.23	1.79	2.44	3.18	4.39

6.2.3 AMMONIA-BASED PROCESSES

In 1839, a German technical chemist named Kuhlman discovered the role of platinum in the oxidation of ammonia. He predicted its commercial application, but at that time there was no commercial necessity. As the demand for fertilizers and explosives increased, the Germans realized that local production of nitric acid was a need. The first plant based on the Ostwald process was allocated in Westphalia, Germany, in 1909. Soon after, in 1918, the first one in England started its production. The process itself consists of five stages:

1. Oxidation of NH_3 to NO (catalyzed with platinum). The high cost of ammonia and platinum requires a high conversion of the reaction.
2. Oxidation of NO to NO_2 (no catalyst). It is a slow reaction.
3. Absorption of NO_x in water for the production of nitric acid. There is also a reduction of NO_2 to NO. In the column both absorption and reaction take place.
4. Tail gas purification. This stream is also used as a cooling agent.
5. Removal of NO_x from the nitric acid, ie, bleaching.

6.2.3.1 Process description

Nitric acid production processes are classified by their operating pressure. Apart from early designs that worked at atmospheric pressure, currently they work at medium (1.7—6.5 bar), high (6.5—13 bar), or dual pressure. These processes try to make the most of the chemical principles governing the operations. Oxidation, as will be explained later, is favored at lower pressure, while the absorption of a gas in water has higher yield at high pressure. Dual process can work at low—medium pressure so that the oxidation occurs below 1.7 bar and the absorption from 1.7 to 6.5, or medium—high pressure, in which case the oxidation occurs from 1.7 to 6.5 bar while the absorption takes place at 6.5—13 bar (Ray and Johnson, 1989, MMAMRM, 2009). Table 6.2 shows typical operating parameters used in industry.

In Table 6.3, examples of industrial process plants recently built are shown with the company that runs them and the main operating conditions. We see that by operating at constant pressure we rarely go beyond a concentration of 65%. Somehow, the advantage achieved by operating at the optimal conditions for one of the stages, ammonia oxidation or NO_2 absorption, mitigates the performance loss in the second one. However, dual pressure optimizes the operation, reaching concentrations around 68%.

Based on the information on real processes presented in the tables above, the different operating modes and the corresponding flowsheets will be described below.

Table 6.2 Typical Operating Parameters of Different Basic Processes (Thyssenkrupp Uhde, 2015)

Plant Type (per t of HNO_3)	Medium Pressure	High Pressure	Dual Pressure
Operating pressure (bar)	5.8	10	4.6/12 bar
Ammonia (kg)	284	286.0	282.0
Electric power (kWh)	9.0	13.0	8.5
Platinum (primary losses (g))	0.15	0.26	0.13
With recovery (g)	0.04	0.08	0.03
Cooling water ($\Delta T = 10$ K) (t)	100	130	105
Process water (t)	0.3	0.3	0.3
LP heating steam, 8 bar, saturated (t)	0.05	0.20	0.05
HP excess steam, 40 bar, 450°C (t)	0.76	0.55	0.65
N_2 yield (%)	95.7	94.5	>96
HNO_3 conc. (%)	Max. 65	Max. 67	>68
Tail gas (ppm)	500	200	150

Table 6.3 New Plant Operating Conditions (Thyssenkrupp Uhde, 2015)

Year	Company	Location	NH_3 Oxidation Pressure (bar)	Absorption Pressure (bar)	Acid Concentration %	Capacity mtpd 100%
2005	Rashtriya Chemicals and Ferlilizers Ltd.	India	7.5	7.5	60	352
2003	Namhae Chemical Corporation	Korea	4.6	12	67	1150
2001	BP Koln GmbG	Germany	4.6	12	68.25	1500
2001	Radice Chimica GmbH	Germany	10	10	65	250
2001	ACE Pressureweld	Singapore	4.4	4.4	60	12
2000	Queensland Nitrates Pty Ltd.	Australia	10	10	60	405
1999	Enaex SA	Chile	10	10	60	925
1999	Namhae Chemical Corporation	Korea	10	10	65	300
1998	SKW Stickstoffwerke Piesteritz GmbH	Germany	5.6	5.6	62	500
1998	CF Industries Inc.	United States	4.6	11	57.5	870

Some *medium-pressure process* examples are Montecatini (1.7−5 atm), Chemico (4−4.5 atm), Stamicarbon (4.5 atm), and Uhde (4.5 atm). The main advantage of these processes is the reduced consumption of catalyst. Furthermore, the energy consumption is lower, as well as the risk that explosive mixtures will be formed. However, the construction materials are expensive, and typically the acid concentration is lower, requiring distillation. Note that the azeotrope water−nitric acid occurs at 68% concentration. This method is appropriate when nitric acid is to be devoted to fertilizer production.

Fig. 6.2 shows Uhde's medium-pressure process flowsheet. Ammonia is evaporated and filtrated. On the other hand, the air to be used in the combustion is purified using a two- or even three-step filtration system. The filters are expected to remove all the particles in the ammonia and air streams to avoid interferences with the catalyst during the oxidation of NH_3. Next, the flow is pressurized. The air stream is split in two. One stream is sent to the catalytic converter, the other to the bleaching section of the absorption column.

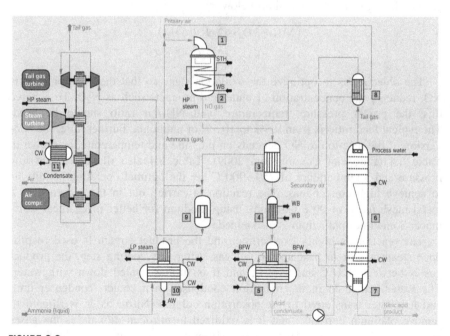

FIGURE 6.2

Medium-pressure process.
1. Reactor; 2. Process gas cooler; 3. Tail gas heater 3; 4. Economizer; 5. Cooler condenser and feed water preheater; 6. Absorption; 7. Bleacher; 8. Tail gas heaters 1 & 2; 9. Tail gas reactor; 10. Ammonia evaporator & superheater; 11. Turbine steam condenser.

Table 6.4 Ammonia Conversion as Function of Pressure and Temperature

Pressure (bar)	Temperature (°C)	% NO
<1.7	810–850	97
1.7–6.5	850–900	96
>6.5	900–940	95

The first air stream is mixed with ammonia in a ratio of 10:1, with careful consideration to avoid an explosive mixture composition. The mixture can be filtered again. Next, NH_3 reacts with air over the catalysts. The reaction, which is in fact an equilibrium that produces nitric oxide and water, is as follows:

$$4NH_3 + 5O_2 \leftrightarrow 4NO + 6H_2O$$

Apart from this main reaction, nitrogen and nitrous oxide (N_2O) can also be produced, as given by the reactions below:

$$4\,NH_3 + 3\,O_2 \leftrightarrow 2\,N_2 + 6\,H_2O$$

$$4\,NH_3 + 4\,O_2 \leftrightarrow 2\,N_2O + 6\,H_2O$$

The objective is to optimize the oxidation stage so that the main product is NO, reducing the concentration of undesired products such as N_2O. To achieve this, the proper pressure, temperature, and NH_3/air ratio should be used. The typical feed ratio is from 9.5% to 10.5% of ammonia. Furthermore, the conversion of the reaction to NO depends on pressure and temperature, as shown in Table 6.4 (European Commission, 2009). Table 6.4 also illustrates that low pressure and a temperature of 750–900°C are the optimal operating conditions to achieve high conversions. The reaction is carried out in the presence of a metal mesh made of 90% platinum, using rhodium for better mechanical resistance. Sometimes palladium is also added.

The reaction is highly exothermic, and the product stream is used to produce steam and/or to heat up the tail gas. After this cooling step, the product gas is between 100°C and 200°C, and it is further cooled down with water. The water generated in the oxidation condenses in a cooler–condenser unit, and it is later transferred to the absorption column. Nitric oxide is filtered to remove platinum particles, and it is oxidized to nitrogen dioxide as the gases are cooled down to 45–50°C in the absence of any catalyst following the reaction:

$$2NO + O_2 \leftrightarrow 2NO_2 \leftrightarrow N_2O_4$$

It is a homogenous process that depends on the working pressure and temperature. High pressure and low temperature are the best operating conditions. The absorption column is fed using demineralized water and condensed steam, or process condensate. The weak acid solution (around 43%) generated in the refrigeration of the product gas is also fed to the column. The NO_2 is put into countercurrent contact with water in the tower, producing HNO_3 and NO:

$$3NO_2 + H_2O \leftrightarrow 2HNO_3 + NO$$

Secondary air is fed to oxidize the NO produced. The oxidation and absorption of NO is carried out in the liquid and gas phases, respectively. In the trays the absorption takes place while the oxidation occurs in-between trays. High pressure and low temperature favor these reactions. The tower is typically 2 m in diameter and 46 m high, with 49 trays.

The process that takes place in the absorption tower is highly exothermic and requires continuous cooling. The nitric acid produced contains dissolved NO_2 and therefore has to be bleached using secondary air. The NO_2 is stripped out from the nitric acid and later reused. Typically the nitric acid produced is a solution with 55–70% acid, depending on the tower design and the operating conditions. The unabsorbed gases, or tail gas (at 20–30°C), exit from the top of the column. They are heated up using the excess energy generated in the oxidation of NO to NO_2. Heat integration is therefore included in process design. The hot tail gas is then processed to reduce the NO_x content. A fraction of the feed of ammonia is used as per the reactions bellow so as to decompose the nitrogen oxides.

$$4NO + 4NH_3 + O_2 \rightarrow 4N_2 + 6H_2O$$

$$3NO_2 + 4NH_3 \rightarrow (7/2)N_2 + 6H_2O$$

Next, tail gas is expanded in a turbine to recover part of the energy–power integration. The tail gas, typically over 100°C to avoid precipitation of ammonium nitrate and ammonium nitrite, is sent to a chimney.

In general, *high-pressure processes* are characterized by high catalyst, energy, and water consumption. The advantage is the smaller size of the equipment and the higher nitric acid concentration. Typically, these processes are better-suited when the acid is used in the production of explosives. Fig. 6.3 shows the flowsheet, which is similar to the previous one. The higher pressure requires the use of interstage cooling between compressors. Some examples are Dupont de Nemour (8–9 atm), Chemico (10–12 atm), Weathey (10–12 atm), and Uhde (12 bar).

Analyzing the principles of operation of various processes, it is clear that ammonia oxidation achieves higher yields at low pressures while the absorption of NO_2 in water is favored at high pressure. Therefore new plants use a *dual-pressure process* by adding a compression stage between the oxidation of the ammonia and the condenser. Typically, an external bleacher is used to strip out the NO_2 dissolved in the nitric acid produced. Fig. 6.4 shows a flowsheet for a dual process.

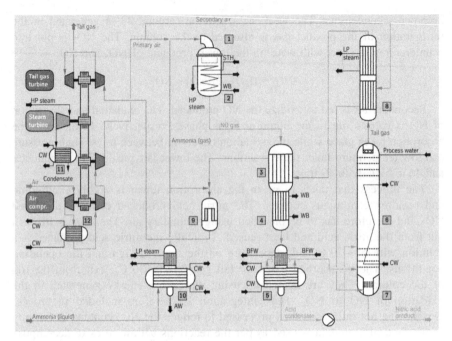

FIGURE 6.3

High-pressure process.
1. Reactor; 2. Process gas cooler; 3. Tail gas heater 3; 4. Economizer; 5. Cooler
condenser and feed water preheater; 6. Absorption; 7. Bleacher; 8. Tail gas heaters 1 & 2;
9. Tail gas reactor; 10. Ammonia evaporator & superheater; 11. Turbine steam condenser;
12. Air intercooler.

*With permission from Thyssenkrupp Uhde, 2015. http://www.thyssenkrupp-industrial-solutions.com/
fileadmin/documents-/brochures/uhde_brochures_pdf_en_4.pdf.*

Following the *production of nitric acid*, its concentration ranges from 60% to
70%, depending on the process conditions. This concentration is not enough for
certain applications. There are *two methods for obtaining a concentrated acid*:
direct and indirect.

Direct methods are based on the production of liquid N_2O_4, which is then
reacted with oxygen and water under pressure to produce HNO_3:

$$2N_2O_4 + O_2 + 2H_2O \rightarrow 4HNO_3$$

To produce N_2O_4, NO from the converter at low pressure is oxidized to NO_2
via oxidation and postoxidation steps. Next, the NO_2 is purified with concentrated nitric acid from the absorption stage, process condensate, and weak nitric
acid from the final absorption. Subsequently, the NO_2 or the N_2O_4 are separated

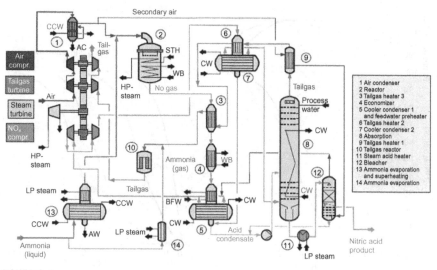

Secondary air
CCW
① AC Tail-gas
② STH
⑥
⑨
CW
WB
Air compr.
Tailgas turbine
Air
HP-steam No gas
Tailgas
Process water
③
Steam turbine
NO_x compr.
HP-steam
⑩
Ammonia (gas) ④
WB
CW
⑧
LP steam
Tailgas
⑬ CCW
BFW
CW
⑫
CW
CCW
AW LP steam
CW
⑤ Acid condensate
Ammonia (liquid)
LP steam
⑭
⑪
Nitric acid product
LP steam

1 Air condenser
2 Reactor
3 Tailgas heater 3
4 Economizer
5 Cooler condenser 1 and feedwater preheater
6 Tailgas heater 2
7 Cooler condenser 2
8 Absorption
9 Tailgas heater 1
10 Tailgas reactor
11 Steam acid heater
12 Bleacher
13 Ammonia evaporation and superheating
14 Ammonia evaporation

FIGURE 6.4

Dual-pressure process.

With permission from Thyssenkrupp Uhde, 2015. http://www.thyssenkrupp-industrial-solutions.com/fileadmin/documents-/brochures/uhde_brochures_pdf_en_4.pdf.

from the concentrated acid by bleaching and are then condensed. The most challenging stage is the production of liquid nitrogen dioxide. It can be either obtained by condensing NO_2 under pressure or by absorption on nitric acid. Then the reaction presented above is carried out, but instead of using water, nitric acid from the final absorption stage is used in a reactor at 50 bar and 75°C. A part is already a product, while the rest is recycled to absorption and final oxidation.

The residual gas from absorption can be discharged to the atmosphere, providing the NO_x concentration is within legal limits. Typically, condensates containing nitric acid are recycled within the process, otherwise they are processed as wastes.

Indirect methods are based on the extractive distillation of weak nitric acid (with sulfuric acid, for example). In Fig. 6.5 a phase diagram of the water–nitric acid system is presented. It shows a maximum boiling point azeotrope at around 68% weight fraction. Therefore, a dehydrating agent such as sulfuric acid or a solution of magnesium nitrate ($Mg(NO_3)_2$) can be used. Fig. 6.6 shows that if a solution of sulfuric acid with a concentration over 50% is used, the azeotrope disappears. Typically, sulfuric acid at 60% is used. Thus the nitric acid with a concentration of 55–65% is distilled with sulfuric acid. The liquid is fed from the top and flows countercurrently to the vapors that

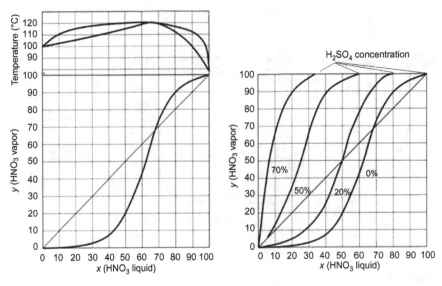

FIGURE 6.5

Phase diagram for nitric acid extractive distillation.

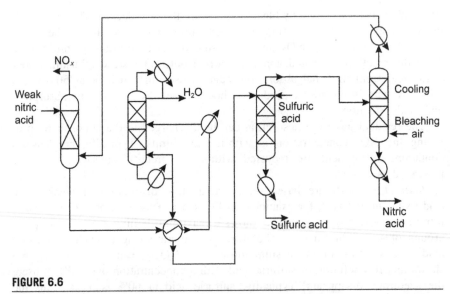

FIGURE 6.6

Extractive distillation scheme for concentrated nitric acid production.

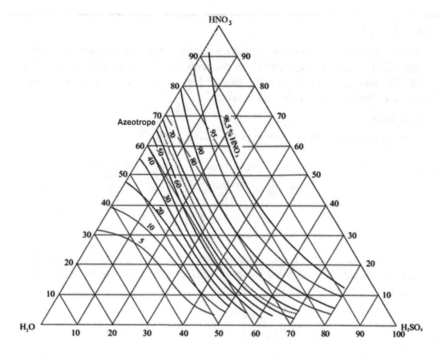

FIGURE 6.7

Triangular diagram nitric acid–sulfuric acid–water system (1 atm).

rise. Nitric acid in the vapor phase exits from the top of the column with a concentration of 99%. The impurities consist of small amounts of NO_2 and O_2. To purify the nitric acid, it is fed into a bleacher system to remove the dissolved NO_x. The vapors are hereby condensed and treated in countercurrent with the air. The air and the NO_x are sent to an absorption column where the NO is oxidized with secondary air to produce NO_2 that is later recovered as a weak acid. The unreacted gases are released to the atmosphere. Tail gases are treated in a wet scrubber to remove the NO_x using alkali solutions, or by means of molecular sieves that catalyze the oxidation of NO to NO_2 and later adsorb NO_2 to be recycled to the absorber. Fig. 6.6 shows a scheme of the flowsheet described above.

In the extractive column an equilibrium among water, nitric acid, and sulfuric acid takes place. Fig. 6.7 shows the triangular diagram for the nitric acid–sulfuric acid–water system including the azeotrope. We illustrate the operation of this column in Example 6.1.

EXAMPLE 6.1

Nitric acid at 65% is produced from a medium-pressure process. Compute the sulfuric acid required (99%) to concentrate 10 kg/s of nitric acid up to 98.5%. Sulfuric acid at 80% by weight leaves the system. Determine the flows of the streams exiting the tower.

Solution

A mass balance to water can be performed as follows:

$$m_{\text{water,HNO}_3\,\text{ini}} + m_{\text{water,H}_2\text{SO}_4\,\text{ini}} = m_{\text{water,HNO}_3,\text{fin}} + m_{\text{water,H}_2\text{SO}_4,\text{fin}}$$

$$0.35 \cdot 10 + \frac{m_{\text{H}_2\text{SO}_4}}{0.99} \cdot 0.01 = \frac{m_{\text{HNO}_3}}{0.985} \cdot (1 - 0.985) + \frac{m_{\text{H}_2\text{SO}_4}}{0.8} \cdot 0.2$$

$$m_{\text{H}_2\text{SO}_4} = 14.2 \text{ kg/s}$$

The composition of the streams that cross in the rectifying section can be computed by aligning the point difference, Δ, to the tangent to the equilibrium line of 98.5% (Fig. 6E1.1):

$$S2 = A(10)\frac{A - S3}{S2 - S3} = 10(6.6/1.8) = 36.7 \text{ kg/s}$$

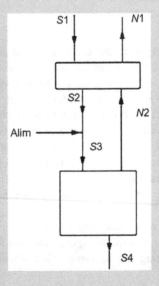

FIGURE 6E1.1

Feed tray scheme.

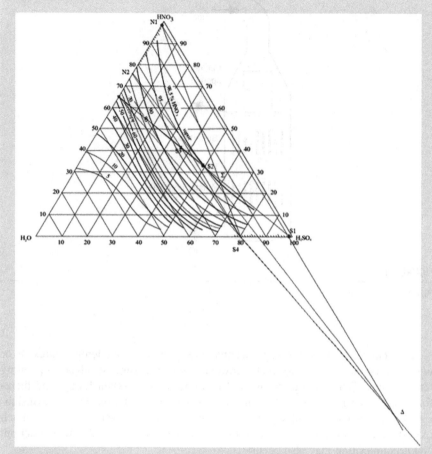

FIGURE 6E1.2

Extractive distillation.

The nitric vapors are computed as follows (Fig. 6E1.2):

$$N2 = S2\frac{S2 - \Delta}{N2 - \Delta} = 36.7(9.3/12.6) = 20.5 \text{ kg/s}$$

6.2.3.2 Process analysis

6.2.3.2.1 Ammonia oxidation

Reaction and equilibrium. Fig. 6.8 shows a scheme of the converter with the sponge of platinum catalyst and heat recovery regions. The reaction that takes place is the following:

$$NH_3 + \frac{5}{4}O_2 \rightarrow NO + \frac{3}{2}H_2O \quad \Delta H = -54 \text{ kcal/mol}$$

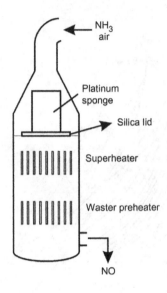

FIGURE 6.8

Ammonia converter.

Le Chatelier's principle suggests that low pressures and temperatures favor the conversion of NH_3 to NO. However, kinetics suggest higher operating temperatures. Temperatures below 400°C do not allow oxygen fixing, and therefore there is no reaction. If the temperature is over 600°C, the rate is diffusion-controlled. The oxygen is fixed to the catalyst and reacts. Over 1000°C, too much oxygen is absorbed and the reaction rate decreases. Apart from the main reaction, NO decomposition may also occur:

$$NO \rightarrow \frac{1}{2}N_2 + \frac{1}{2}O_2$$

The effect of temperature, pressure, and feed composition on the conversion can be seen in Fig. 6.9. The lower the pressure and the higher the oxygen-to-ammonia ratio, the higher the conversion. However, an optimum can be found due to the fact that low temperatures do not allow oxygen fixing. A higher operating temperature produces more nitrogen and N_2O. Over 950°C, the catalyst losses increase, mainly due to vaporization. N_2O is unstable at 850−950°C, and it is partially reduced to nitrogen and oxygen. Larger contact times and higher temperatures favor this reaction. The reaction temperature is directly related to the ratio of ammonia-to-air so that a 1% increment in ammonia results in an increase of 68°C. Finally, the use of air instead of oxygen improves the temperature control.

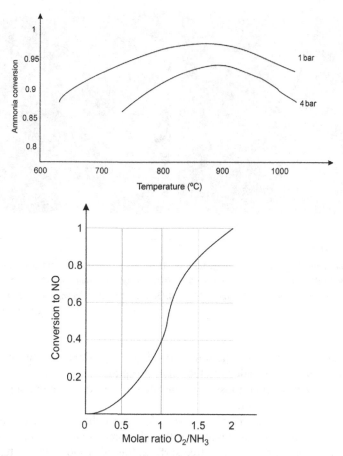

FIGURE 6.9

Effect of pressure and temperature and feed composition on ammonia conversion and NO formation.

Factors that affect catalyst losses

- *Temperature and thermal shocks*: Catalyst losses increase exponentially with temperature from 0.2% at 720°C to 1% at 880°C. At the same temperature, the richer the gas in ammonia, the lower the catalyst losses. Start-up and shutdown create temperature gradients that may result in catalyst structural breakage.

- *Catalyst composition*: The addition of rhodium to the catalyst (up to 10%) reduces catalyst losses with temperature. While at 720°C 0.2% losses are reported, at 880°C no more than 0.5% losses are expected. Above 10% of rhodium, the catalyst becomes fragile, increasing the losses. Typical platinum losses are on the order of 0.12 g/t of nitric acid produced, and the ammonia

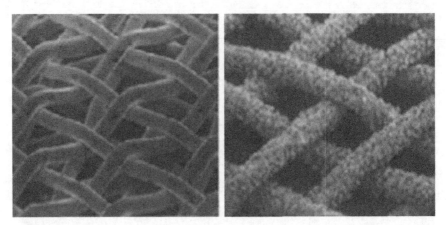

FIGURE 6.10

Platinum meshes.

From Thiemann, M., Scheibler, E., Wiegand, K.W., 1998. Nitric acid, nitrous acid, and nitrogen oxides.
ULLMANN'S "Encyclopedia of Industrial Chemistry," Wiley-VCH, Weinheim, Thiemann et al. (1998)
with permission. http//dx.doi.org/10.1002/14356007.a17_293d Copyright © 2002 by Wiley-VCH
Verlag GmbH & Co. KGaA.

conversion reaches a maximum for around 10−20% rhodium concentration. Palladium can also be added to the catalyst (up to 4%).

- *Pressure effect*: Platinum losses increase five times when the pressure increases from 1 to 8 atm.
- *Number of meshes and relative position*: The most characteristic parameters that define a mesh are the number of threads, thread diameters, meshes per square centimeter, surface, and free surface. Fig. 6.10 shows the pictures taken with an electronic microscope of the initial stage and when it was activated. The threads have 0.06−0.09 mm of diameter, and there are typically 1024 meshes per square centimeter. Meshes allocated to the exit are the most affected. Currently, for structural purposes, some meshes are substituted by stainless steel or alloys, but the yield of the converter decreases.

Explosion limit for ammonia−oxygen mixtures. Ammonia and air can produce explosive mixtures for certain ammonia concentrations. The value is sensitive to pressure. Fig. 6.11 shows the mixtures with explosive composition at 900°C, and Table 6.5 presents the typical mixtures used and the explosion limits at different pressures. For security reasons no more than 11% ammonia in air is fed to the converter. Typically for a low-pressure operation, a 10.8% volume is set. The ratio of O_2 to ammonia should be around 2.

Reaction kinetics and mechanism. In the reactor, a number of reactions take place, from the decomposition of the ammonia to produce nitrogen and hydrogen, to the production of water if the ammonia reacts with the NO. NO can also

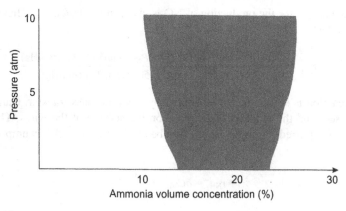

FIGURE 6.11

Explosion limits.

Table 6.5 Explosion Limits and Typical Mixture Compositions

Pressure (atm)	% NH₃ Air	(O₂/NH₃)ₘₒₗₑₛ
1.0	13.5	1.35
5.0	12.0	1.54
7.0	11.5	1.62
10.0	10.8	1.73

decompose to nitrogen and hydrogen, and water is formed from the hydrogen and oxygen in the mixture:

$$NH_3 \rightarrow \frac{1}{2}N_2 + \frac{3}{2}H_2$$

$$NH_3 + \frac{5}{4}O_2 \rightarrow NO + \frac{3}{2}H_2O \rightarrow \begin{cases} O_2 \rightarrow 2O^* \\ NH_3 + O^* \rightarrow NH_2OH \\ NH_2OH \rightarrow NH^* + H_2O \\ NH^* + O_2 \rightarrow HNO_2 \\ HNO_2 \rightarrow NO + OH^* \\ 2OH^* \rightarrow H_2O + O^* \end{cases}$$

$$NH_3 + \frac{3}{2}NO \rightarrow \frac{5}{4}N_2 + \frac{3}{2}H_2O$$

$$NO \rightarrow \frac{1}{2}N_2 + \frac{1}{2}O_2$$

$$H_2 + \frac{1}{2}O_2 \rightarrow H_2O \tag{6.1}$$

The actual rate for the production of NO is given in (moles/cm^2 s) (Hickman and Schmidt, 1991):

$$-r_{NH_3} = \frac{2.1 \cdot 10^{16} \cdot \exp(10850/T(K))P_{NH_3}(\text{mmHg}) \cdot P_{O_2}(\text{mmHg})}{1 + 4.0 \cdot 10^{-5} \cdot \exp(12750/T(K))P_{NH_3}(\text{mmHg})} \quad (6.2)$$

The reaction is so fast that it is limited by external mass transfer. Gauze reactors are used, and the reaction takes place on the surface of the wires. The kinetics can be simplified as proposed by Apelbaum, Temkim, and Viwump (Wallas, 1959):

$$-r_{NH_3} = qC_{NH_3} = \frac{k\beta C_{NH_3}}{k + \beta} \quad (6.3)$$

For temperatures below 500°C, $\beta >>> k$, $-r = kC_{NH_3}$ $\quad (6.4)$

For temperatures above 600°C, $k >>> \beta$, $-r = \beta C_{NH_3}$ $\quad (6.5)$

As presented before, temperatures from 600°C to 900°C are common in industrial practice, and therefore the second approximation holds. The design of such reactors is carried out as presented below (Harriott, 2003). Assuming that the gas flow does not change much along the reactor:

$$-r_{NH_3} = \beta C_{NH_3} = u_o \frac{dC_{NH_3}}{da} = u_o \frac{dC_{NH_3}}{a' \cdot dn} \quad (6.6)$$

n is the number of gauzes, a' is the external area of one gauze per unit cross section, and u_o is the velocity of the gas. For a gauze with square openings:

$$a' = 2\pi Nd \quad (6.7)$$

where N is the number of wires per unit length and d is the wire diameter. The conversion, X, of the reactor becomes:

$$\ln\left(\frac{C_{NH_{3,0}}}{C_{NH_3}}\right) = \ln\left(\frac{1}{1-X}\right) = \frac{\beta a' n}{u_o} \quad (6.8)$$

$$\beta = j_{D,\gamma} \frac{u_o}{\gamma} \frac{1}{Sc^{2/3}} = 0.644(Re)^{-0.57} \frac{u_o}{(1-Nd)^2} \frac{1}{Sc^{2/3}}$$

$$Re = \frac{d \cdot u_o \cdot \rho}{(1-Nd)^2 \mu} = \frac{d \cdot u_o \cdot \rho}{(1-Nd)^2 \mu} \quad (6.9)$$

$$Sc = \frac{\mu}{\rho D_{NH_3}}$$

where μ and ρ, are the viscosity and density of the gas, respectively, and D_{NH_3} is the diffusivity.

EXAMPLE 6.2

Determine the temperature of the gases exiting the converter. Assume that the ammonia mixture is fed at 65.5°C, the reactor operates adiabatically, and 90% conversion is reached. Neglect secondary reactions.

a. Stoichiometric air is used.
b. 10% air excess with respect to the stoichiometric one is used.
c. Pure oxygen is fed.

Solution

The reaction that takes place is as follows:

$$NH_3 + \frac{5}{4}O_2 \Rightarrow NO + \frac{3}{2}H_2O$$

In this example the data related to heat capacities and formation enthalpies are presented in the appendix.

a. For *stoichiometric air* we formulate a mass and energy balance assuming a reference temperature of 25°C:

$$Q_{in(flow)} + Q_{generated} = Q_{out(flow)}$$

$$\sum m_i \int_{T_{ref}}^{T_{in}} c_{p,i} dT + Q_{generated} = \sum m_i \int_{T_{ref}}^{T_{out}} c_{p,i} dT$$

We need to solve a four-degree polynomial on temperature. Table 6E2.1 show the results of the mass and energy balance. The outlet temperature is 1164 K.

Table 6E2.1 Results for Case Study A

	In			Out			
	kmol	kg	Q$_f$(kcal)	kmol	kg	Q$_f$(kcal)	Q$_r$(kcal)
N$_2$	4.7	131.6	1331.68547	4.7	131.6	30,243.7232	
O$_2$	1.25	40	366.917421	0.125	4	1143.13708	
H$_2$O	0	0		1.35	24.3	10,642.232	−78,027.3
NO	0	0		0.9	27	6016.39115	19,440
NH$_3$	1	17	352.082501	0.1	4.6	2728.50194	9864
Temp	338.6			1163.97265			
Q			2050.68539			50,773.9854	−48,723.3
		$Q_{in} + Q_{generated}$ = 50,773.9854		Q_{out} = 50,773.9854			

Table 6E2.2 Results for Case Study B

	In			Out			
	kmol	kg	Q_f(kcal)	kmol	kg	Q_f(kcal)	Q_r(kcal)
N_2	5.17	144.76	1464.85401	5.17	144.76	30,927.8436	
O_2	1.375	44	403.609163	0.25	8	2060.48012	
H_2O	0	0	0	1.35	24.3	9854.66725	−78,027.3
NO	0	0	0	0.9	27	5590.22948	19,440
NH_3	1	17	352.082501	0.1	4.6	2510.62525	9864
Temp	338.6			1107			
Q			2220.54568			50,943.8457	−48,723.3
		$Q_{in} + Q_{generated}$ $= 50,943.8457$		$Q_{out} =$ 50,943.8457			

b. For *10% excess of air* the gas exit temperature is 1107 K, see Table 6E2.2, lower than before since the air has to be heated up, and therefore the gas absorbs part of the generated energy.

c. For the *pure oxygen* case, no nitrogen is present and all the energy produced in the reaction is carried by the product gas. Thus the exit temperature would be 2096 K.

6.2.3.2.2 NO oxidation to NO_2

The reaction is an *equilibrium* that is exothermic towards NO_2 and very slow. While the NO is colorless, the NO_2 is yellow.

$$2NO + O_2 \leftrightarrow 2NO_2 \quad \Delta H = -27.3 \text{ kcal/mol}$$

The equilibrium constant is given as Eq. (6.10):

$$P_{NO_2} = y_{NO_2}P$$

$$P_{NO} = y_{NO}P \Rightarrow K_p P_T = \frac{y_{NO_2}^2}{y_{O_2}y_{NO}^2} \quad (6.10)$$

$$P_{O_2} = y_{O_2}P$$

In Table 6.6, some representative values of the equilibrium constant can be found. For low temperatures (below 200°C), almost 100% conversion can be reached. High pressure is also recommended.

The *reaction kinetics* is given by the following expression:

$$r = k_1 \left(P_{NO}^2 P_{O_2} - \frac{1}{K_p} P_{NO_2}^2 \right) \quad (6.11a)$$

Table 6.6 Equilibrium Constant Values

T (°C)	38	200	400	600	800	1000
K_p	1.0×10^{11}	1.0×10^5	1.0×10^1	1.0×10^{-1}	1.7×10^{-3}	1.2×10^{-3}

At low temperatures, based on the equilibrium shown above we have:

$$r = k_1(P_{NO}^2 P_{O_2})$$ (6.11b)

where the kinetic constant, according to Bodenstein (1922), is the following:

$$\log_{10} k_1 = \frac{641}{T} - 0.725$$ (6.12)

To illustrate the rate of the reaction, assuming a batch reactor for the sake of argument, we compute the time required to achieve a certain conversion. Example 6.6 provides a plug flow reactor example. For now, assuming ideal gases:

$$PV = nRT \Rightarrow P = n\frac{RT}{V}$$ (6.13)

For constant temperature and volume, P is directly proportional to the number of moles:

$$(n_{O_2})_{ini} = (n_{O_2})_{fin} + 0.5(n_{NO_2})_{fin}$$ (6.14)

$$(P_{O_2})_{ini} = (P_{O_2})_{fin} + 0.5(P_{NO_2})_{fin} \Rightarrow (P_{O_2})_{fin} = (P_{O_2})_{ini} - 0.5(P_{NO_2})_{fin}$$ (6.15)

$$(P_{NO})_{ini} = (P_{NO})_{fin} + (P_{NO_2})_{fin} \Rightarrow (P_{NO})_{fin} = (P_{NO})_{ini} - (P_{NO_2})_{fin}$$ (6.16)

We define the conversion as Eq. (6.17):

$$X = (P_{NO_2})_t / (P_{NO})_{ini}$$ (6.17)

Thus the reaction kinetics are as follows:

$$r = \frac{dP_{NO_2}}{dt} = \frac{dP_c}{dt} = k_1((P_{NO})_{ini} - (P_{NO_2})_t)^2 \left[(P_{O_2})_{ini} - 0.5(P_{NO_2})_t\right]$$

$$r = (P_{NO})_{ini}\frac{dX}{dt} = k_1((P_{NO})_{ini} - X(P_{NO})_{ini})^2 \left[(P_{O_2})_{ini} - 0.5X(P_{NO})_{ini}\right]$$ (6.18)

$$\frac{dx}{dt} = k_1(P_{NO})_{ini}(1-X)^2 \left[(P_{O_2})_{ini} - 0.5X(P_{NO})_{ini}\right]$$

$$t = \int \frac{dX}{k_1(P_{NO})_{ini}(1-X)^2((P_{O_2})_{ini} - 0.5X(P_{NO})_{ini})}$$

$$t = \frac{2}{k_1(2(P_{O_2})_{ini} - (P_{NO})_{ini})} \left[\frac{X}{(P_{NO})_{ini}(1-X)} - \frac{1}{2(P_{O_2})_{ini} - (P_{NO})_{ini}} \ln \frac{2(P_{O_2})_{ini} - (P_{NO})_{ini}X}{2(P_{O_2})_{ini}(1-X)}\right]$$

(6.19)

EXAMPLE 6.3

Using the equations developed for the reactor kinetics below 200°C, determine the time required to reach 90% conversion for a reaction mixture of 10% NO, 7% O_2, and 83% inert gases operating at:

a. 45°C and 1 atm.
b. 45°C and 10 atm.

Solution

a. $P_T = 1.0$ atm

$$T = 45°C \rightarrow 318\ K$$
$$P_a = 0.1 \cdot P_T = 0.1\ \text{atm}$$
$$P_b = 0.07 \cdot P_T = 0.07\ \text{atm}$$
$$x = 0.9$$

$t = 148$ s

b. $P_T = 10.0$ atm

$$T = 45°C \rightarrow 318\ K$$
$$P_a = 0.1 \cdot P_T = 1\ \text{atm}$$
$$P_b = 0.07 \cdot P_T = 0.7\ \text{atm}$$
$$x = 0.9$$

$t = 1.48$ s

The time required for the reaction decreases by two orders of magnitude when pressure increases by one order of magnitude.

EXAMPLE 6.4

Compute the conversion from NO to NO_2 as a function of pressure and temperature for the following cases using the equilibrium constants shown in Table 6.6:

a. Only NO and O_2 are present and oxygen is in stoichiometric proportion.
b. Assume that oxygen is not in stoichiometric proportions.
c. Assume that there are NO and O_2 (but not in stoichiometric proportions), and that there are inert gases.
d. Plot the equilibrium constant for various pressures from 1 to 8 atm. Assume that the initial stream consists of 10% NO, 6% O_2, and inert.

Solution

Instead of developing each case, a general formulation of the problem is provided. Next, the general expression is simplified for each case. For the sake of simplicity, let b be the moles of oxygen, $2a$ the number of moles of NO, n the moles of inert gas, and X the conversion of the following reaction:

$$2NO + O_2 \leftrightarrow 2NO_2$$

Table 6E4.1 Mass Balance to the NO Oxidation Equilibrium

	Initial	Reacts	Equilibrium
NO	$2a$	$-2aX$	$2 \cdot a - 2 \cdot a \cdot X$
O_2	b	$-aX$	$b - a \cdot X$
NO_2		$2aX$	$2 \cdot a \cdot X$
Inert	n		n
Total	$2 \cdot a + b + n$	$-aX$	$2 \cdot a + b + n - a \cdot X$

Table 6E4.1 presents the mass balance to the species involved in the equilibrium.

The equilibrium constant can be written as follows:

$$K_p = \frac{P_{NO_2}^2}{P_{NO}^2 \cdot P_{O_2}} = \frac{\left(\dfrac{2 \cdot a \cdot X}{2 \cdot a + b + n - a \cdot X} \cdot P_T\right)^2}{\left(\dfrac{b - a \cdot X}{2 \cdot a + b + n - a \cdot X} \cdot P_T\right)\left(\dfrac{2 \cdot a - 2 \cdot a \cdot X}{2 \cdot a + b + n - a \cdot X} \cdot P_T\right)^2}$$

$$K_p = \frac{(2 \cdot a \cdot X)^2(2 \cdot a + b + n - a \cdot X)}{(b - a \cdot X)(2 \cdot a - 2 \cdot a \cdot X)^2 P_T} = \frac{X^2[a \cdot (2 - X) + b + n]}{(b - a \cdot X)(1 - X)^2 P_T}$$

a. Only reactants and no inert $(n = 0)$. Oxygen in stoichiometric proportion $(a = b)$. The values for the equilibrium constant in Table 6.6 are correlated to the equation shown below:

$$K_p = \frac{X^2[a \cdot (2 - X) + b]}{(b - a \cdot X)(1 - X)^2 P_T}$$

$$K_p = \frac{X^2[(2 - X) + 1]}{(1 - X)(1 - X)^2 P_T} = 10^{(-1.086 \cdot 10^{-08} \cdot T^3 + 3.459 \cdot 10^{-05} T^2 - 3.896 \cdot 10^{-02} T + 1.218 \cdot 10)}$$

For $P = 1$ atm, see the results in Fig. 6E4.1.

b. Only reactants and no inert $(n = 0)$, but the oxygen is not in stoichiometric proportion $(a \neq b)$.

$$K_p = \frac{X^2[a \cdot (2 - X) + b]}{(b - a \cdot X)(1 - X)^2 P_T} = 10^{(-1.086 \cdot 10^{-08} \cdot T^3 + 3.459 \cdot 10^{-05} T^2 - 3.896 \cdot 10^{-02} T + 1.218 \cdot 10)}$$

c. Reactants and inert (this is the general case).

$$K_p = \frac{P_{NO_2}^2}{P_{NO}^2 \cdot P_{O_2}} = \frac{X^2[a \cdot (2 - X) + b + n]}{(b - a \cdot X)(1 - X)^2 P_T}$$

$$= 10^{(-1.086 \cdot 10^{-08} \cdot T^3 + 3.459 \cdot 10^{-05} T^2 - 3.896 \cdot 10^{-02} T + 1.218 \cdot 10)}$$

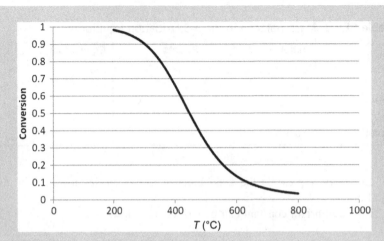

FIGURE 6E4.1

Equilibrium curve case a.

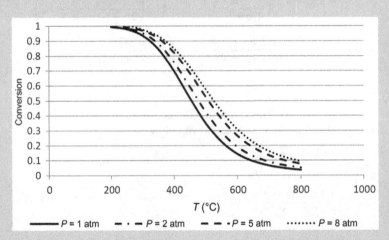

FIGURE 6E4.2

Effect of pressure and temperature on the equilibrium NO to NO_2.

d. Particular example with 10% NO, 6% O_2, and the rest inert. Compute the conversion for pressures ranging from 1 to 8 atm. Thus:

$$2a = 10; \quad n = 84; \quad b = 6$$

Fig. 6E4.2 shows the effect of pressure and temperature in the conversion. Low temperatures and high pressures are recommended for high conversions.

6.2.3.2.3 Nitrogen peroxide equilibrium

Although it lacks peroxo bonding, it has typically received that name. It is a quick reaction, so the equilibrium is assumed to be reached instantaneously:

$$2NO_2 \leftrightarrow N_2O_4 \quad -13.7 \text{ kcal/mol of } N_2O_4$$

The *kinetics of the reaction* are given by the following expression (Thiemann et al., 1998):

$$r = \frac{k_r}{RT}\left(p_{NO_2}^2 - \frac{p_{N_2O_4}}{K_p} \right) \tag{6.20}$$

where the kinetic rate, K_p, is almost independent of the temperature, and equal to $5.7 \times 10^5 \text{ atm}^{-1} \text{ s}^{-1}$.

The *equilibrium* constant is computed as given by Eq. (6.21):

$$K_p = \frac{P_{N_2O_4}}{P_{NO_2}^2} \tag{6.21}$$

Several expressions can be found for its dependence on temperature, for instance (Thiemann et al., 1998):

$$\log_{10}\left(\frac{1}{K_p}\right) = \frac{-2692}{T} + 1.75 \log_{10} T + 0.00483T - 7.144 \cdot 10^{-6} T^2 + 3.062 \tag{6.22}$$

$$K_p = 0.698 \cdot 10^{-9} \exp\left(\frac{6866}{T}\right) \tag{6.23}$$

The equilibrium is driven to the right at low temperature and high pressure. N_2O_4 can easily be liquefied, a useful property for its transport. In most cases, the term *equivalent NO_2* is used in order to avoid including the dimer in the balances. It is defined as the NO_2 that would be available if all the dimer were decomposed into NO_2:

$$NO_2^{\text{Equivalent}} = N_{NO_2,\text{eq}} = N_{NO_2} + 2N_{N_2O_4} \tag{6.24}$$

It is also possible that N_2O_3 is produced. The kinetic reaction and equilibrium constant are given as:

$$NO + NO_2 \leftrightarrow N_2O_3 \quad -40 \text{ kJ/mol}$$

$$r = \frac{k_r}{RT}\left(p_{NO} \cdot p_{NO_2} - \frac{p_{N_2O_3}}{K_p} \right) \tag{6.25}$$

where (Thiemann et al., 1998):

$$K_p = \frac{P_{N_2O_3}}{P_{NO} \cdot P_{NO_2}} = 65.3 \cdot 10^{-9} \cdot \exp\left(\frac{4740}{T}\right) \tag{6.26}$$

EXAMPLE 6.5

Plot the figure for the decomposition of N_2O_4 into NO_2 as a function of pressure and temperature. Develop the equation for the equilibrium assuming the presence of inert in the gas stream.

Solution

The equilibrium constant is as follows:

$$2NO_2 \leftrightarrow N_2O_4 \quad K_p = \frac{P_{N_2O_4}}{P_{NO_2}^2}$$

where:

$$\log_{10}\left(\frac{1}{K_p}\right) = -\frac{2692}{T} + 1.75 \log_{10} T + 0.00483 \cdot T - 7.144 \cdot 10^{-6} T^2 + 3.062$$

We use as a base for our computations 1 kmol of N_2O_4.

The mass balance to the equilibrium is presented below. d represents the dissociation degree for the dimer.

	2NO$_2 \leftrightarrow$ N$_2$O$_4$		Inters
Initial	0	1	a
Equilibrium	$2d$	$1-d$	a

The molar fraction of each species and the partial pressure are computed as follows:

$$y_{N_2O_4}^{eq} = \frac{1-d}{1-d+2d+a} = \frac{1-d}{1+d+a} \Rightarrow P_{N_2O_4}^{eq} = \frac{1-d}{1+d+a} P_T$$

$$y_{NO_2}^{eq} = \frac{2d}{1-d+2d+a} = \frac{2d}{1+d+a} \Rightarrow P_{NO_2}^{eq} = \frac{2d}{1+d+a} P_T$$

Substituting both into the equilibrium constant equation it becomes:

$$K_p = \frac{P_{N_2O_4}}{P_{NO_2}^2} = \frac{\dfrac{1-d}{1+d+a} P_T}{\left[\dfrac{2d}{1+d+a} P_T\right]^2} = \frac{(1-d)(1+d+a)}{4 \cdot d^2 \cdot P_T}$$

$$d = \frac{-a + \sqrt{a^2 + 4 \cdot (1+a)(1 + 4 \cdot P_T \cdot K_p)}}{2 \cdot (1 + 4 \cdot P_T \cdot K_p)}$$

From the definition of equivalent NO_2:

$$\%NO_2Eq = \frac{2}{a+2} 100 \Rightarrow a = \frac{2 \cdot (100 - \%NO_2Eq)}{\%NO_2Eq}$$

$$NO_2Eq = NO_2 + 2N_2O_4 = 2d + 2(1-d) = 2$$

For a total pressure of 1 atm, and different equivalent NO_2 compositions and temperatures, we compute d. In Fig. 6E5.1 we see that dissociation increases with temperature and with the amount of equivalent NO_2. In Fig. 6E5.2 we present the effect of pressure. N_2O_4 dissociation decreases with pressure, as expected based on Le Chatelier's principle.

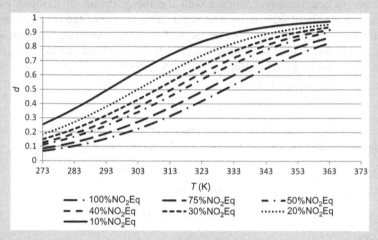

FIGURE 6E5.1

Effect of temperature on dissociation fractions for different equivalent NO_2 compositions.

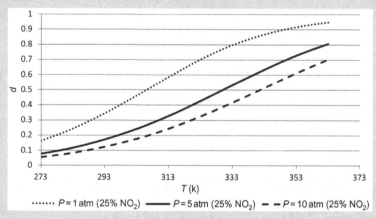

FIGURE 6E5.2

Effect of temperature and pressure on dissociation fractions.

6.2.3.2.4 Absorption of NO$_x$

Absorption—oxidation towers are used to completely oxidize NO to NO$_2$. The design of such columns depends on the kinetics. The global reaction that takes place is the reaction between the NO$_2$ and N$_2$O$_4$ with water to produce nitric acid and NO. There are reactions in both phases: liquid and gas.

The unit is a tray reactive column with intercooling. Water is fed from the top and is put in countercurrent contact with the rising stream of nitrogen oxides. Typically, a stream of 60% nitric acid is produced and exits the column from the bottom. NO$_2$ is soluble in the product acid. Therefore, a stripping column, working as a bleacher, is used to remove the NO$_2$ from the liquid product. The model for the tower can be found in Holma and Sohlo (1979).

Reactions in the liquid phase:

$$3NO_2 + H_2O \Leftrightarrow 2HNO_3 + NO$$

The gas—liquid equilibrium for a particular tray is computed based on the equilibrium conditions and the reaction stoichiometries as follows:

$$3NO_2 + H_2O \Leftrightarrow 2HNO_3 + NO$$

$$2NO_2 \Leftrightarrow N_2O_4$$

$$NO + NO_2 + H_2O \Leftrightarrow 2HNO_2$$

where:

$5 \leq w(\text{wt}\%) \leq 65$

$$K_1(\text{bar}^{-2}) = 30.086 - 0.0693T - (0.1917 - 3.27 \cdot 10^{-4}T)w - 1.227\left(\frac{100-w}{w^2}\right) \tag{6.27}$$

$w(\text{wt}\%) \leq 5$

$$K_1(\text{bar}^{-2}) = 31.96 - 0.0693T - (3.27 \cdot 10^{-4}T - 0.4193)w$$

$$K_3 = \frac{P_{N_2O_4}}{P_{NO_2}^2} = e^{\frac{6891.61}{T} - 21.244} \ (\text{bar}^{-1}) \tag{6.28}$$

$$K_4 = \frac{P_{HNO_2}^2}{P_{NO} \cdot P_{NO_2} \cdot P_{H_2O}} = 0.185 \cdot 10^{-6} e^{\frac{4323}{T}} \ (\text{bar}^{-1}) \tag{6.29}$$

As a result:

$$3\frac{V_n}{V_{n-1}}\frac{P_{n-1}}{P_n}K_1 p_{NO_2,n}^3 + \frac{V_n}{V_{n-1}}\frac{P_{n-1}}{P_n}\left(2K_3 + \frac{1}{2}\sqrt{K_1 K_4 P_{H_2O,n}}\right)p_{NO_2,n}^2$$
$$+ \frac{V_n}{V_{n-1}}\frac{P_{n-1}}{P_n}P_{NO_2,n} - (3P_{NO} + P_{NO_2} + 2 \cdot K_3 \cdot P_{NO_2}^2)_{n-1} = 0 \tag{6.30}$$

Reactions between trays:

$$NO + \frac{1}{2}O_2 \Leftrightarrow NO_2$$

$$2NO_2 \Leftrightarrow N_2O_4$$

These reactions have been analyzed in a previous section. The first reaction is slow, and almost irreversible below 150°C. Dimerization is quick; therefore the equilibrium concentrations between the NO_2 and the dimer can be computed with the equilibrium constant. In this case, inside the reaction tower, the following expression is used:

$$K_3 = \frac{P_{N_2O_4}}{P_{NO_2}^2} = e^{\frac{6891.61}{T} - 21.244} \, (bar^{-1}) \tag{6.31}$$

When the oxides rise (as the conversion increases), the partial pressures of the species decrease, and therefore the residence time should be larger, resulting in a larger separation between trays. We can assume that space between trays behaves as a plug flow reactor. The kinetics of the reaction are given by the following equation:

$$r_{NO} = \frac{dP_{NO}}{dt} = -0.2166 \exp(1399/T)\frac{P_{NO}^2}{P_{O_2}} (bar/s) \tag{6.32}$$

As a result, the global reaction is the following:

$$3NO_2(g) + H_2O(l) \leftarrow \rightarrow 2HNO_3(l) + NO(g) \quad \Delta H = -72.8 \, kJ/mol$$

with an equilibrium constant of this:

$$K_p = \frac{P_{NO}P_{HNO_3}^2}{P_{NO_2}^3 P_{H_2O}} \Rightarrow K_p = K_1 \cdot K_2 = \frac{P_{NO}}{P_{NO_2}^3} \cdot \frac{P_{HNO_3}^2}{P_{H_2O}} \tag{6.33}$$

where the pressure for liquids is their vapor pressure; for gases it corresponds to their partial pressure. Fig. 6.12 shows the vapor pressures of water and nitric acid as functions of the acid concentration. The value for the equilibrium constant is provided in the literature, either referred to NO_2 or to equivalent NO_2:

$$\log_{10}(K_p) = -8.88 + \frac{1900}{T} \quad \text{(Referred to } NO_2) \tag{6.34}$$

$$\log_{10}(K_p) = -4.71 + \frac{787}{T} \quad \text{(Referred to equivalent } NO_2) \tag{6.35}$$

Another one:

$$\log(K_p) = -7.35 + \frac{2.64}{T} \quad \text{(Referred to equivalent } NO_2) \tag{6.36}$$

These constants are given referred to gas phase. As the gases react and their partial pressures decrease, the tray distance required needs to increase. As a result, the product gases still have nitrogen oxides. Among them, NO_2 provides a yellow color to the nitric acid. Furthermore, if released to the atmosphere, it is responsible for acid rain. Fig. 6.13 shows the mechanisms of the reactions taking place inside the column from the gas bulk to the liquid. The dissociation of the dimer decreases with the concentration of the acid.

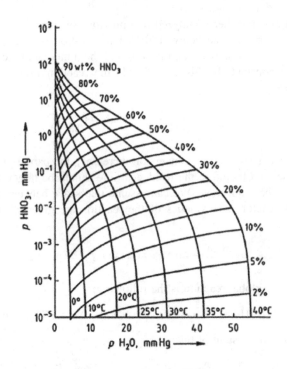

FIGURE 6.12

Vapor pressures of nitric acid and water as a function of acid concentration and temperature.

Nonhebek Gas Purification Processes for Air Pollution Control (1972) with permission.

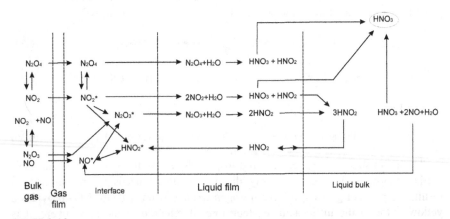

FIGURE 6.13

Mechanism of nitric acid production in the absorption tower.

Nonhebek Gas Purification Processes for Air Pollution Control (1972) with permission.

EXAMPLE 6.6

Determine the distance between the first two trays of an absorption column assuming that the kinetics are given as:

$$r_{NO} = -0.2166 \exp(1399/T)\frac{P_{NO}^2}{P_{O_2}} \text{ (bar/s)}$$

Solution

Assuming an isothermal plug flow reactor between trays we have:

$$F_{NO}\frac{dX}{dV} = -r_{NO}$$

$$NO + \frac{1}{2}O_2 + N_2 \leftrightarrow NO_2 + N_2$$

$$F_{A_o} \quad F_{B_o} \quad F_{C_o}$$

$$F_{A_o}(1-X) \quad F_{B_o} - \frac{1}{2}F_{A_o}X \quad F_{C_o} \quad F_{A_o}X$$

$$\Theta_i = \frac{F_{i_o}}{F_{A_o}}$$

$$F_T = F_{A_o}\left[(1-X) + \Theta_{O_2} - \frac{1}{2}X + \Theta_{N_2} + \Theta_{NO_2} + X\right] = F_{A_o}(1 + \Theta_{O_2} + \Theta_{N_2} + \Theta_{NO_2}) + F_{A_o}\delta X$$

$$\delta = -\frac{1}{2}$$

$$F_T = F_{T_o} + F_{A_o}\delta X$$

Assuming ideal gases:

$$C_T = \frac{F_T}{v} = \frac{P}{RT}$$

$$C_{T_o} = \frac{F_{T_o}}{v_o} = \frac{P_o}{RT_o}$$

Dividing both:

$$v = v_o\frac{P_o}{P}\frac{T}{T_o}\frac{F_T}{F_{T_o}} = v_o\frac{P_o}{P}\frac{T}{T_o}\frac{F_{T_o} + F_{A_o}\delta X}{F_{T_o}} = v_o\frac{P_o}{P}\frac{T}{T_o}\left(1 + \frac{F_{A_o}}{F_{T_o}}\delta X\right)$$

Then, the concentration of the species i is as follows:

$$C_i = \frac{F_i}{v} = \frac{F_i}{v_o\frac{P_o}{P}\frac{T}{T_o}\frac{F_T}{F_{T_o}}} = C_{T_o}\frac{F_i}{F_T}\frac{P}{P_o}\frac{T_o}{T}$$

$$C_i = C_{T_o}\frac{F_i}{F_{T_o} + F_{A_o}\delta X}\frac{P}{P_o}\frac{T_o}{T} = C_{T_o}\frac{F_{A_o}(\Theta_i + v_i X)}{F_{T_o} + F_{A_o}\delta X}\frac{P}{P_o}\frac{T_o}{T}$$

$$C_i = C_{T_o} \frac{F_{A_o}}{F_{T_o}} \frac{(\Theta_i + v_i X)}{1 + (F_{A_o}/F_{T_o})\delta X} \frac{P}{P_o} \frac{T_o}{T} = C_{A_o} \frac{(\Theta_i + v_i X)}{1 + \varepsilon X} \frac{P}{P_o} \frac{T_o}{T}$$

$$\varepsilon = y_{A_o}\delta = (F_{A_o}/F_{T_o})\delta$$

Thus the partial pressure of component i is calculated as follows:

$$P_i = C_i RT = C_{A_o} \frac{T_o}{P_o} \frac{(\Theta_i + v_i X)}{1 + \varepsilon X} R \frac{T}{T} \cdot P = P_{A_o} \frac{(\Theta_i + v_i X)}{1 + \varepsilon X} \frac{P}{P_o}$$

$$P_{A_o} = C_{A_o} R T_o$$

$$-r_{NO} = -\frac{dP_{NO}}{dV} = 0.2166 \exp(1399/T) \frac{P_{NO}^2}{P_{O_2}}$$

$$F_{NO} \frac{dX}{dV} = 0.2166 \exp(1399/T) \frac{\left(P_{NO} \dfrac{(1-X)}{1 + (-(F_{NO}/F_{T_o})0.5)X}\right)^2}{P_{NO} \dfrac{(P_{O_2}/P_{NO} - 0.5X)}{1 + (-(F_{NO}/F_{T_o})0.5)X}}$$

$$= 0.2166 \exp(1399/T) \frac{P_{NO}(1-X)^2}{(P_{O_2}/P_{NO} - 0.5X) \cdot (1 + (-(F_{NO}/F_{T_o})0.5)X)}$$

The MATLAB code for solving the problem is shown below. For details on how to use MATLAB for reactor modeling we refer the reader to Martín (2014).

```
[a,b] = ode45('TorreOxidaNO',[0 0.3],[0]);

plot(a,b)
xlabel('Height (Z)')
ylabel('x')

function Reactor = TorreOxidaNO(t,x)
%UNTITLED Summary of this function goes here
%    Detailed explanation goes here

Temp = 30 + 273.15;
Ptotal = 15;
k1 = 0.2166*exp(1399/Temp);
PO2ini = 0.06*Ptotal;
PNOini = 0.1*Ptotal;
FNO = 1; %kmol/s
A = 1;%m2
Reactor(1,1) = A*(k1/FNO)*PNOini*(1−x)^2/( (PO2ini/PNOini−0.5*x)*
  (1-0.5*x*(PNOini/Ptotal)));

End
```

At 15 bar we get more than 90% conversion after 10 cm (Fig. 6E6.1).

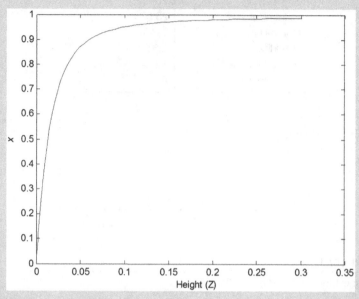

FIGURE 6E6.1

Conversion as a function of the tray separation.

EXAMPLE 6.7

Perform the mass and energy balances to a dual-pressure plant of nitric acid. The feedstock consists of liquid ammonia and enriched air with oxygen, both at 1.5 atm and 25°C. The ammonia is evaporated and mixed before entering the converter. Table 6E7.1 shows the feed information. A two-stage compression with intercooling is used to reach 10 atm. The compressed gas is cooled down to 30°C from the temperature exiting the converter. Water condenses, producing nitric acid 3.5% w/w. This stream is mixed with weak acid to feed the absorption column. The mass ratio between the liquid and gas phases is 0.25, and the composition of the liquid is 2.5% in nitric acid w/w before the condensation. Assume isothermal operation of the columns. The product is nitric acid 60% w/w (Fig. 6E7.1).

Table 6E7.1 Feed Information

T (°C)	25
P_v (mmHg)	23.57
P_t (mmHg)	1140
Humidity	0.35
Rel Air/O_2 added	12
Rel O_2 total/NH_3	2

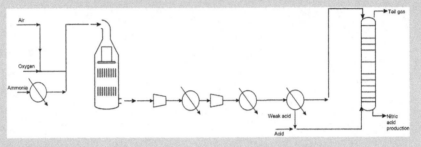

FIGURE 6E7.1

Flowsheet for nitric acid production.

Solution

$$1 \text{ kmol}_{\text{air}} = n_{O_2} + 3.76 n_{O_2} + 4.76 n_{O_2} \left(\frac{P_v}{P_T - P_v} \right)$$

$$\frac{n_{\text{air}}}{12} = n_{O_2,\text{added}}$$

$$\frac{n_{O_2,\text{total}}}{2} = n_{NH_3}$$

To evaporate the ammonia, the energy provided to the heat exchanger is as follows:

$$Q(\text{vaporization}) = \lambda \, m_{NH_3}$$

The mixture enters the reactor at the same pressure and temperature, 1.5 bar and 25°C. The summary of the mass and energy balances is presented in Fig. 6E7.2.

The reaction taking place in the converter is as follows:

$$NH_3 + \frac{5}{4}O_2 \rightarrow NO + \frac{3}{2}H_2O$$

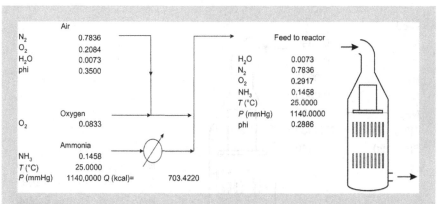

FIGURE 6E7.2

Results of the feed composition.

As presented in the description of the process, ammonia is expensive, and so is the platinum. Therefore, a conversion of 100% of ammonia is assumed. Based on the stoichiometry, we compute the flows of NO, H_2O, oxygen, and nitrogen:

$$n_{NH_3}\big|_{out} = n_{NH_3}\big|_{in}(1-X)$$

$$n_{NO}\big|_{out} = n_{NH_3}\big|_{in}(X)$$

$$n_{O_2}\big|_{out} = n_{O_2}\big|_{in} - \frac{5}{4}n_{NH_3}\big|_{in}(X)$$

$$n_{H_2O}\big|_{out} = n_{H_2O}\big|_{in} + \frac{3}{2}n_{NH_3}\big|_{in}(X)$$

$$n_{N_2}\big|_{out} = n_{N_2}\big|_{in}$$

The temperature of the gases is computed by an adiabatic energy balance to the reactor where:

$$Q_{in(flow)} + Q_{reaction} = Q_{out(flow)}$$

$$\sum m_i \int_{T_{ref}}^{T_{in}} c_{p,i}dT + Q_{reaction} = \sum m_i \int_{T_{ref}}^{T_{out}} c_{p,i}dT$$

The summary of the results can be seen in Fig. 6E7.3.

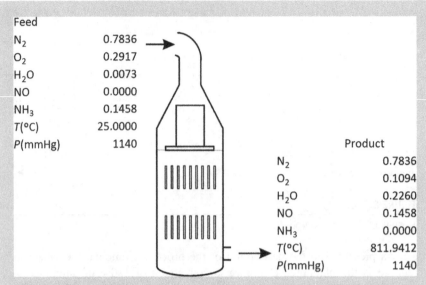

Feed	
N_2	0.7836
O_2	0.2917
H_2O	0.0073
NO	0.0000
NH_3	0.1458
$T(°C)$	25.0000
P(mmHg)	1140

Product	
N_2	0.7836
O_2	0.1094
H_2O	0.2260
NO	0.1458
NH_3	0.0000
$T(°C)$	811.9412
P(mmHg)	1140

FIGURE 6E7.3

Results of the converter operation.

The gases are compressed up to 10 atm using a two-stage compression system with intercooling. The pressure ratio at each compressor is $(P_2/P_1)^{0.5} = 2.58$. Assuming polytropic behavior of the compressors, the temperature and energy required are computed as follows:

$$T_2 = T_1 \left(\frac{P_2}{P_1}\right)^m$$

$$m = \frac{1}{2}\frac{1.4-1}{1.4}$$

$$W = 2T_1 R \frac{1.4}{1.4-1}\left[\left(\frac{P_2}{P_1}\right)^m - 1\right]$$

Furthermore, the cooling requirements in between the compression stages so that each compressor receives the gas at the same initial temperature are as follows:

$$Q = -2\sum_i m_i \int_{T_i}^{T_{out}} c_p dt$$

During the cooling to 30°C, we assume that the conversion of NO to NO_2 is 100% since it is favored at high pressure and low temperature. In this condensation, we assume that nitric acid 3.5% by weight is produced with the water that condenses. Furthermore, there is an equilibrium between the NO_2 and the N_2O_4. The composition of the gas exiting the cooler is computed in three stages:

1. Determine the gas composition after the oxidation:

$$NO + \frac{1}{2}O_2 \rightarrow NO_2$$

$$n_{NO}\Big|_{out} = n_{NO}\Big|_{in}(1-X)$$

$$n_{NO_2}\Big|_{out} = n_{NO}\Big|_{in}(X)$$

$$n_{O_2}\Big|_{out} = n_{O_2}\Big|_{in} - \frac{1}{2}n_{NO}\Big|_{in}(X)$$

$$n_{H_2O}\Big|_{out} = n_{H_2O}\Big|_{in}$$

$$n_{N_2}\Big|_{out} = n_{N_2}\Big|_{in}$$

2. Compute the amount of HNO_3 produced. Although the moment we have NO_2 the equilibrium to N_2O_4 can be established (as long as nitric acid is produced), the N_2O_4 will decompose to provide for the NO_2 that has been consumed. Thus, we first compute the acid production. The stoichiometry of the reaction is as follows:

$$2NO_2 + H_2O + \frac{1}{2}O_2 \rightarrow 2HNO_3 \quad \Delta H_r = -127.28 \text{ kJ}$$

During the cooling, there is also water condensation. This water is the one that is forming the weak nitric acid solution. Thus, the water remaining in the gas is the humidity that can be handled at 30°C. The rest condenses. Therefore, the acid produced is computed using the stoichiometry of the reaction as follows:

$$\%_{acid} = \frac{n_{HNO_3} \cdot M_{HNO_3}}{n_{HNO_3} \cdot M_{HNO_3} + M_{H_2O}\left(n_{H_2O} - y_{sat} \cdot GAS - \frac{1}{2}n_{HNO_3}\right)}$$

$$GAS = n_{NO_2,fin} + n_{N_2,fin} + n_{O_2,fin}$$

where

$$n_{NO_2,fin} = n_{NO_2,ini} - n_{HNO_3}$$

$$n_{O_2,fin} = n_{O_2} - \frac{1}{4}n_{HNO_3}$$

$$n_{N_2,fin} = n_{N_2,ini}$$

Thus, the amount of water that accompanies the gas is determined as follows:

$$m_{H_2O} = \left(\frac{100 - \%_{acid}}{\%_{acid}}\right) m_{HNO_3}$$

The energy involved in the water phase change is included in the global energy balance as:

$$Q_{cond} = -m_{H_2O}\lambda$$

3. Finally, with the gas phase computed in the previous stage, we determine the equilibrium concentration as the final gas phase to be sent to the absorption tower,

$$2NO_2 \leftrightarrow N_2O_4 \quad K_p = \frac{P_{N_2O_4}}{P_{NO_2}^2}$$

where:

$$\log_{10}\left(\frac{1}{K_p}\right) = -\frac{2692}{T} + 1.75\log_{10}T + 0.00483 \cdot T - 7.144 \cdot 10^{-6}T^2 + 3.062$$

For the sake of argument, allow y to be the moles of N_2O_4 and x the moles of NO_2 at equilibrium. Thus the equilibrium constant is given as follows:

$$K_p = \frac{\frac{y}{n_T}}{P_T\left(\frac{x}{n_T}\right)^2} = \frac{y \cdot n_T}{P_T x^2}$$

$$2y + x = NO_{2,eq} = cte$$

$$K_p(P_{T,atm})x^2 = \left(\frac{NO_{2,eq} - x}{2}\right)\left(n_{H_2} + n_{O_2} + n_{N_2} + n_{H_2O} + x + \left(\frac{NO_{2,eq} - x}{2}\right)\right)$$

$$n_{H_2} + n_{O_2} + n_{N_2} + n_{H_2O} = N_{ini}$$

$$x = \frac{-0.5N_{ini} + \sqrt{(0.5N_{ini})^2 - 4K_p(P_{T,atm} + 0.25)(0.25NO_{2,eq}^2 + 0.5NO_{2,eq}N_{ini})}}{2K_p(P_{T,atm} + 0.25)}$$

The energy involved is computed as (Houghen et al., 1959):

$$Q_{ref} = \sum_i m_i\left(\Delta H_f^{25} + \int_{T_{ref}}^{T_{out}} c_{p,i}dT\right) - \sum_i m_i\left(\Delta H_f^{25} + \int_{T_{ref}}^{T_{in}} c_{p,i}dT\right)$$

where

$$\Delta H_{f,HNO_3}^{25} = \Delta H_f^{25}\Big|_{pure} + \Delta H_{solution}$$

$$\sum \Delta H_{f,products}(25°C) = -15,941.257 \text{ kcal}$$

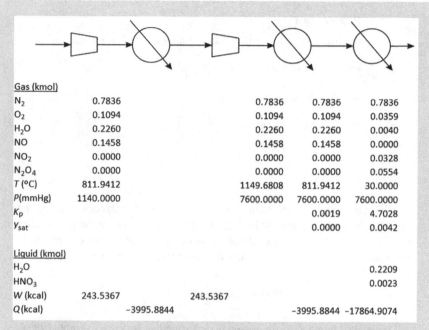

FIGURE 6E7.4

Results of the compression stage and condensation.

For the products, remember that nitric acid is produced and diluted and that energy is also involved in that stage. Water leaves partially as liquid and partially by saturating the gas, thus we have both formation energies for the different aggregation stages:

$$\sum \Delta H_{f,reactants}(25°C) = -9914.745 \text{ kcal}$$
$$Q_{in} \text{ flow} = 11,891 \text{ kcal}$$
$$Q_{out} \text{ flow} = 52.5 \text{ kcal}.$$

The summary of the compression and cooling stages is presented in Fig. 6E7.4.

The exit gases are fed to the absorption column. The tower operates at 30°C and 10 atm. The liquid-to-gas mass ratio of operation is 0.25 and the nitric acid produced has a concentration of 60% by weight. To compute the mass of liquid added, the gas phase fed to the tower is used as follows (the liquid fed to the column is nitric acid 2.5%):

Balance HNO_3

$$n_{HNO_3,total} \cdot M_{HNO_3} = 0.60 \cdot Liq_{out} = (n_{HNO_3,ini} + n_{HNO_3,prod})M_{HNO_3}$$

$$n_{HNO_3,prod} = \frac{0.60 \cdot Liq}{M_{HNO_3}} - n_{HNO_3,ini}$$

Balance water

$$(n_{H_2O,vap} + n_{H_2O,liq})_{ini} = 0.40 \cdot Liq_{out} + n_{H_2O,vap} + 0.5 \cdot n_{HNO_3,prod}$$

$$n_{H_2O,vap} = (n_{N_2} + n_{O_2} + n_{NO_2,eq}) \frac{P_v}{P_T - P_v}$$

$$n_{NO_2,out} = n_{NO_2,ini} - n_{HNO_3,prod}$$

$$n_{O_2,out} = n_{O_2,ini} - \frac{1}{4} n_{HNO_3,prod}$$

$$n_{N_2,out} = n_{N_2,ini}$$

$$Liq_{ini} = 0.25 V_{ini}$$

The gas phase is determined from the coupled balances. Again, the equilibrium between the NO_2 and the N_2O_4 is to be recomputed after the production of nitric acid at the pressure and temperature of operation. The reaction is also highly exothermic, and therefore requires cooling.

$$2NO_2 + H_2O + \frac{1}{2}O_2 \rightarrow 2HNO_3$$

An energy balance to the tower is performed as follows (Sinnot, 1999):

$$Q_{ref} = \sum_i m_i \left(\Delta H_f^{25} + \int_{T_{ref}}^{T_{out}} c_{p,i} dT \right) - \sum_i m_i \left(\Delta H_f^{25} + \int_{T_{ref}}^{T_{in}} c_{p,i} dT \right)$$

$$\Delta H_{f,HNO_3}^{25} = \Delta H_f^{25} \Big|_{pure} + \Delta H_{solution}$$

When computing the energy balance, beware of the different aggregation stages of water, fact that the solution of the concentrated acid has a specific heat capacity while the liquid in flow is mostly water and dilution energies as a function of the concentration (Houghen et al., 1959, Sinnot, 1999):

$$\sum \Delta H_{f,products}(25°C) = -30097 \text{ kcal}$$

$$\sum \Delta H_{f,reactants}(25°C) = -28513 \text{ kcal}$$

$$Q_{in}, \text{ flow} = 65 \text{ kcal}$$

$$Q_{out}, \text{ flow} = 29 \text{ kcal}.$$

The summary of the mass and energy balances is given in Fig. 6E7.5.

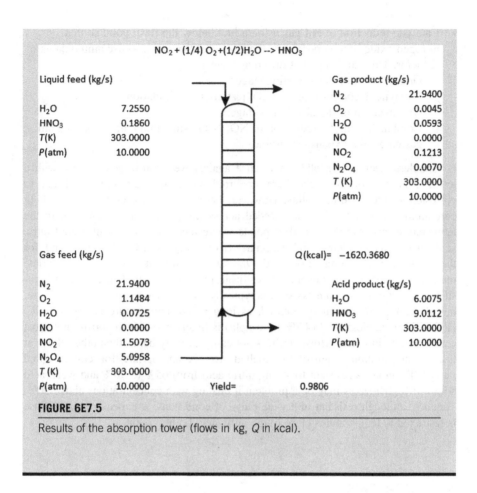

$$NO_2 + (1/4) O_2 + (1/2)H_2O \longrightarrow HNO_3$$

Liquid feed (kg/s)

H_2O	7.2550
HNO_3	0.1860
$T(K)$	303.0000
$P(atm)$	10.0000

Gas product (kg/s)

N_2	21.9400
O_2	0.0045
H_2O	0.0593
NO	0.0000
NO_2	0.1213
N_2O_4	0.0070
$T(K)$	303.0000
$P(atm)$	10.0000

Gas feed (kg/s)

N_2	21.9400
O_2	1.1484
H_2O	0.0725
NO	0.0000
NO_2	1.5073
N_2O_4	5.0958
$T(K)$	303.0000
$P(atm)$	10.0000

$Q(kcal) = -1620.3680$

Acid product (kg/s)

H_2O	6.0075
HNO_3	9.0112
$T(K)$	303.0000
$P(atm)$	10.0000

Yield=   0.9806

FIGURE 6E7.5

Results of the absorption tower (flows in kg, Q in kcal).

6.3 EMISSION CONTROL AND BUILDING ISSUES

NO_x species are a hazardous waste responsible for acid rain. Therefore, tail gas must be properly processed to avoid or reduce those emissions. There are four typical approaches that nitric acid plants use:

- Improving column performance by careful control of the operating pressure and temperature.
- Chemical washing using ammonia, lime, or hydrogen peroxide.
- Adsorption in molecular sieves.

- Catalytic reduction of tail gas using a fuel below the flash point so that nitrogen oxides decompose over their surface. We can also use ammonia as reductor. This can happen at different stages:
  - Optimization of the oxidation stage.
  - Catalytic decomposition of N_2O, just after the oxidation.
  - Optimization of the absorption stage.
  - Combined catalytic reduction of NO_x/N_2O using ammonia before the tail gas enters the expansion turbine.

Stainless steel has excellent resistance against weak nitric acid. Those with a lower concentration of carbon are preferred since they are more resistant to corrosion in the welding points. However, better quality materials such as high chromium alloys (20–27%) are needed across the process in sections where the temperature reaches 900°C, and in points where there is contact with liquid nitric acid, such as the condenser and the preheater of the tail gas. On the other hand, for the production of strong nitric acid (98–99% by weight) the most appropriate materials are aluminum, tantalus, borosilicate glass, silicon molted iron, and high-silicon stainless steel. Stainless steel is typically used for storage up to 95% weight, but above that aluminum is required, as long as the temperature is below 38°C. High-silicon stainless steel (4.7% by weight) is highly resistant to corrosion against concentrated nitric acid. However, it is not comparatively better than other alloys at lower concentration. High-silicon melted iron is appropriate for columns and pumps. Titanium is resistant to strong nitric acid from 65% to 90% and weak acid below 10%, but stress may be a problem if dealing with concentrations above 90%. Tantalus, although resistant in a wide range of conditions, is expensive, and its use is restricted to temperatures close to the nitric acid boiling point.

6.4 PROBLEMS

P6.1. A gas mixture contains nitrogen peroxide at 1 atm. The equivalent NO_2 is 30% and the dissociated fraction of N_2O_4 is equal to 0.4. Determine the value for K_p in the equilibrium

$$2NO_2 \longleftrightarrow N_2O_4$$

and the temperature of operation. The equation for the equilibrium constant is given as follows:

$$\log_{10}\left(\frac{1}{K_p}\right) = -\frac{2692}{T} + 1.75\log_{10}T + 0.00483\cdot T - 7.144\cdot10^{-6}T^2 + 3.062 \quad T[=]K$$

P6.2. In the production of nitric acid, the oxygen proportion of the air is increased by mixing atmospheric (humid) air with oxygen so that the molar ratio between both is 4:1.

This stream is at 760 mmHg, and before mixing it with ammonia, the dew point is 11.24°C and the relative humidity is 0.42.

Ammonia is fed to the converter at the same temperature as that of the rich air. The oxygen fed to the converter is in 20% excess with respect to the stoichiometry of the reaction:

$$NH_3 + \frac{5}{4}O_2 \Rightarrow NO + \frac{3}{2}H_2O$$

Determine:

a. Composition and temperature of the product gases.

b. Maximum concentration of the acid produced if there is not stream separation and the required oxygen for the oxidation from NO to NO_2 is provided.

P6.3. Determine the exit temperature for the product gases of an ammonia converter. The feed to the converter is 65.6°C, and the conversion is a function of the final temperature as given in Fig. P6.3. Assume no secondary reaction, adiabatic operation, and an excess of 10% oxygen with respect to the stoichiometric one.

$$NH_3 + \frac{5}{4}O_2 \Rightarrow NO + \frac{3}{2}H_2O$$

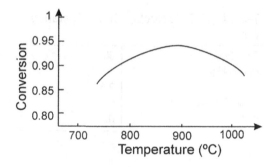

FIGURE P6.3

Conversion versus temperature for NH_3 oxidation.

P6.4. Perform the mass and energy balances for the production of HNO_3 via catalytic oxidation of NH_3.

Atmospheric air (see Table P6.4.1) is mixed with oxygen so that the ratio of oxygen to ammonia becomes 2.15. The ratio of air to added oxygen is 11.5. Assume 1 kmol of atmospheric air for the calculations.

Table P6.4.1 Feed Data

T air (°C)	15
P_{total} (mmHg)	699
Relative humidity	0.25
Ratio Air/O_2 added	11.5
Ratio O_2 total/NH_3	2.15

The converter is fed at 36.5°C and 699 mmHg. Assume 100% conversion. *Compute the temperature and composition of the gases exiting the converter assuming adiabatic operation.* The gases are cooled down to 25°C at 699 mmHg. In the process, NO is oxidized to NO_2 with a conversion of 15%. Nitric acid is also produced with a composition of 3.5%. The gas is next fed to an oxidation tower. *Determine the temperature of the gases exiting the tower assuming adiabatic operation.* The gases are cooled down to 30°C and fed to the absorption–oxidation tower that operates isothermally. A solution of nitric acid 2.5% is put into contact with the gas. The *L/V* mass ratio is 0.2. The product gas is expected to reach 60% concentration. *Compute the flow of acid solution required* (Table P6.4.2).

Table P6.4.2 Absorption Tower Operating Data

T in tower (°C)	30
P_t (mmHg)	699
Conc liq abs (%)	2.5
Conc HNO_3 prod (%)	60
L/V	0.2

Data:

$$Pv_{Water}(mmHg) = EXP\left(18.3036 - \frac{3816.44}{227.02 + T(C)}\right)$$

$$2NO_2 \leftrightarrow N_2O_4 \quad K_p = \frac{P_{N_2O_4}}{P_{NO_2}^2}$$

$$\log_{10}\left(\frac{1}{K_p}\right) = -\frac{2692}{T} + 1.75 \log_{10}T + 0.00483 \cdot T$$
$$- 7.144 \cdot 10^{-6}T^2 + 3.062 \quad T[=]K, \quad K_p[=]atm^{-1}$$

P6.5. A sphere of 10L is filled with a mixture consisting of 25% NO and the rest dry air at 20°C and 1 atm. We allow equilibrium to be reached. Determine:
   a. The total final pressure and the partial pressure of the species in the equilibrium.
   b. The total final pressure and the partial pressure of the species in the equilibrium if the temperature becomes 10°C or 40°C (A.F. Tena).

P6.6. Compute the exit temperature and the gas composition of an oxidation tower. Assume 100% conversion of NO to NO_2. The operating pressure is 10 atm and the feed properties are given in Table P6.6.

Table P6.6 Feed to the Oxidation Tower

	In	
	kmol	kg
N_2	0.78357048	21.9399735
O_2	0.10798747	3.45559896
H_2O	0.00244144	0.04394591
NO	0.11752368	3.52571029
NO_2	0.0076061	0.12930368
N_2O_4	0.00546848	0.34451412
T (K)	293.6	
P (atm)	10	

P6.7. Determine the exit temperature for the product gases of an ammonia converter. The feed to the converter is 65.6°C, and the conversion is a function of the final temperature as given in Fig. P6.3. Assume no secondary reaction, and the excess of air to reach 90% conversion:

$$NH_3 + \frac{5}{4}O_2 \Rightarrow NO + \frac{3}{2}H_2O$$

P6.8. Determine the time it takes to reach 90% conversion in the production of NO_2 from NO for a range of temperatures from 30°C to 100°C and 1 atm (Table P6.8):

Table P6.8 Data for Equilibrium Constant as a Function of the Temperature

T (°C)	38	200	400	600	800	1000
K_p	1E+11	100,000	100	0.1	0.0017	0.0012

$$2NO^{\text{colorless}} + O_2 \rightarrow 2NO_2^{\text{yellow}}$$

$$r = k_1 \left(P_{NO}^2 P_{O_2} - \frac{1}{K_p} P_{NO_2}^2 \right)$$

$$\text{Ec. Bodenstein} \Rightarrow \log_{10} k_1 = \frac{641}{T} - 0.725 \quad T[=]K$$

P6.9. Compute the composition of the HNO_3 solution fed to an absorption–oxidation tower to obtain a concentrated acid at 65%. The mass ratio L/V is 0.25 and the conversion of the equivalent NO_2 is 90%. The gas stream is fed to the tower at 30°C and 10 atm and the composition is given in Table P6.9.

Table P6.9 Inlet Gas Composition of an Absorption Tower

GAS	kmol
N_2	1
O_2	0.1
H_2O	0.005
NO	0
NO_2	0.12
N_2O_4	0.12
Total	1.345
T (K)	303
P (atm)	10

Determine the composition of the product gas assuming isothermal operation.

P6.10. Determine the tray separation required in an absorption-oxidation tower to reach a conversion of 85% assuming the following reaction rate for an operating pressure of 15 atm and 30°C. The composition of the feed is 6% molar of oxygen and 10% molar of NO, and the rest nitrogen. Assume a flow of NO of 1 kmol/s and an area of 1 m^2.

$$r_{NO} = \frac{dP_{NO}}{dV} = -0.2166 \exp(1399/T) \frac{P_{NO}^2}{P_{O_2}} (bar/s)$$

REFERENCES

Bodenstein, M., 1922. Bildung und Zersetzung der Hoheren Stickoxyde. Z. Phys. Chem. 100, 68–123.

Clarke, S.I., Mazzafro, W.J., 1993. Nitric acid. Kirk-Othmer Encyclopedia of Chemical Technology. Wiley.

European Commission, 2009. Integrated pollution prevention and control reference document on best available techniques for the manufacture of large volume inorganic chemicals—ammonia, acids and fertilisers. August, 2007.

Harriott, P., 2003. Chemical Reactor Design. CRC Press. Taylor and Francis, Boca Ratón, FL.

Hickman, D.A., Schmidt, L.D., 1991. Modeling catalytic gauze rectors: ammonia oxidation. Ind. Eng. Chem. Res. 20 (1), 50–55.

Holma, H., Sohlo, J., 1979. A mathematical mode of an absorption tower of nitrogen oxides in nitric acid production. Comp. Chem. Eng. 3, 135–141.

Houghen, O.A., Watson, K.M., Ragatz, R.A., 1959. Chemical Process Principles. vol. 1. Material and Energy Balances. Wiley, New York, NY.

Martín, M., 2014. Introduction to Software for Chemical Engineers. CRC Press, Boca Raton, FL.

MMAMRM, 2009. <http://www.prtr-es.es/data/images/LVIC-AAF-FINAL.pdf.>.

Nonhebek, E., 1972. Gas Purification Processes for Air Pollution Control. Elsevier, Oxford, UK.

Ortuño, A.V., 1999. Introducción a la Química Industrial. Reverté, Barcelona, Spain.

Ray, M.S., Johnson, D.W., 1989. Chemical Engineering Design Project. A Case Study Approach. Topics in Chemical Engineering Volume 6. Gordon and Breach Science Publishers, London, <http://www.prtr-es.es/data/images/LVIC-AAF-FINAL.pdf>.

Sinnot, R.K., 1999. Coulson & Richardson Chemical Engineering. Vol. 6. Elsevier, Singapore.

Thiemann, M., Scheibler, E., Wiegand, K.W., 1998. Nitric acid, nitrous acid, and nitrogen oxides. ULLMANN'S "Encyclopedia of Industrial Chemistry". Wiley-VCH, Weinheim.

Thyssenkrupp Uhde, 2015. <http://www.thyssenkrupp-industrial-solutions.com/fileadmin/documents-/brochures/uhde_brochures_pdf_en_4.pdf>.

Wallas, S.M., 1959. Reactor Kinetics for Chemical Engineers. McGraw-Hill, New York, NY.
<www.uhde.com> Nitric acid.

Sulfuric acid

7.1 INTRODUCTION

History: The discovery of sulfuric acid is attributed to the Persian alchemist Abu-Beskv-Ahhases in the 15th century. However, it was already known long before, back in the era of Albertus Magnus (1193–1280). During the 15th century in Germany, sulfuric acid was produced by heating iron(II) sulfate heptahydrated (green vitriol), which inspired the name "oil of vitriol." The process was reported in the 16th century. The properties of the oil were first described in 1570 by Pornaeius. Later, Johann Van Helmont (c.1580) obtained it via dry distillation of green vitriol (hydrated iron(II) sulfate, $FeSO_4 \cdot 7H_2O$) and blue vitriol (hydrated copper(II) sulfate, $CuSO_4 \cdot 5\ H_2O$). The chemical basis for the lead chamber process was introduced by Andrea Libavius in 1595:

$$S + O_2 \rightarrow SO_2; \quad SO_2 + H_2O \rightarrow H_2SO_4$$

The reactor for running this process was developed in Birmingham, England, in 1746 by John Roebuck. The breakthrough in the production process consisted of substituting the glass units used before, and increasing the conversion up to 78%. For years this process was improved by recovering the catalyst using Gay-Lussac and Glover Towers, and increasing the concentration of the product. By 1831, a patent was issued that described the catalytic oxidation of SO_2 over platinum, the contact process. This process made it possible to increase the reaction yield from 70% to 95%. Although the initial catalyst was platinum, others were tested. In 1913 BASF was granted a patent for the use of vanadium (V) oxide (V_2O_5), which became the catalyst of choice due to its lower cost and resistance to poisonous species. In 1960 a further evolution was patented by Bayer for a process involving a double absorption (double catalyst) process. For years sulfuric acid production has been used as a metric for the industrial development of a country; it has a global production volume above 150 Mt/yr.

Raw materials: Sulfur-based species are typically used as raw materials for sulfuric acid production, from sulfur, copper, iron, and zinc sulfides, to industrial gases like SO_2 and H_2S. The Claus process can transform these gases into elementary sulfur as a raw material for sulfuric acid production (see Chapter 5: Syngas). Sulfides are raw materials for the metals, while sulfur is a byproduct. However, for iron(II) sulfide, both elements need to be considered as equally important for the profitability of the process. Pyrite complexes are species that are found together

Industrial Chemical Process Analysis and Design. DOI: http://dx.doi.org/10.1016/B978-0-08-101093-8.00007-0

Table 7.1 Pyrite and Complex Compositions

Species	% by Weight	
	Pyrite	Complex
S	46.0–50.0	44.0–48.0
Fe	41.0–45.0	36.0–40.0
Cu	0.4–1.0	0.4–1.3
Zn	0.2–2.3	3.0–8.0
Pb	0.5–1.3	2.0–6.0
Ag	0.3–0.6	0.3–0.7

with iron sulfide and other sulfides such as those from copper, zinc, and/or lead. Furthermore, other elements such as copper, selenium, gold, and silver can be found in small amounts. Table 7.1 shows the typical mass composition of pyrites and complexes. Finally, other sulfur-based species from refineries can also be used. In particular, pyrites are an interesting raw material.

Applications: The industrial applications for sulfuric acid are based on two properties:

- It is a strong acid when in water solution; concentrations below 5% by weight are needed for the second proton to be dissociated.
- It is a dehydration agent when concentrated over 80% by weight (and above 95% is preferable).

Furthermore, it is relatively cheap. It is typically commercialized in long-term contracts from large production and consumer companies. The most important consumers are phosphoric acid producers, who typically allocate their facilities in the vicinity of the sulfuric acid production plants. Moreover, it is a reagent used in the production of numerous other chemicals such as explosives, synthetic fibers, fertilizers, etc., and in metallurgy for purification processes for sugar dehydration or concentrated nitric acid production.

Health and safety issues: Sulfuric acid is a strong acid when in dilute solution. It attacks the skin, eyes, and mucoses. Furthermore, when in contact with metals, a large volume of hydrogen can be generated. It reacts exothermally with water, dehydrates and even burns organics, and reacts with other reductants as well. Oleum and sulfur trioxide (SO_3) also react energetically with water.

Global production: The production of sulfuric acid has increased from 130 Mt to over 220 Mt in the last 40 years due to the increased consumption of phosphate and sulfate fertilizers. It is typically sold as solutions of 98–99% by weight as produced, but it can be commercialized as solutions of SO_3 in sulfuric acid, also known as oleum. If water is added, additional sulfuric acid can be produced. It is typically shipped in 200,000 t container ships. It is stored at room temperature and in carbon steel tanks; the tanks are cylindrical with a conic roof, and are in

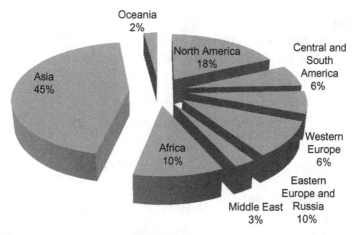

FIGURE 7.1

Distribution of sulfuric acid production across regions.

concrete structures re-covered with stoneware bricks. They need to have a pipe to alleviate pressure and allow any generated hydrogen to leave, and for loading and unloading purposes. The price is usually stable (around $50/t), but in 2008 it reached $400/t. Current prices are around $100/t. Fig. 7.1 shows the distribution of production by region.

Chemical properties: The water–sulfuric acid system has a maximum boiling point azeotrope at 98.5% and 336°C that can easily be overcome in the production process. While diluted it is a strong acid capable of attacking all non-noble metals but lead, which can handle 80% acid at 90°C; concentrated solutions have a small acid constant and can be stored in carbon steel. There is a relationship between the concentration and the density up to 97%. Commercial concentration is close to freezing point.

EXAMPLE 7.1

Fill in Table 7E1.1.

Table 7E1.1 Data Table

% H_2SO_4	°*Be*	%SO_3 Free	%SO_3 Total	Name
65				Chamber acid
	60			
		20		
			93.6	

Solution

Definitions ($°Be$ = degrees Baumé):

$$°Be = 145 - \frac{145}{\rho_{rel}\big|_{15.6°C}}$$

$$SO_3 \text{ free} = \text{monohydrate} + SO_3 \equiv \text{oleum}$$

$$SO_3 \text{ total} = \text{free} + \text{bonded with water}$$

1. 65%

   Using the tables in Perry and Green (1997) we compute the density at 15.6°C.

   15°C: 1.5578 g/cm^3
   20°C: 1.5533 g/cm^3
   So, for 15.6°C we have $\rho = 1.5573$ g/cm^3.

   $$\rho_{rel}\big|_{15.6°C} = \frac{\rho_{H_2SO_4}}{\rho_{H_2O}}\bigg|_{15.6°C} = \frac{1.5573}{0.999} = 1.5589$$

   $$°Be = 145 - \frac{145}{1.5589} = 51.99$$

   %SO$_3$ free: NONE

   $$\%SO_3\big|_{total} = 65\% \ H_2SO_4 + 35\% \ H_2O$$

   From the 65% only a fraction $\dfrac{Mw_{SO_3}}{Mw_{H_2SO_4}}$ is SO$_3$

   $$\%SO_3\big|_{total} = 65 \cdot \frac{80}{98} = 53.06$$

2. 60°Be

   Using the definition above we compute the relative density and the density of the sulfuric acid:

   $$°Be = 145 - \frac{145}{\rho_{rel}} = 60 \Rightarrow \rho_{rel}\big|_{15.6°C} = 1.7059$$

   $$\rho_{rel}\big|_{15.6°C} = \frac{\rho_{H_2SO_4}}{\rho_{H_2O}}\bigg|_{15.6°C} \Rightarrow \rho_{H_2SO_4} = 1.7041 \ g/cm^3$$

   We interpolate that value in density tables as seen in Table 7E1.2.

Table 7E1.2 Sulfuric Acid Density

g/cm^3 (%)	15°C	20°C	15.6°C
77	1.6976	1.6927	1.6970
78	1.7093	1.7043	1.7087

Table 7E1.3 Results Table

% H_2SO_4	°Be	%SO_3 Free	%SO_3 Total	Name
65	52	NA	53.06	Chamber acid
77.6	60	NA	63.30	Glover acid
104.50	NA	20	85.31	Oleum 20%
114.67	NA	65.16	93.6	Oleum 65.16%

Thus, with 1.7041 g/cm³, the mass fraction is 77.6%:

$$\% \ SO_3 \ \text{free} = 0$$

$$\% \ SO_3 \Big|_{\text{total}} = 77.6 \cdot \frac{80}{98} = 63.30$$

3. We cannot longer used °*Be* units. In 100 kg of mixture we have:
   H_2SO_4 80kg → 100%
   SO_3 20 kg → 20 × 98/80 = 24.5 kg of acid can be produced out of it. Thus, 104.5 kg of acid are potentially available in total, although only 80 kg are real. To compute the total SO_3, we have the SO_3 free, plus that in the acid.

$$\% SO_3 \Big|_{\text{total}} = 20 + 80 \cdot \frac{80}{98} = 85.31$$

4. $\% SO_3 \Big|_{\text{total}} = SO_{3,\text{free}} + (100 - SO_{3,\text{free}}) \cdot \dfrac{80}{98} = 93.6 \Rightarrow SO_{3,\text{free}} = 65.16 \ \text{kg}$

   Therefore, in 100 kg, 65.16 kg are SO_3, and the rest (34.84 kg) sulfuric acid. Thus the potential sulfuric acid in the solution is computed as (Table 7E1.3):

$$34.84 + 65.16 \frac{98}{80} = 114.67$$

7.2 PYRITE ROASTING

In this analysis it will be assumed that the mineral consists of sulfur and iron alone in the form of pyrite (FeS_2) or pyrrhotite (Fe_7S_8). The reactions that take place are the following:

1. $7FeS_2 + 6O_2 \rightarrow Fe_7S_8 + 6SO_2$
2. $Fe_7S_8 + O_2 \rightarrow 7FeS + SO_2$
3. $3FeS + 5O_2 \rightarrow Fe_3O_4 + 3SO_2$
4. $4Fe_3O_4 + O_2 \rightarrow 6Fe_2O_3$

The global reaction is computed as:

$$(1) + (2) + (3) \cdot 7/3 + (4) \cdot 7/6$$

This results in:

$$FeS_2 + \frac{11}{4}O_2 \rightarrow \frac{1}{2}Fe_2O_3 + 2SO_2$$

Pyrites lose moisture up to $100-110°C$, but from $415°C$ onwards, one of the labile sulfur atoms is detached and it burns, as in reactions (5) and (6). Therefore with oxygen available, we produce FeS as slag:

5. $FeS_2 \rightarrow \frac{1}{2}S_2 + FeS$

$$\rightarrow FeS_2 + O_2 \rightarrow FeS + SO_2$$

6. $\frac{1}{2}S_2 + O_2 \rightarrow SO_2$

If the combustion of the mineral is not complete, the FeS_2 transforms into FeS as in the reactions above. If all of it is burned, the FeS_2 produces Fe_3O_4 as in reaction 3. The mechanism is as follows:

$$FeS + 2O_2 \rightarrow FeSO_4$$
$$FeSO_4 \rightarrow FeO + SO_3$$
$$FeO + \frac{1}{2}O_2 \rightarrow Fe_3O_4 \Rightarrow FeS + \frac{10}{6}O_2 \rightarrow SO_2 + \frac{1}{3}Fe_3O_4$$

$$SO_3 \rightarrow SO_2 + \frac{1}{2}O_2$$

For the combustion to be efficient, the lower the temperature, the more oxygen required. We need 150% excess at $680°C$, but only 115% excess at $700°C$. The sulfates produced at different stages decompose to oxides over $900°C$. The type of mineral determines the melting temperature. Furthermore, SO_3 in small amounts can be produced. It is typically neglected in the analysis. The production is due to reactions such as:

$$SO_2 + \frac{1}{2}O_2 \rightarrow SO_3$$

Both are slow and favored at lower temperatures. The SO_3 can generate SO_4^{2-} (Table 7.2).

$$\text{Pyrite roasting: } SO_3 + Fe_2O_3 \rightarrow Fe_2(SO_4)_3$$

Gas cleaning: The product gas does not only contain the SO_2, which is useful for sulfuric acid production, but also dust, that must be eliminated, arsenic and moisture. Dust is typically removed by gravity using colliders, cyclones, centrifuges, filters, or electrostatic precipitators. Moisture may become a problem since it generates sulfuric acid before the converter; this will corrode the catalysts and the pipes. Finally, arsenic oxides must be eliminated. CaO can be used to absorb

Table 7.2 Pyrite Roasting Furnaces

Rotary and mechanical furnaces can process particle sizes up to 10–12 mm.

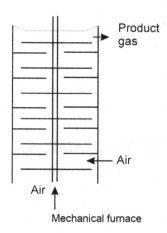

Mechanical furnace

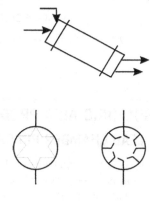

Rotary furnace

Flash Furnaces: The pyrite is fed from the top and the air from the bottom. Small particle sizes are required. The high operating temperatures result in the *vitrification* of the ash. Arsenic remains inside and iron cannot be recovered.

Fluidized beds: These allow processing of different types of particles with good heat recovery. Arsenic remains in the ash.

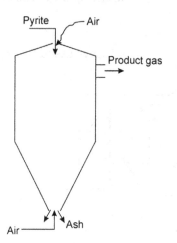

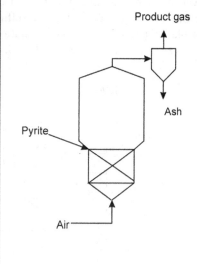

arsenic, generating $Ca_3(AsO_4)_2$. Alternatively, we can eliminate the arsenic oxides using a Tyndall box, where $Fe(OH)_3$ is sprayed over the gases to remove the arsenic oxides as per the following reaction:

$$As_2O_3 + 2Fe(OH)_3 \rightarrow 2FeAsO_3 + 3H_2O.$$

7.3 SULFURIC ACID PRODUCTION

7.3.1 LEAD CHAMBER PROCESS: HOMOGENEOUS CATALYST

This process is no longer in use. However, it was the first one on the market and its evolution is a good example of process design. The first facility in the United Kingdom was installed in 1749 in Scotland, and it consisted of only the lead chambers. By 1772–77, based on Lavoisier's discovery of the sulfuric acid formula, it was shown that niter was not a component of the acid; it was thought to be a catalyst. In 1827 Gay-Lussac introduced the tower that was named after him; it was designed to recover the nitrogen oxides that were lost with the gases. However, this tower was not used until 1859, when Glover introduced a new tower (a denitrification tower) to recover the catalyst from the liquid produced in the Gay-Lussac Tower, so that the sulfuric acid is denitrified. This process is only capable of producing low-concentration sulfuric acid. The process consists of three stages, as seen in Fig. 7.2 (Lloyd, 2011).

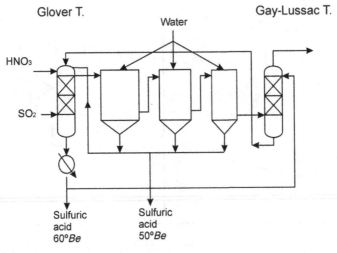

FIGURE 7.2

Flowsheet for the production of sulfuric acid using the lead chamber process.

In plants up to 1920, niter pots were used to produce in situ the HNO_3 used as catalyst. The gas was passed through a niter oven where $NaNO_3$ and concentrated H_2SO_4 were heated. Later, an ammonia oxidation unit was used (see Chapter 6: Nitric acid, for nitric acid production from ammonia).

1. The SO_2, impure with other gases, is processed through a dust chamber before being fed to the Glover Tower for denitrification. The tower is a cylinder of $12-14$ m of diameter, whose walls are covered by Volvic rock, volcanic rock, resistant to acids. The outside surface is covered by a thick lead layer. From 1 m to 1.5 m from the base up to three fourths of its height, it is filled with a porous acid-proof packing. The lower limit is hydraulically sealed with the acid to avoid gas leaks. The hot SO_2 is fed to the lower part of the Glover Tower. From the top, a liquid mixture of diluted sulfuric acid ($53°Be$) rich in nitrous gases ($NO + NO_2$) from the Gay-Lussac Tower is sprayed inside, the nitrous vitriol (sulfuric acid with NO and NO_2 dissolved). Nitric acid is also fed in as makeup. Part of the SO_2 is oxidized into SO_3, and it is dissolved into the acid solution to obtain the so-called Glover acid, a solution of 78% H_2SO_4. The process is described by the following reactions:

$$HSO_4\ NO + HNO_3 \rightarrow H_2SO_4 + 2\ NO_2 \quad (H_2SO_4 \text{ denitrification})$$
$$2HSO_4\ NO + H_2O \rightarrow 2\ H_2SO_4 + NO + NO_2 \quad (NO_x \text{ release})$$
$$SO_2 + NO_2 \rightarrow NO + SO_3 \quad (\text{Partial oxidation of } SO_2)$$
$$SO_3 + H_2O \rightarrow H_2SO_4 (\text{Glover acid}) \quad (\text{Acid concentration})$$

Thus, the sulfuric acid is denitrified in the tower. The nitrous vapors rise, acting as catalysts for the oxidation of SO_2. The acid, as it descends, gives up the water and cools down the gases from the furnace. These gases mix with the steam and the nitrous vapors and are sent to the first lead chamber through a lead pipe located at the top of the tower.

2. From the Glover Tower, a mix of gases including sulfur dioxide and sulfur trioxide, nitrogen oxides, nitrogen, oxygen is transferred to a chamber internally covered by lead, where it is sprayed with water. The lead chambers are large tanks in the form of a truncated cone (100 m long, 15 m high, and 25 m wide) whose walls are made of lead supported by a wood structure. The walls and the bottom constitute a hydraulic seal where the produced sulfuric acid is collected. From the roof of the chambers, a fine spray of water is fed to dissolve the SO_3 obtained in the chambers. The sulfuric acid, formed in the series of reactions presented below, condenses on the walls and is gathered at the bottom of the chambers:

$$SO_2 + H_2O \rightarrow H_2SO_3$$
$$H_2SO_3 + NO_2 \rightarrow H_2SO_4\ NO$$
$$H_2SO_4\ NO \rightarrow H_2SO_4 + NO$$
$$NO + 0.5O_2 \rightarrow NO_2$$
$$SO_3 + H_2O \rightarrow H_2SO_4 \quad (\text{Production of } H_2SO_4)$$

The unconverted gases are fed to the second chamber and so on. Each chamber is smaller than the previous one. There can be from three to six chambers. The sulfuric acid produced has a concentration ranging from 62% to 68% H_2SO_4. It is called chamber acid. Finally, the gases are fed to the Gay-Lussac Tower.

3. The unconverted gases are washed with cooled concentrated acid from the Glover Tower in the Gay-Lussac Tower. The nitrogen oxides and the sulfur dioxide are dissolved in the acid, regenerating the nitrous vitriol that is sent to the Glover Tower. The unabsorbed gases are released to the atmosphere.

$$H_2SO_4 + N_2O_4 + N_2O_3 \rightarrow HSO_4 \, NO + HNO_3 + H_2O \quad \text{(Absorption of NO}_x\text{)}$$

N_2O_3 is formed from NO and NO_2. The catalyst losses are small: 5–8 kg of HNO_3 per t of H_2SO_4.

The acid produced using this method can be concentrated from 78% (the typical concentration exiting the Glover Tower) using a cascade of silica dishes or by using the Gaillard process. The first method concentrates the sulfuric acid step-by-step in each dish. The Gaillard process consists of spraying the sulfuric acid inside a Gaillard Tower (diameter: 6–15 m; height: 10–20 m). The spray is put in countercurrent with hot gases from a coke furnace so that water is evaporated, concentrating the acid up to 92%. The hot gases are further processed in another tower where dilute acid is also sprayed. Next, the gases are fed to a precipitator where lead bars are used to maintain a high potential so that the mist of acid droplets is recovered.

7.3.2 INTENSIVE METHOD

This method is based on the fact that if the concentration of sulfuric acid is above 70%, the use of lead chambers is no longer needed, and Fe is enough to handle the streams (which reduces the capital costs). The acid from the chambers is recycled and sprayed to increase the concentration. A number of modifications are carried out based on the lead chamber process (see Fig. 7.3):

- The production from the chambers is mixed with the stream that sends the nitrous vitriol from the Gay-Lussac Tower to the Glover Tower.
- To increase the concentration, the acid from the chambers is recycled to them.

As a result, the intensive method produces sulfuric acid with a concentration of 78–92%. Throughout the following examples we analyze the process for the production of sulfuric acid based on the lead chambers. The case study is based on the one in Houghen et al. (1959). Fig. 7E2.1 shows the flowsheet. In Example 7.2 we analyze the pyrite roasting process, in Example 7.3 the Glover Tower, and in Example 7.4 the lead chambers and the Gay-Lussac Tower.

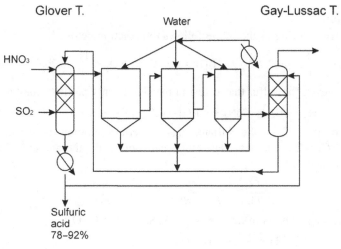

FIGURE 7.3

Intensive method for sulfuric acid production.

EXAMPLE 7.2

Mineral (100 kg/s) consisting of 85% pyrite (FeS_2), 13% inerts, and 2% H_2O is roasted in a furnace to produce SO_2. Air at 25°C and 760 mmHg with a relative humidity of 30% is used. The gas product exits the furnace at 450°C with a composition of 8.50% SO_2, 10.00% O_2, and 81.50% N_2. Only 0.5% of the initial sulfur goes with the slag. The slag exits the furnace at 450°C. Compute the air needed for the process and the heat losses in the furnace. (Fig. 7E2.1)

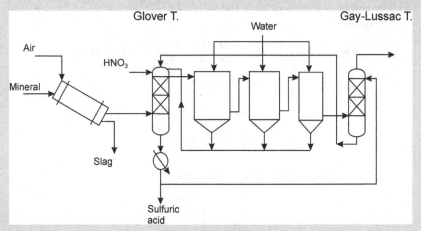

FIGURE 7E2.1

Flow diagram for Examples 7.2–7.4.

Solution

Complete roasting of the pyrite follows the reaction below:

$$FeS_2 + \frac{11}{4}(O_2 + 3.76N_2 + yH_2O) \rightarrow \frac{1}{2}Fe_2O_3 + 2SO_2 + \frac{11}{4} \cdot (3.76N_2 + yH_2O)$$

However, the sulfur that is lost in the slag is only partially roasted:

$$FeS_2 + (O_2 + 3.76N_2 + yH_2O) \rightarrow FeS + SO_2 + (3.76N_2 + yH_2O)$$

The air fed to the furnace has a certain amount of water, and $P_v(25°C) = 23.76$ mmHg. Therefore, per mol of air, the amount of water is computed as:

$$\frac{H_2O_{vap}}{Mol_{air}} = \frac{\varphi P_v^o}{P_T - \varphi P_v^o} = 9.98 \cdot 10^{-3} \frac{mol_{H_2O}}{mol_{dryair}}$$

The slag consists of the inerts, the FeS, and the Fe_2O_3:

$$Slag = Inert + FeS + Fe_2O_3$$
$$Inerts = 13 \text{ kg}/s$$
$$FeS = \frac{Mw_{FeS}}{Mw_S} S_{slag} = \frac{87.9}{32}\left[0.005 \cdot 85 \cdot \frac{32 \cdot 2}{119.9}\right] = 0.625 \text{ kg}/s$$

$$Fe_2O_3 = \frac{1}{2}\frac{Mw_{Fe_2O_3}}{Mw_{FeS_2}} FeS_{2,roasted} = 56.075 \text{ kg}/s$$

$$FeS_{2,roasted} = 85 - \frac{Mw_{FeS_2}}{Mw_{FeS}} FeS = 85 - \frac{119.9}{87.9}0.625 = 84.15 \text{ kg}/s$$

$$Slag = 13 + 0.625 + \frac{1}{2}\frac{159.8}{119.9}(85 - 0.853) = 69.700 \text{ kg}/s$$

The slag composition is given in Table 7E2.1.

Using the stoichiometry of the reaction and the amount of Fe_2O_3 and FeS produced, we can compute the sulfur dioxide, and with it, the dry gas composition:

$$n_{SO_2} = 4n_{Fe_2O_3} + n_{FeS} = 4\frac{56.075}{159.8} + \frac{0.625}{87.9} = 1.411 \text{ kmol}/s$$

$$n_{dry,gas} = \frac{n_{SO_2}}{0.085} = 16.560 \text{ kmol}/s$$

$$n_{O_2} = n_{dry,gas} \cdot 0.1 = 1.656 \text{ kmol}/s$$

$$n_{N_2} = n_{dry,gas} \cdot 0.815 = 13.526 \text{ kmol}/s$$

Table 7E2.1 Slag Composition

	%w
Fe_2O_3	80.45
FeS	0.90
Inerts	18.65

Table 7E2.2 Results for Pyrite Roasting Process

	Furnace Inlet			Gas outlet			Reaction	Slag	
	kmol/s	kg/s	Q (flow) (kcal/s)	kmol/s	kg/s	Q(flow) (kca	Reaction	kg/s	Q flow (kcal/s)
N2	13.526449	378.740572	0	13.526449	378.740572	40964.7239			
O2	3.59563834	115.060427	0	1.659686994	53.10998381	5835.20057			
H2O	0.28096222	5.05731992	0	0.280962218	5.057319916	1016.2259	64.95067635		
SO2			0	1.410733945	90.28697248	6662.0562	−100253.234		
SO3			0						
FeS2	0.70833333	85	0				30078.74801		
FeS		0	0					0.62534362	42.52336614
FesO3		0	0				−68691.3928	56.07460634	3813.073231
Slag		13	0					13	884
Temp									
Total	17.4030496								
	298		0	723		54478.2066	−138800.928	723	4739.596597
				138800.9281	59217.80316	79583.125			
				Q in + Q gen =	Q out	Losses			

The nitrogen comes directly from the air fed. Thus, the air fed to the system is as follows:

$$n_{\text{dry,air}} = n_{\text{N}_2}\frac{1}{0.79} = 17.12 \text{ kmol/s}$$

$$n_{\text{O}_2} = n_{\text{N}_2}\frac{0.21}{0.79} = 3.596 \text{ kmol/s}$$

$$n_{\text{H}_2\text{O}} = n_{\text{dry,air}}\cdot y = 0.170 \text{ kmol/s}$$

Together with the mineral (100 kg/s), a flowrate of 496.9 kg/s of air is fed to the furnace.

The moisture that accompanies the gas product is given by that in the air plus that in the mineral:

$$n_{\text{H}_2\text{O}} = n_{\text{dry,air}}\cdot y + \frac{2}{18} = 0.17 + 0.111 = 0.281 \text{ kmol/s}$$

The total flow of product gases is 527.2 kg/s, the slag is 69.7 kg/s, and the total product flow is 596.9 kg/s. The energy balance is given as follows:

$$Q_{\text{in(flow)}} + Q_{\text{reaction}} = Q_{\text{out(flow)}}$$

$$\sum m_i \int_{T_{\text{ref}}}^{T_{\text{in}}} c_{p,i}\,dT + Q_{\text{reaction}} = \sum m_i \int_{T_{\text{ref}}}^{T_{\text{out}}} c_{p,i}\,dT + Q_{\text{losses}}$$

Assume that the slag has a heat capacity of 0.16 kcal/kg. Table 7E2.2 summarizes the operation of the furnace.

EXAMPLE 7.3

The product gas from the furnace is fed into the Glover Tower. Fifteen percent (15%) of the SO_2 is transformed into H_2SO_4. The Glover Tower also receives a flow rate of 600 kg/s from the Gay-Lussac Tower; it consists of 77% H_2SO_4, 22% H_2O, and 1% N_2O_3. A stream of 1.5 kg of HNO_3 with a composition of 36% is fed as makeup, and from the lead chambers a stream of 175 kg/s of H_2SO_4 with a composition of 64% is also fed to the Glover Tower. The product from the Glover Tower is sulfuric acid 78%. Compute the composition of the gas that will be sent to the lead chambers and the cooling needs so that the product acid is at 25°C. The gases from and to the lead chambers are at 91°C.

Solution

The conversion of SO_2 to H_2SO_4 is 15%. Assume that the reaction taking place is as follows:

$$SO_2 + \frac{1}{2}O_2 + H_2O \rightarrow H_2SO_4$$

We compute the gas products and the sulfuric acid produced:

$$n_{H_2SO_4,produced} = n_{SO_2,converted} = 0.15 \cdot n_{SO_2,ini} = 0.212 \text{ kmol}/s$$

$$n_{H_2O,consumed} = n_{H_2SO_4,produced} = 0.212 \text{ kmol}/s$$

$$n_{O_2,consumed} = \frac{1}{2}n_{H_2SO_4,produced} = 0.105 \text{ kmol}/s$$

Products

$$n_{SO_2} = n_{SO_2,ini} - n_{SO_2,converted} = 1.411 - 0.212 = 1.199 \text{ kmol}/s$$

The sulfuric acid that exits the tower is computed using a mass balance:

$$\text{Gay-Lussac} + \text{Chambers} + \text{Generated} =$$
$$0.77 \cdot 600 + 0.64 \cdot 175 + 0.212 \cdot 98 = 594.78 \text{ kg}/s$$

Out of the total liquid stream, sulfuric acid represents 78% by weight; thus, the water that accompanies it can be computed as:

$$\text{Acidflow} = \frac{m_{H_2SO_4}}{0.78} = 762.48 \text{ kg}/s$$

$$\text{Waterflow} = 0.22 \cdot \text{Acidflow} = 167.75 \text{ kg}/s$$

Next, a balance to water is performed. The reactions of the decomposition of the nitric acid must be accounted for. Furthermore, to compute the gas product, the decomposition of the N_2O_3 is also considered:

$$2HNO_3 \rightarrow 2NO + H_2O + \frac{3}{2}O_2$$

$$N_2O_3 \rightarrow 2NO + \frac{1}{2}O_2$$

We assume that the conversions of these reactions are 100% because nitric acid is not exiting with the products:

$$n_{HNO_3} = \frac{m_{HNO_3}}{Mw_{HNO_3}} = 8.57 \times 10^{-3} \text{ kmol}/s$$

$$n_{N_2O_3} = \frac{m_{N_2O_3}}{Mw_{N_2O_3}} = 0.0789 \text{ kmol}/s$$

$$n_{NO,produced} = n_{HNO_3} + 2 \cdot n_{N_2O_3} = 8.57 \cdot 10^{-3} + 2 \cdot 0.0789 = 0.166 \text{ kmol}/s$$

$$n_{H_2O,produced} = \frac{1}{2} n_{HNO_3} = 4.285 \cdot 10^{-3} \text{ kmol}/s$$

$$n_{O_2,produced} = \frac{3}{4} n_{HNO_3} + \frac{1}{2} n_{N_2O_3} = \frac{3}{4} 8.57 \cdot 10^{-3} + \frac{1}{2} 0.0789 = 0.0459 \text{ kmol}/s$$

Performing a global mass balance to the tower we have the amount of gas products:

$$F_{Gay\text{-}Lussac} + F_{Chambers} + F_{Gas} + F_{Nitric} = F_{Product\ acid} + F_{Gases}$$
$$600 + 175 + 527.2 + 1.5 = 762.48 + F_{Gases}$$
$$F_{Gases} = 541.21 \text{ kg}/s$$

Next, performing a mass balance to water, we compute the water that leaves with the gases (Table 7E3.1):

$$Water_{Gay\text{-}Lussac} + Water_{Chambers} + Water_{Gas} + Water_{Nitric} + Water_{Produced,HNO_3}$$

$$- Water_{consumed,H_2SO_4} = Water_{Productacid} + Water_{Gases}$$

$$0.22 \cdot 600 + 0.36 \cdot 175 + 5.064 + 0.64 \cdot 1.5 + 0.077 - 3.809 = 0.22 \cdot 762.48 + Water_{Gases}$$

$$Water_{Gases} = 29.53 \text{ kg}/s$$

Table 7E3.1 Gas Composition

	kmol/s	kg/s
N_2	13.526	378.741
O_2	1.599	51.193
H_2O	1.641	29.539
SO_2	1.199	76.744
NO	0.166	4.994

Table 7E3.2 Results for Glover Tower

			Furnace			Gay Lussac			Chambers			Catalyst			Product gas			Acid product			
		Hf (kcal/kmc	Entra												Salida						
			kmol/s	kg/s	Qf (kcal/s)	kmol/s	kg/s	Qf (kcal/s)	kmol/s	kg/s	Qf (kcal/s)	kmol/s	kg/s	Qf (kcal/s)	kmol/s	kg/s	Qf (kcal/s)	kmol/s	kg/s	Qf (kcal/s)	Q react(kcal/s)
N2	28	0	13.52645	378.7406	40964.72										13.526449	378.74057	6231.332				
O2	32	0	1.653687	53.10998	5835.201										1.5997842	51.193095	768.2515				
H2O	18	−57798	0.280962	5.05732	1016.226	7.33333333	132		3.5	63		0.053333	0.96		1.6410514	29.538925	878.619	187.7466			28470.56
SO2	64	−71064.6	1.410734	90.28697	6662.056										1.1991239	76.743927	777.3684				15037.93
SO3	80	−94624.4																			
H2SO4	98	−193900				4.71428571	462		1.1428571	112								594.7378			−42724.1
HNO3	63	−41650										0.008571	0.54								357.9524
FeS2	120	−42464.1																			0
FeS	88																				
Fe2O3	160	−196000																			0
N2O3	76	139300				0.07894737	6														−10387.4
NO	30	21600													0.1664682	4.993985	78.72795				3595.669
Escorias																					
Qdis								−43295.9			−13485.71			−57.5538					−52191.3		
Total			16.87783	527.1946	54478.21	12.1285964	600	0	4.6428571	175	7700	0.061905	1.5	0	18.132875	541.2105	8734.239	762.4843	33549.31	0	−6259.28
Temp			723			298			384			298			384			338			
																			0		
HHNO3 (kJ/kg)														−176.4							
mol Wa/mol ac							1.555556			3.0625								1.535613			
									Qin=	5339.0678		Qgen=	−6259.28		Qout=	−3907.665	Qref=	21506.01			

The energy balance is performed considering the formation of solutions of sulfuric acid and NO_x:

$$\sum_i m_i \left(\Delta H_f^{25} + \int_{T_{ref}}^{T_{in}} c_{p,i} dT \right) = \sum_i m_i \left(\Delta H_f^{25} + \int_{T_{ref}}^{T_{out}} c_{p,i} dT \right) + Q_{losses}$$

For H_2SO_4

$$\Delta H_f^{25} = \Delta H_f^{25} \Big|_{pure} + \Delta H_{solution}$$

For N_2O_3

$$\Delta H_f^{25} = \Delta H_f^{25} \Big|_{gas} + \Delta H_{solution}(-28,900 \text{ kcal/kmol})$$

Table 7E3.2 summarizes the results of the mass and energy balances.

EXAMPLE 7.4

The acid leaving the Glover Tower, free from nitrogen oxides, is partially sent to the Gay-Lussac Tower. The rest is collected as product. The gases from the lead chambers are sent to the Gay-Lussac Tower. Determine the acid produced in the facility, the water added to the lead chambers, and the composition of the residual gases from the Gay-Lussac Tower.

Solution

The total amount of sulfuric acid produced in the facility is the sum of that produced in the lead chambers and that produced in the Glover Tower.

This is the one that must leave the facility to avoid buildup. The acid produced in the Glover Towers was determined in Example 7.3. The acid produced in the lead chambers is in the data from the problem. Furthermore, the acid produced from the facility has the same composition as that from the Glover Tower. Thus:

$$m_{H_2SO_4}{}^{glover} = 0.212 \cdot Mw_{H_2SO_4} = 20.78 \text{ kg/s}$$

$$m_{H_2SO_4}{}^{chambers} = n_{SO_2,consumed} \cdot Mw_{H_2SO_4} = 175 \cdot 0.64 = 112 \text{ kg/s}$$

$$m_{H_2SO_4}{}^{glover} + m_{H_2SO_4}{}^{chambers} = 132.776 \text{ kg/s}$$

$$m_{H_2SO_4}{}^{Glover,out} = 762.67 \text{ kg}$$

$$\text{Recycle} = 762.67 \text{ kg} - \frac{132.776 \text{ kg}}{0.78} = 592.26 \text{ kg/s}$$

What follows is the *analysis of the lead chamber operation.*

To determine the water added to the chambers, we perform a molar balance to the entire process:

$$\text{Water}_{air} + \text{Water}_{mineral} + \text{Water}_{HNO_3} + \text{Reacts}_{nitric} - \text{Reacts}_{sulfuric} + \text{Water}_{Chambers}$$
$$= \text{Water}_{product}$$

$$0.17 + 0.11 + \frac{1.5 \cdot 0.36}{18} + (0.00428) + \text{Water}_{chambers} =$$

$$\frac{H_2SO_4^{Glover} + H_2SO_4^{Chambers}}{98} + \frac{1}{18}\left(H_2SO_4^{Glover} + H_2SO_4^{chambers}\right)\frac{0.22}{0.78}$$

$$\text{Water}_{chambers} = 3.146 \text{ kmol/s} \rightarrow 56.619 \text{ kg/s}$$

This is the reaction that takes place in the lead chambers:

$$SO_2 + \frac{1}{2}O_2 + H_2O \rightarrow H_2SO_4$$

The product is a flowrate of 175 kg/s with a composition of 64% by weight. Thus, the composition of the exiting gas is computed based on the stoichiometry in the chambers:

$$n_{SO_2,consumed} = \frac{m_{T,acid} \cdot \%_{H_2SO_4}}{M_{w,H_2SO_4}} = \frac{175 \cdot 0.64}{98} = 1.143 \text{ kmol/s}$$

$$n_{O_2,consumed} = \frac{1}{2} n_{SO_2,consumed} = 0.571 \text{ kmol/s}$$

$$n_{H_2O,consumed} = n_{SO_2,consumed} = 1.143 \text{ kmol}$$

The molar balance to water that exits with the gases is calculated as follows:

$$n_{\text{chambers}} + n_{\text{gases}} - n_{\text{reacts}} - n_{\text{prod,acid}} = 3.146 + 1.640 - 1.143 - \frac{0.36 \cdot 175}{18}$$

$$= 0.143 \text{ kmol}/s$$

The energy balance to the lead chambers is as follows:

$$\sum_i m_i \left(\Delta H_f^{25} + \int_{T_{\text{ref}}}^{T_{\text{in}}} c_{p,i} dT \right) = \sum_i m_i \left(\Delta H_f^{25} + \int_{T_{\text{ref}}}^{T_{\text{out}}} c_{p,i} dT \right) + Q_{\text{losses}}$$

where

$$\Delta H_{f,\text{H}_2\text{SO}_4}^{25} = \Delta H_f^{25} \Big|_{\text{pure}} + \Delta H_{\text{solution}}$$

The results are summarized in Table 7E4.1.

Now we look at the *analysis of the Gay-Lussac Tower.*

The gas exiting the chambers is fed to the Glover Tower. It is dry gas since the acid is a well-known dehydrating agent. Thus, the gas product from the tower is computed as shown below.

The SO_2 exiting the tower is that which has not produced sulfuric acid:

$$n_{\text{SO}_2} = 1.411 - \frac{\text{H}_2\text{SO}_4^{\text{Glover}} + \text{H}_2\text{SO}_4^{\text{Chamber}}}{98} = 0.037 \text{ kmol}/s$$

Table 7E4.1 Results for Lead Chambers

	Hf (kcal/kmol)	Water fed to chambers In			Gas from Glover In			Acdi produced at chambers Out			Gas product to Gay Lussac In		
		kmol/s	kg/s	Qf(kcal/s)	kmol/s	kg/s	Qf(kcal/s)	kmol/s	kg/s	Qf(kcal/s)	kmol/s	kg/s	Qf(kcal/s)
N2	28	0			13.526449	378.740572	6251.3328				13.526449	378.74057	6136.8543
O2	32	0			1.599784204	51.1930945	768.25393				1.0283556	32.90798	486.23127
H2O	18	-57798	3.1455009	56.619015	1.641051403	29.5389258	878.61903	3.5	63		0.1436951	2.5865121	75.759904
SO2	64	-71064.5933			1.199123853	76.7439266	786.14598				0.0562667	3.6010695	36.3111
SO3	80	-94624.40191											
H2SO4	98	-193900						1.14285714	112				
HNO3	63	-41650											
FeS2	120	-42464.11483											
FeS	88												
FesO3	160	-196000											
N2O3	76	139300											
NO	30	21600			0.166466165	4.99398496	78.727954				0.1664662	4.993985	77.529601
Slag													
Qdis										-13485.71429			
Total		3.1455009	56.619015	0	18.13287463	541.210503	8743.0792	4.64285714	175	0	14.921233	422.82952	6812.6862
Temp		298			364			298			383		
mol Wa/mol Ac								3.0625					
Qdis					Qin=	8743.07919					Qout=	6812.6862	
											Qin GL=	6812.6862	
								Losses=	15416.1073				

Table 7E4.2 Results for Lead Chambers

The *oxygen* in the outlet gases is computed as follows. Oxygen enters with the air and is produced in the decomposition of HNO_3. Moreover, it is consumed in the furnance and in the production of sulfuric acid:

$$n_{O_2} = 1.641 - \frac{H_2SO_4^{Glover} + H_2SO_4^{Chambers}}{98}\frac{1}{2} + \frac{1.5 \cdot 0.36}{63}\frac{3}{4} = 0.979 \, \text{kmol}/s$$

Alternatively, we can determine the oxygen in the gases from the total amount exiting the chambers minus that needed for the regeneration of the catalyst.

The NO in the gas is that entering with the HNO_3. Alternatively, we can compute it as that from the lead chambers minus that needed in the regeneration of the catalyst:

$$2NO + \frac{1}{2}O_2 \rightarrow N_2O_3$$

Finally, we perform an energy balance to the Gay-Lussac Tower as before:

$$\sum_i m_i \left(\Delta H_f^{25} + \int_{T_{ref}}^{T_{in}} c_{p,i} dT \right) = \sum_i m_i \left(\Delta H_f^{25} + \int_{T_{ref}}^{T_{out}} c_{p,i} dT \right) + Q_{losses}$$

where

$$\Delta H_{f,H_2SO_4}^{25} = \Delta H_f^{25}\Big|_{pure} + \Delta H_{solution}$$

for N_2O_3

$$\Delta H_f^{25} = \Delta H_f^{25}\Big|_{gas} + \Delta H_{solution}$$

The results are summarized in Table 7E4.2.

7.3.3 CONTACT METHOD: HETEROGENEOUS CATALYSIS

7.3.3.1 History

This method was first described by Peregrine Phillips in 1831 when the patent was issued. It consisted of the oxidation of sulfur dioxide into sulfur trioxide and its absorption in water. This method was not widely used for over 40 years due to the lack of demand and the slow progress of technology necessary to work with pure gases. A few years later, in 1875 Clemens and Winkler showed that the SO_2 and the O_2 should be put into contact with the catalyst in stoichiometric proportions. By 1898 BASF had already evaluated several catalysts, from the original platinum to V_2O_5, which is the catalyst of choice today. The process allows high-purity sulfuric acid production. The first facility in the United States did not use water directly to absorb and hydrate the SO_3, but sulfuric acid (King et al., 2006).

7.3.3.2 Process description

The process consists of three stages: sulfur burning, catalytic oxidation of SO_2, and SO_3 hydration. Fig. 7.4 shows a scheme of the layout of the plant.

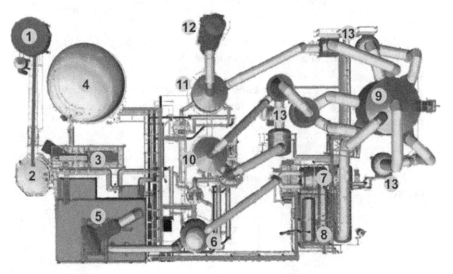

FIGURE 7.4

Layout of a sulfuric acid plant based on the contact method.
1. Solid sulfur storage; 2. Sulfur melting; 3. Liquid sulfur filtration; 4. Liquid sulfur storage; 5. Air filtration and silencer; 6. Air dryer; 7. Sulfur combustion with two distinct burners with individual air supplies; 8. Steam, water storage, boiler; 9. Converter; 10. Intermediate absorber; 11. Final absorber; 12. Chimney; 13. Heat exchangers, economizers, heaters.

Courtesy of Outotec OYJ with permission.

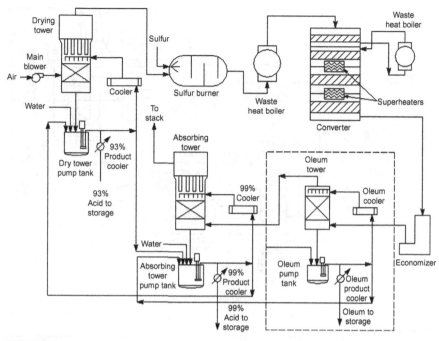

FIGURE 7.5

Single-stage absorption process for the production of sulfuric acid.

Reproduced with permission from Muller, T., 1993. Sulfuric acid and sulfur trioxide. Kirk-Othmer Encyclopedia of Chemical Technology. vol. 23, Wiley, Muller (1993). Copyright Kirk Othmer Encyclopedia of Chemical Technology.

Note the size of the units and the pipes that connect them. There are two main alternatives: single-stage or double-stage absorption. Figs. 7.5 and 7.6 show each of them, respectively. We can use sulfur or pyrite as raw material in any of the processes. The disadvantage of using pyrite—in addition to the fact that it contains arsenic—is that the dust, arsenic, and antimony must be treated, which reduces the global yield of the plant. Furthermore, the dust can never be completely removed. Thus, the energy balance is better for sulfur-based plants.

7.3.3.2.1 Sulfur combustion

The sulfur is received as a solid. It is melted and clarified before storage. Next, the clean sulfur is burned with air to produce sulfur dioxide. Fig. 7.7 shows a photograph of a typical burner. This air needs to have been previously treated to remove dust and humidity. The air is filtered and sent to a drying tower where it is washed with sulfuric acid (95–96 wt%) to remove the humidity—air moisture

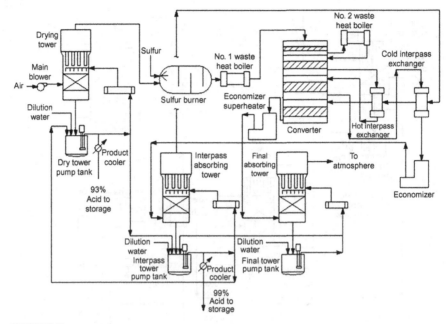

FIGURE 7.6

Double-stage absorption process for the production of sulfuric acid.

Reproduced with permission from Muller, T., 1993. Sulfuric acid and sulfur trioxide. Kirk-Othmer Encyclopedia of Chemical Technology. vol. 23, Wiley, Muller (1993). Copyright Kirk Othmer Encyclopedia of Chemical Technology.

FIGURE 7.7

Sulfur burner.

With permission http://www.chemithon.co.in/burner.php.

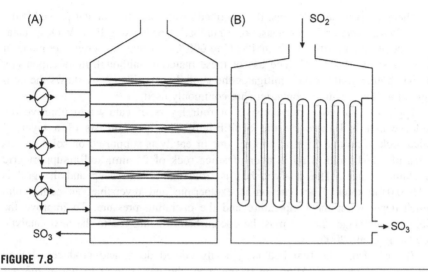

FIGURE 7.8

SO_2 converter configurations. (A) Multibed reactor and (B) Tubular packed bed.

is responsible for corrosion in pipes and towers. Typically the dry air has a composition of 21% oxygen and 79% nitrogen. The reaction that takes place is the following:

$$S + O_2 \rightarrow SO_2$$

The reaction is highly exothermic, so the exiting gas is at a high temperature. The final temperature depends not only on the ratio of oxygen to sulfur, but also on the heat loss due to radiation. Only a fraction of the oxygen is used to burn the sulfur. The rest is needed at a later stage in the converter. The gas composition from the sulfur burner depends on the oxygen-to-sulfur ratio.

7.3.3.2.2 Catalytic oxidation of SO_2

For the reaction to progress, the feed to the converter has to be cooled down to 420−450°C. This is achieved using a boiler to produce steam. The gas stream containing the SO_2 is filtered to remove the dust so that dry gases containing 7−10% SO_2 are fed to the converter. The lower bound corresponds to pyrite-based facilities and the upper bound is common for sulfur-based ones. The stream composition with regards to oxygen is 11−14%. Thus, the typical ratio between O_2 and SO_2 is 1.7. In the converter, SO_2 reacts with the oxygen already in the gas phase to produce sulfur trioxide:

$$2SO_2 + O_2 \rightarrow 2SO_3$$
$$\text{Mechanism: } SO_2 + V_2O_5 \rightarrow V_2O_4 + SO_3; \quad V_2O_4 + \tfrac{1}{2} O_2 \rightarrow V_2O_5$$

There are two basic designs: the multibed reactor and the tubular packed bed.

Multibed reactor: This consists of a number of beds using Pt or V_2O_5 as catalyst, operating adiabatically from 400°C to 600°C. Current converters are made of stainless steel 304 or 321. The use of these materials, although more expensive, allows thinner walls, which mitigates the capital costs. Fig. 7.8A shows the configuration. Two or more converters are commonly used.

Typically, the converter consists of four layers of catalyst to improve the yield, which is limited by the equilibrium. The bed consists of a layer of silica rock of around 25 mm, another one of catalysts supported on solid porous silica of 10−12 mm, a third layer of silica rock of 25 mm, and a support grid of stainless steel. The catalytic bed is from 0.5 to 1 m deep and the grid is 0.02−0.04 m thick. The reaction is exothermic and reversible, an equilibrium which depends on the temperature and the operating pressure. To improve the yield at each stage, the gas must be cooled before being fed to the next catalytic bed (King et al., 2006).

The gas from the first bed is typically cooled down and produces stream; see Figs. 7.5 and 7.6. After the second bed, the gas is used to reheat the gas coming from the absorption tower where SO_3 has been absorbed into sulfuric acid. The gas that has reacted in the third bed is cooled down to generate steam and is then sent to a tower to remove the selenium. Subsequently, the gas is fed to the intermediate absorption column where the SO_3 is removed from the stream by absorption on sulfuric acid 98%. This gas is reheated using the hot product gases from the second and third catalytic beds. The remaining SO_2 is converted into SO_3 in the fourth bed. The final product gas is cooled down in an economizer and fed to the final absorption tower where the remaining SO_3 is absorbed into sulfuric acid 98%.

The catalysts used have evolved over the years. They are quite sensitive to impurities in the gas phase, and thus require proper purification. Initially, platinum catalysts were proposed and tested in 1831 by Phillips. Snhnuder in 1847 suggested the use of lead. A year later, Leming used Pb with 1% MnO_2. In 1853 pyrite cinders were used by Robb. The same year, Hunt patented the use of silica as a catalyst support. By 1894, Mannheim suggested a mixture of iron (III) oxide, but in the same year vanadic acid was also tested. From 1900 on, V_2O_5 has been used as the active component (containing 4−9%), together with alkali metal sulfates as promoters (Lloyd, 2011).

Tubular packed beds: The feed gas is heated up to the proper reaction temperature and fed to the reactor. In this case, the catalyst is packed in tubes, and the tubes are put in heat exchangers where they will be cooled by a boiling liquid. The outside diameter of the tubes is a compromise between the heat transfer to cool down the process and the number of tubes needed. Severe radial temperature gradients have been observed in oxidation systems, although these systems had platinum catalysts and greatly different operating conditions than those being considered here. Fig. 7.8B shows an example.

The conversion of SO_2 to SO_3 is between 96% and 97% since the design efficiency drops with the operation. Pyrite-based facilities also suffer this efficiency reduction problem to a large extent. If their arsenic content is high (which poisons the catalyst), the yield decreases to 95%. Typically the operating temperature in the first converter is from 400°C to 600°C, while in a second converter a temperature range from 500°C to 600°C is used for reaching an optimal conversion at a reduced cost. The residence time in the converters is short: 2−4 s (King et al., 2006).

7.3.3.2.3 SO_3 hydration

The SO_3 produced in the converter, once cooled to 100°C, is absorbed into a sulfuric acid solution (98−99%). The gas is combined with the water in the acid in a two-step procedure after the third and fourth beds, as described in the stage above. These are the reactions:

$$SO_3 + H_2SO_4 \rightarrow H_2S_2O_7$$
$$H_2S_2O_7 + H_2O \rightarrow 2H_2SO_4$$

The reactions take place in towers of 7−9 m diameter, 18−25 m height, a packed bed height of 2.5−6 m, and typically, ceramic saddles of 5−7 cm. The acid is therefore recycled to the dryer where the moisture of the air dilutes it. A plug filter is allocated at the top of the final absorption tower. Therefore, while it is diluted with the air, it is concentrated in the absorption. Fig. 7.9 shows a photograph of an absorption tower.

Water is added to the reservoir tanks of the absorption towers and dryer, increasing the volume and maintaining the acid concentration to 98−99%. The double absorption process allows higher production yields with smaller

FIGURE 7.9

Absorption tower.

Table 7.3 Operation Data for Sulfuric Plants (Outotec, 2012)

	Single Absorption Process	Double Absorption Process
Capacity mtpd Mh	50–7900	50–7900
Conversion %	98–99	99.9
Specific steam production t/mtMh		1.4
Specific cooling	15–25 kWh per t Mh	2 GJ
Specific power consumption	1–1.56 GJ per t Mh	40–60 kWh/mtMh
Specific catalyst quantity L per t Mh	200–260	150–200

Mh, Monohydrate sulfuric acid.

equipment. Table 7.3 shows the comparison between the processes that use single and double absorption.

7.3.3.2.4 Selenium removal

As mentioned before, the stream containing the SO_3 from the third bed still contains selenium from the sulfur. It has to be removed so that the sulfuric acid reaches the proper concentration.

7.3.3.2.5 Oleum production

Oleum is a solution of SO_3 in sulfuric acid, which is used in sulfonation processes. It is also known as Nordhausen acid. The chemical reactions are as follows:

$$(H_2SO_4)_x + H_2O + SO_3 \rightarrow (H_2SO_4)_{x+1} \quad + Heat$$
$$|\text{---acid } 98\% \quad \text{----} \quad | \text{ acid } 100\%$$
$$(H_2SO_4)y + (SO_3)z \quad \rightarrow (H_2SO_4) \cdot SO_3 + Heat$$

Note that x, y, and z are arbitrary numbers of molecules.

The process consists of an absorption tower filled with ceramic material, a recycle pump, a cooling system, and a recycle tank. The absorption of SO_3 has a low yield so that the gas has to be recycled to the sulfuric acid plant to avoid environmental contamination. The final product contains 22–24% of SO_3 dissolved in sulfuric acid, and the theoretical purity is 105% since it is possible to generate more sulfuric acid just by adding water. The product can extract water from organic material as well as from human skin, and careful temperature control is required. Selenium is typically removed at this stage.

7.3.3.2.6 Secondary products

Several stages produce energy. Out of this energy, 57–64% can be used to produce steam that is later used to power the compressor for the air.

7.3.3.2.7 Heat integration

Both the excess energy and the produced energy are reused within the process. In particular, the multibed design allows efficient energy reuse.

7.3.3.2.8 Construction materials

The corrosive nature of sulfuric acid requires the use of different alloys in industrial plants. Mostly metals such as Ni, Cr, Mo, Cu, and Si are used to reduce the corrosion in sulfuric acid plants. The corrosivity of the streams varies across the process since it is a function of the acid concentration, the operating temperature, and the flow velocity.

7.3.3.2.9 Emissions

The emissions of SO_3 or H_2SO_4 mists are due to inefficient absorption. They can be minimized by strict control of sulfur impurities, and the proper dry process for air. Candle filters can be used to remove the mists, but they are not effective for removing the excess SO_3.

7.3.3.3 Process analysis

7.3.3.3.1 Oxidation thermodynamics

The reaction taking place in the converter is an exothermic (26,700 kcal/kmol) equilibrium:

$$SO_2 + 0.5\ O_2 \longleftrightarrow SO_3$$

The equilibrium is driven to products at lower temperatures by increasing the pressure and/or increasing the concentration of the reactants. The corrosive nature of the species involved prevents the use of high pressures or a higher concentration of reactants. Therefore, temperature control is used to improve the conversion of the reaction. The equilibrium constant is given as follows:

$$K_p = \frac{P_{SO_3}}{P_{SO_2} P_{O_2}^{0.5}} \tag{7.1}$$

The values for the equilibrium constant as a function of the temperature can be found elsewhere. Two correlations are presented below (Duecker and West, 1975; Müller, 1998; Ortuño, 1999):

$$\log_{10}(K_p) = \frac{4956}{T} - 4.6780 \tag{7.2}$$

$$\log_{10}(K_p) = \frac{5186.5}{T} + 0.61 \log_{10}(T) - 6.7496 \tag{7.3}$$

As the reaction progresses, the temperature increases, which reduces the maximum equilibrium conversion that can be achieved. Thus, refrigeration is required to obtain high conversions.

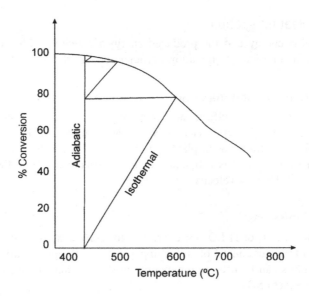

FIGURE 7.10

SO$_2$ converter operation.

    Ideally, isothermal operation would be the best option (see the vertical line in Fig. 7.10), but the difficulties in removing the generated energy suggest the use of intercooling between the catalytic beds (see the step-based path in the same figure). In the following examples we compute the equilibrium line as a function of the operating fed composition, pressure, and temperature for different feedstocks (Examples 7.5 and 7.6), and also design a multibed reactor (Example 7.7).

EXAMPLE 7.5

Determine the equilibrium conversion from SO$_2$ to SO$_3$ from sulfur. Plot the equilibrium curve for different pressures (0.8 atm, 1 atm, 1.5 atm) and initial concentrations of SO$_2$ (6%, 8%, 12%).

Solution

Consider the reaction stoichiometry and the equilibrium constant, given as follows:

$$SO_2 + \frac{1}{2}O_2 \Leftrightarrow SO_3$$

$$K_p = \frac{P_{SO_3}}{P_{SO_2} \cdot P_{O_2}^{0.5}}$$

$$\log_{10}(K_p) = \frac{4956}{T} - 4.678$$

Table 7E5.1 Balances to Sulfur Burning and Equilibrium Stages

| | $S \rightarrow SO_2$ | | $SO_2 \rightarrow SO_3$ | |
	Initial	Final	Initial	Equilibrium
SO_2	–	a	a	$a \cdot (1 - X)$
N_2	0.79	0.79	0.79	0.79
O_2	0.21	$0.21 - a$	$0.21 - a$	$0.21 - a - 0.5 \cdot a \cdot X$
SO_3	–			$a \cdot X$

From sulfur and 1 kmol of dry air, we perform the mass balances to the process. Let X be the conversion and a the fraction of SO_2 in the feed to the converter (Table 7E5.1).

The total number of moles in the equilibrium is the summation of the last column of the table:

$$n_T = a - a \cdot X + 0.79 + 0.21 - a - 0.5 \cdot a \cdot X + a \cdot X = 1 - 0.5 \cdot a \cdot X$$

Thus, the partial pressures for the species involved are as follows:

$$P_{SO_2} = \frac{a \cdot (1 - X)}{1 - \frac{a \cdot X}{2}} P_T; \quad P_{SO_3} = \frac{a \cdot (X)}{1 - \frac{a \cdot X}{2}} P_T; \quad P_{O_2} = \frac{0.21 - a - \frac{a \cdot X}{2}}{1 - \frac{a \cdot X}{2}} P_T$$

Substituting in the expression and reordering:

$$K_p = \frac{P_{SO_3}}{P_{SO_2} \cdot P_{O_2}^{0.5}} = \frac{\left[\dfrac{a \cdot (X)}{1 - \frac{a \cdot X}{2}} P_T \right]}{\left[\dfrac{a \cdot (1 - X)}{1 - \frac{a \cdot X}{2}} P_T \right] \left[\dfrac{0.21 - a - \frac{a \cdot X}{2}}{1 - \frac{a \cdot X}{2}} P_T \right]^{0.5}} = \frac{X \cdot \left[1 - \frac{a \cdot X}{2} \right]^{0.5}}{(1 - X) \left[0.21 - a - \frac{a \cdot X}{2} \right]^{0.5} P_T^{0.5}}$$

Table 7E5.2 shows the conversion for different pressures and temperatures for $a = 0.06$. Fig. 7E5.1 shows the results repeating the procedure for the rest.

Table 7E5.2 Conversion as a Function of Pressure and Temperature

K_p (T)	T	P = 0.8 atm X	P = 1 atm X	P = 1.5 atm X
3819.442708	600	0.99916852	0.99925623	0.99939263
884.3320899	650	0.9964198	0.99679643	0.99738259
252.3480772	700	0.98757686	0.98887226	0.99089366
85.11380382	750	0.96413344	0.96778569	0.97352465
32.88516309	800	0.91261604	0.92104806	0.93449999
14.20980883	850	0.82004997	0.83570639	0.86136331
6.740105069	900	0.68664695	0.7096612	0.74884936
1.896705921	1000	0.38827411	0.41448564	0.46323565

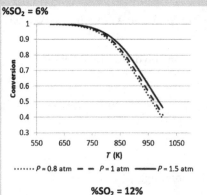

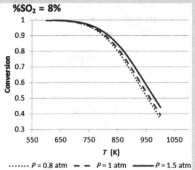

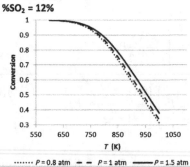

FIGURE 7E5.1

Effect of pressure, temperature, and SO_2 feed on conversion to SO_3 (sulfur burning).

EXAMPLE 7.6

Starting from pyrite, plot the equilibrium curve for various pressures (0.8 atm, 1 atm, 1.2 atm) and initial concentrations of SO_2 to the converter (6%, 8%, 10%).

Solution

Consider the reaction stoichiometry and the equilibrium constant for the oxidation of SO_2:

$$SO_2 + \frac{1}{2}O_2 \Leftrightarrow SO_3$$

$$K_p = \frac{P_{SO_3}}{P_{SO_2} \cdot P_{O_2}^{0.5}}$$

$$\log_{10}(K_p) = \frac{4956}{T} - 4.678$$

Assume 1 kmol of dry air. The mass balances for the pyrite roasting and the sulfur dioxide conversion are as follows (Table 7E6.1):

The fraction of SO_2 in the gas that enters the converter is given by:

$$\%SO_2 = \frac{2a}{1 - 0.75a} \rightarrow a = \frac{\%SO_2}{2 + 0.75 \cdot \%SO_2}$$

In the equilibrium, the total number of moles is given by the equation below:

$$n_T = 2 \cdot a \cdot (1 - X) + 0.79 + 0.21 - 2.75a - a \cdot X + 2 \cdot a \cdot X = 1 - 0.75 \cdot a - a \cdot X$$

Therefore, the partial pressures for the species involved are the following:

$$P_{SO_2} = \frac{2 \cdot a \cdot (1 - X)}{1 - 0.75a - a \cdot X} P_T$$

$$P_{SO_3} = \frac{2 \cdot a \cdot (X)}{1 - 0.75a - a \cdot X} P_T$$

$$P_{O_2} = \frac{0.21 - 2.75 \cdot a - a \cdot X}{1 - 0.75a - a \cdot X} P_T$$

Table 7E6.1 Balances to Pyrite Roasting and Equilibrium Stages

	$FeS_2 + (11/4)O_2 \rightarrow 0.5Fe_2O_3 + 2SO_2$		$SO_2 \rightarrow SO_3$	
	Initial	**Final**	**Initial**	**Equilibrium**
SO_2	–	$2a$	$2a$	$2 \cdot a \cdot (1 - X)$
N_2	0.79	0.79	0.79	0.79
O_2	0.21	$0.21 - 2.75a$	$0.21 - 2.75a$	$0.21 - 2.75a - a \cdot X$
SO_3	–			$2 \cdot a \cdot X$

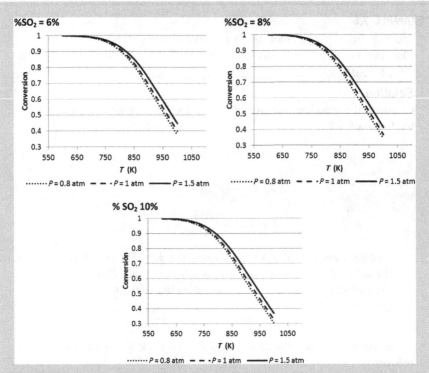

FIGURE 7E6.1

Effect of pressure, temperature, and SO_2 feed on conversion to SO_3 (pyrite roasting).

The equilibrium constant is given by this equation:

$$K_p = \frac{P_{SO_3}}{P_{SO_2} \cdot P_{O_2}^{0.5}} = \frac{\left[\dfrac{2 \cdot a \cdot (X)}{1 - 0.75a - a \cdot X} P_T \right]}{\left[\dfrac{2 \cdot a \cdot (1 - X)}{1 - 0.75a - a \cdot X} P_T \right] \left[\dfrac{0.21 - 2.75 \cdot a - a \cdot X}{1 - 0.75a - a \cdot X} P_T \right]^{0.5}}$$

$$= \frac{X \cdot [1 - 0.75a - a \cdot X]^{0.5}}{(1 - X)[0.21 - 2.75 \cdot a - a \cdot X]^{0.5} P_T^{0.5}}$$

Fig. 7E6.1 shows the effect of feed composition, operating pressure, and temperature on the conversion.

EXAMPLE 7.7

Multibed reactor design: A production facility operates a four bed reactor. The feedstock enters the reactor at 400°C with an SO_2 content of 10% from burning sulfur. The total pressure is 1 atm. The feed temperatures before the second, third, and fourth beds are 450°C, 450°C, and 430°C, respectively. Each catalytic bed reaches 97% of the equilibrium conversion and operates adiabatically. Determine the gas composition after each bed.

Solution

Using the results presented in Example 7.5, with $a = 0.1$, we compute the equilibrium line. As before, from sulfur, and considering 1 kmol of dry air, we have the mass balances as seen in Table 7E7.1.

The moles in the equilibrium are as follows:

$$n_T = a - a \cdot X + 0.79 + 0.21 - a - 0.5 \cdot a \cdot X + a \cdot X = 1 - 0.5 \cdot a \cdot X$$

The equilibrium constant can be computed as below (and plotted in Fig. 7E7.1):

$$K_p = \frac{P_{SO_3}}{P_{SO_2} \cdot P_{O_2}^{0.5}} = \frac{\left[\dfrac{a \cdot (X)}{1 - \dfrac{a \cdot X}{2}} P_T\right]}{\left[\dfrac{a \cdot (1-X)}{1 - \dfrac{a \cdot X}{2}} P_T\right]\left[\dfrac{0.21 - a - \dfrac{a \cdot X}{2}}{1 - \dfrac{a \cdot X}{2}} P_T\right]^{0.5}}$$

$$= \frac{X \cdot \left[1 - \dfrac{a \cdot X}{2}\right]^{0.5}}{(1 - X)\left[0.21 - a - \dfrac{a \cdot X}{2}\right]^{0.5} P_T^{0.5}}$$

Table 7E7.1 Balances to Sulfur Burning and Equilibrium Stages

	$S \rightarrow SO_2$		$SO_2 \rightarrow SO_3$	
	Initial	Final	Initial	Equilibrium
SO_2	–	a	a	$a \cdot (1 - X)$
N_2	0.79	0.79	0.79	0.79
O_2	0.21	$0.21 - a$	$0.21 - a$	$0.21 - a - 0.5 \cdot a \cdot X$
SO_3	–			$a \cdot X$

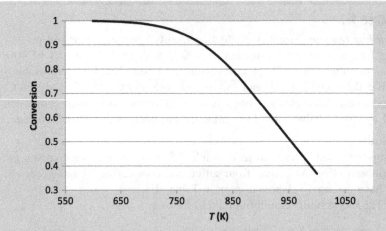

FIGURE 7E7.1

Equilibrium curve.

For the first bed, we perform an energy balance:

$$\sum_{i \in \text{out}} n_i \cdot \int_{T_{\text{ref}}}^{T_{\text{out}}} c_p dT - \sum_{i \in \text{in}} n_i \cdot \int_{T_{\text{ref}}}^{T_{\text{in}}} c_p dT = \Delta H_r$$

$$n_{SO_3} \int_{T_{\text{ref}}}^{T_{\text{out}}} c_p dT + n_{SO_2} \int_{T_{\text{ref}}}^{T_{\text{out}}} c_p dT + n_{N_2} \int_{T_{\text{ref}}}^{T_{\text{out}}} c_p dT + n_{O_2} \int_{T_{\text{ref}}}^{T_{\text{out}}} c_p dT \bigg|_{\text{out}} -$$

$$n_{SO_3} \int_{T_{\text{ref}}}^{T_{\text{in}}} c_p dT + n_{SO_2} \int_{T_{\text{ref}}}^{T_{\text{in}}} c_p dT + n_{N_2} \int_{T_{\text{ref}}}^{T_{\text{in}}} c_p dT + n_{O_2} \int_{T_{\text{ref}}}^{T_{\text{in}}} c_p dT \bigg|_{\text{in}} = \Delta H_r$$

$$n_{SO_2} X \int_{T_{\text{in}}}^{T_{\text{out}}} c_p dT + n_{SO_2}(1 - X) \int_{T_{\text{ref}}}^{T_{\text{in}}} c_p dT + n_{N_2} \int_{T_{\text{in}}}^{T_{\text{out}}} c_p dT + (n_{O_2} - n_{SO_2} X) \int_{T_{\text{in}}}^{T_{\text{out}}} c_p dT = \Delta H_r$$

Assuming constant heat capacity for the mixture and the fact that the moles of gas are almost constant in the bed, the balance becomes linear:

$$n_{T,1} \cdot c_{p,\text{mix}}(T - T_o) = (1 - 0.5 \cdot a \cdot X) \cdot c_{p,\text{mix}}(T - T_o) = a \cdot X \cdot 23,200$$

$$c_{p,\text{mix}}(T - T_o) = a \cdot X \cdot 23,200$$

$$T = \frac{1}{c_{p,\text{mix}}} a \cdot X \cdot 23,200 + T_o = (400 + 273) + 293.7X$$

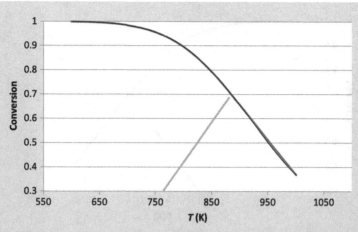

FIGURE 7E7.2

First bed operation.

This straight line reaches the equilibrium line. For the sake of argument, we fit the equilibrium curve to a polynomial as a function of temperature in order to solve the conversion at each bed. X is the conversion, and T is the temperature in K:

$$X = -4.7106 \cdot 10^{-16}T^6 + 2.3838 \cdot 10^{-12}T^5 - 4.9151 \cdot 10^{-09}T^4 + 5.2914 \cdot 10^{-06}T^3$$
$$- 3.1449 \cdot 10^{-03}T^2 + 9.8094 \cdot 10^{-01}T - 1.2473 \cdot 10^2$$

By solving the energy balance and the equilibrium line simultaneously, we have $X = 0.7$ and $T = 881K$; see Fig. 7E7.2. However, the conversion is only 97% of that: 0.686.

For the second bed, the gas is cooled with no progress in the conversion; see the horizontal line in Fig. 7E7.3. The energy balance uses the flow rate exiting the first bed as a starting point and assumes that there is not a large change in the flow across the bed. Note that the equilibrium is computed with respect to total conversion and therefore X is the global conversion, not the one reached at each bed. Thus, the energy balance becomes (Fig. 7E7.3):

$$n_{T,2ini} \cdot c_{p,mix}(T - T_o) = a \cdot (X - X_1) \cdot 23,200$$
$$0.965\, c_{p,mix}(T - T_o) = a \cdot (X - 0.686) \cdot 23,200$$
$$T = \frac{1}{0.965 \cdot c_{p,mix}} a \cdot (X - 0.686) \cdot 23,200 + T_o = (450 + 273) + 304.1(X - 0.686)$$

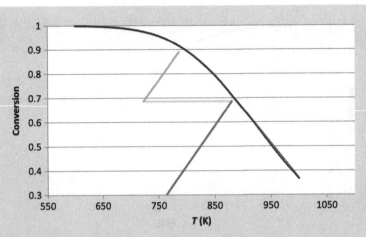

FIGURE 7E7.3

Second bed operation.

By solving the equilibrium and the energy balance we have $X = 0.91$ at $T = 787\text{K}$. Again, only 97% of the equilibrium is reached; thus, $X = 0.89$. For the third bed, the procedure is repeated. The energy balance is as follows:

$$n_{T,3\text{ini}} \cdot c_{p,\text{mix}}(T - T_o) = a \cdot (X - X_2) \cdot 23,200$$

$$0.956\, c_{p,\text{mix}}(T - T_o) = a \cdot (X - 0.89) \cdot 23,200$$

$$T = \frac{1}{0.956 \cdot c_{p,\text{mix}}} a \cdot (X - 0.89) \cdot 23200 + T_o = (450 + 273) + 307(X - 0.89)$$

The conversion becomes 0.97 at 739K. Thus 0.94 is the actual conversion (Fig. 7E7.4).

For the last bed we have the following:

$$n_{T,4\text{ini}} \cdot c_{p,\text{mix}}(T - T_o) = a \cdot (X - X_3) \cdot 23,200$$

$$0.955\, c_{p,\text{mix}}(T - T_o) = a \cdot (X - 0.94) \cdot 23,200$$

$$T = \frac{1}{0.955 \cdot c_{p,\text{mix}}} a \cdot (X - 0.94) \cdot 23,200 + T_o = (430 + 273) + 307.4(X - 0.94)$$

The computed conversion is 0.98 at 722K. The actual one is 0.95; see Fig. 7E7.5.

The summary of the mass balances for each of the beds is presented below. Note that the total molar flow is the one used as $n_{T,\text{ini}}$ for each bed (Table 7E7.2).

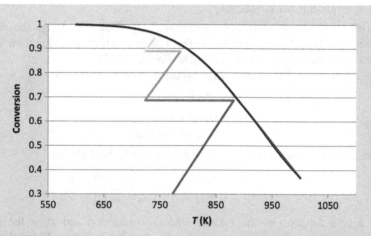

FIGURE 7E7.4

Third bed operation.

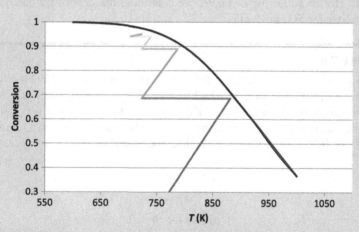

FIGURE 7E7.5

Fourth bed operation.

Table 7E7.2 Gas Composition After Each Bed

	Feed	1st Bed	2nd Bed	3rd Bed	4th Bed
SO_2	0.1	0.0314	0.0111	0.0058	0.0049
N_2	0.79	0.7900	0.7900	0.7900	0.7900
O_2	0.11	0.0757	0.0655	0.0629	0.0624
SO_3	0	0.0686	0.0889	0.0942	0.0995
n_T	1	0.9657	0.9555	0.9529	0.9526

7.3.3.3.2 Oxidation kinetics

In the previous section, only the equilibrium is evaluated to design the reactor. However, the bed depth depends on the kinetics, which is a function of the catalysts—platinum, iron oxide, and V_2O_5 (Fogler, 1997).

The mechanism of the reaction is suggested to be the following:

$$SO_2 + Cat\text{-}O \rightarrow SO_3 + Cat$$
$$Cat + O_2 \rightarrow Cat\text{-}O$$
$$SO_3 + Cat \rightarrow SO_2 + Cat\text{-}O$$

The kinetics of the reaction were established by Eklund (1956):

$$-r'_{SO_2} = k\sqrt{\frac{P_{SO_2}}{P_{SO_3}}}\left[P_{O_2} - \left(\frac{P_{SO_3}}{K_p \cdot P_{SO_2}}\right)^2\right] \tag{7.4}$$

where k is a function of the catalyst and the geometry, and P_i is the partial pressure of each species. For the example, based on the data from Eklund's paper, the catalyst is V_2O_5 supported on volcanic rock with a volumetric density of 33.8 lb/ft^3. In the following example, we determine the depth of a catalytic bed in a converter for SO_3 production.

EXAMPLE 7.8

The first bed of a converter is fed with a flow rate of 467.1292 mol/s at 2 atm and 750K. Out of that, 51.42 mol/s is SO_2. The aim is to reach 50% conversion at the end of the bed. The diameter of the converter is 6.650 m, and the catalyst properties, porosity, particle diameter, and apparent density are given below. Determine the depth of the catalytic bed.

$$\phi = 0.45$$
$$D_p = 0.00457 \text{ m}$$
$$A_c = 34.715 \text{ m}^2$$
$$\rho_b = 542 \text{ kg/m}^3$$

Solution

We assume a plug flow reactor where the sulfur dioxide comes from sulfur burning. With the information provided, and $a = 0.11$, the mass balance for the sulfur burner and the first bed is as shown in Table 7E8.1.

Kinetics of the reactor:

$$F_{A_o}\frac{dX}{dW} = -r'_A$$

If the conversion is below 5%, the reaction rate does not depend on the conversion, and is therefore given as:

$$x \leq 0.05$$
$$-r_{SO_2} = -r_{SO_2}(X = 0.05)$$

Table 7E8.1 Mass Balance for SO_3 Production From Sulfur

	$S + O_2 \rightarrow SO_2$		$SO_2 + (1/2)O_2 \leftarrow \rightarrow SO_3$
	Initial	Final	Initial
SO_2	–	a	a
N_2	0.79	0.79	0.79
O_2	0.21	$0.21 - a$	$0.21 - a$
SO_3	–		

The reaction rate is given by:

$$-r'_{SO_2} = k\sqrt{\frac{P_{SO_2}}{P_{SO_3}}}\left[P_{O_2} - \left(\frac{P_{SO_3}}{K_p P_{SO_2}}\right)^2\right]$$

The equilibrium constant is given as:

$$K_p = 0.0031415 \exp\left(\frac{42.311}{1.987 \cdot (1.8 \cdot T - 273.15) + 491.67} - 11.24\right)$$

$K_p [=] Pa^{-0.5}$, $T [=] K$

For the rate constant (based on Eklund's (1956) results) we have:

$k = 9.86 \cdot 10^{-6}$

$$\exp\left[\frac{-176008}{((1.8 \cdot T - 273.15) + 491.67)} - 110.1\ln(((1.8 \cdot T - 273.15) + 491.67)) + 912.8\right]$$

k is in mol of SO_2/kg cat $S \cdot Pa$, while T is in K.
The stoichiometrics of the reaction are as follows:

$$SO_2 + \frac{1}{2}O_2 + N_2 \leftrightarrow SO_3 + N_2$$

F_{A_o} F_{B_o} F_{C_o}

$F_{A_o}(1 - X)$ $F_{B_o} - \frac{1}{2}F_{A_o}X$ F_{C_o} $F_{A_o}X$

$$\Theta_i = \frac{F_{i_o}}{F_{A_o}}$$

$$F_T = F_{A_o}\left[(1 - x) + \Theta_{O_2} - \frac{1}{2}X + \Theta_{N_2} + \Theta_{SO_3} + X\right] = F_{A_o}(1 + \Theta_{O_2} + \Theta_{N_2} + \Theta_{SO_3}) + F_{A_o}\delta X$$

$$\delta = -\frac{1}{2}$$

$$F_T = F_{T_o} + F_{A_o}\delta X$$

Assuming ideal gases:

$$C_T = \frac{F_T}{v} = \frac{P}{RT}$$

$$C_{T_o} = \frac{F_{T_o}}{v_o} = \frac{P_o}{RT_o}$$

Dividing both:

$$v = v_o \frac{P_o}{P}\frac{T}{T_o}\frac{F_T}{F_{T_o}} = v_o \frac{P_o}{P}\frac{T}{T_o}\frac{F_{T_o} + F_{A_o}\delta X}{F_{T_o}} = v_o \frac{P_o}{P}\frac{T}{T_o}\left(1 + \frac{F_{A_o}}{F_{T_o}}\delta X\right)$$

The concentration of the species i is as follows:

$$C_i = \frac{F_i}{v} = \frac{F_i}{v_o \dfrac{P_o}{P}\dfrac{T}{T_o}\dfrac{F_T}{F_{T_o}}} = C_{T_o}\frac{F_i}{F_T}\frac{P}{P_o}\frac{T_o}{T}$$

$$C_i = C_{T_o}\frac{F_i}{F_{T_o} + F_{A_o}\delta X}\frac{P}{P_o}\frac{T_o}{T} = C_{T_o}\frac{F_{A_o}(\Theta_i + v_i X)}{F_{T_o} + F_{A_o}\delta X}\frac{P}{P_o}\frac{T_o}{T}$$

Dividing both:

$$C_i = C_{T_o}\frac{F_{A_o}}{F_{T_o}}\frac{(\Theta_i + v_i X)}{1 + (F_{A_o}/F_{T_o})\delta X}\frac{P}{P_o}\frac{T_o}{T} = C_{A_o}\frac{(\Theta_i + v_i X)}{1 + \varepsilon X}\frac{P}{P_o}\frac{T_o}{T}$$

$$\varepsilon = y_{A_o}\delta = (F_{A_o}/F_{T_o})\delta$$

Thus, the partial pressure of component i is the following:

$$P_i = C_i RT = C_{A_o}\frac{T_o}{P_o}\frac{(\Theta_i + v_i X)}{1 + \varepsilon X}R \cdot P = P_{A_o}\frac{(\Theta_i + v_i X)}{1 + \varepsilon X}\frac{P}{P_o}$$

$$P_{A_o} = C_{A_o}RT_o$$

Therefore, the kinetics of the reaction become:

$$-r'_{SO_2} = k\sqrt{\frac{P_{SO_2}}{P_{SO_3}}}\left[P_{O_2} - \left(\frac{P_{SO_3}}{K_p \cdot P_{SO_2}}\right)^2\right]$$

$$\frac{dx}{dW} = \frac{-r'_{SO_2}}{F_{A_o}} = \frac{k}{F_{A_o}} \sqrt{\frac{P_{SO_2,o}\left(\frac{1-X}{1+\varepsilon X}\right)\frac{P}{P_o}}{P_{SO_2,o}\left(\frac{0+X}{1+\varepsilon X}\right)\frac{P}{P_o}}}$$

$$\left[P_{SO_2,o}\left(\frac{\Theta_{O_2} - \frac{1}{2}X}{1+\varepsilon X}\right)\frac{P}{P_o} - \left(\frac{P_{SO_2,o}\left(\frac{0+X}{1+\varepsilon X}\right)\frac{P}{P_o}}{K_p P_{SO_2,o}\left(\frac{1-X}{1+\varepsilon X}\right)\frac{P}{P_o}}\right)^2 \right]$$

$$= \frac{k}{F_{A_o}} \sqrt{\frac{1-X}{(X)}} \left[P_{SO_2,o}\left(\frac{\Theta_{O_2} - \frac{1}{2}X}{1+\varepsilon X}\right)\frac{P}{P_o} - \left(\frac{X}{K_p(1-X)}\right)^2 \right]$$

$$\varepsilon = \delta y_{A_o} = -\frac{1}{2}\frac{0.10}{0.1+0.11+0.79} = -0.05$$

$$P_{SO_2,o} = P_T y_{SO_2,o} = 202650 \cdot 0.1 = 20265 \text{ Pa}$$

$$\Theta_{O_2} = \frac{0.11}{0.1} = 1.1$$

$$\Theta_{N_2} = \frac{0.79}{0.1} = 7.9$$

$$F_{Ao} = 51.42 \text{ mol/s}$$

$$\frac{dX}{dW} = \frac{-r'_{SO_2}}{F_{A_o}} = \frac{k}{0.00237} \sqrt{\frac{1-X}{(X)}} \left[20,265\left(\frac{1.1 - \frac{1}{2}X}{1 - 0.05X}\right)\frac{P}{P_o} - \left(\frac{X}{K_p(1-X)}\right)^2 \right]$$

The energy balance is as follows:

$$\dot{Q} - W_s - F_{A_o}\sum_{i=1}^{n}\int_{T_{io}}^{T}\Theta_i c_{p,i} dT - \left[\Delta H_R(T_R) + \int_{T_R}^{T}\Delta c_p dT\right]F_{A_o}X = 0$$

$$\frac{d\dot{Q}}{dV} - F_{A_o}\left(\sum_{i=1}^{n}\Theta_i c_{p,i} + X\Delta c_p\right)\frac{dT}{dV} - \left[\Delta H_R(T_R) + \int_{T_R}^{T}\Delta c_p dT\right]F_{A_o}\frac{dX}{dV} = 0$$

$$-r_A = F_{A_o}\frac{dX}{dV}$$

Assuming adiabatic operation of the bed:

$$\frac{dT}{dW} = \frac{(-r_A)\left[\Delta H_R(T_R) + \int_{T_R}^{T}\Delta c_p dT\right]}{F_{A_o}\left(\sum_{i=1}^{n}\Theta_i c_{p,i} + X\Delta c_p\right)}$$

Thus the heat of reaction is computed as:

$$\Delta H_R(298K) = -98480 \, J/mol \, SO_2$$
$$c_{p,SO_2} = 23.852 + 66.989 \cdot 10^{-3}T - 4.961 \cdot 10^{-5}T^2 + 13.281 \cdot 10^{-9}T^3$$
$$c_{p,O_2} = 28.106 - 3.680 \cdot 10^{-6}T + 17.459 \cdot 10^{-6}T^2 - 1.065 \cdot 10^{-8}T^3$$
$$c_{p,SO_3} = 16.370 + 14.591 \cdot 10^{-2}T - 1.120 \cdot 10^{-4}T^2 + 32.324 \cdot 10^{-9}T^3$$
$$c_{p,N_2} = 31.150 - 1.357 \cdot 10^{-2}T + 26.796 \cdot 10^{-6}T^2 - 1.168 \cdot 10^{-8}T^3$$

where c_p is in J/mol K, and temperature (T) is in K (Sinnot, 1999):

$$\Delta H_R = \Delta H_R(T_R) + \int_{T_R}^{T} \Delta c_p dT$$

$$= \Delta H_R(T_R) + \Delta\alpha(T - T_R) + \frac{\Delta\beta}{2}(T^2 - T_R^2) + \frac{\Delta\gamma}{3}(T^3 - T_R^3) + \frac{\Delta\xi}{4}(T^4 - T_R^4)$$

$$\Delta\alpha = \alpha_{SO_3} - \frac{1}{2}\alpha_{O_2} - \alpha_{SO_2} = 16.370 - 0.5(28.106) - 23.852 = -21.535$$

$$\Delta\beta = 0.07892$$
$$\Delta\gamma = -7112 \cdot 10^{-5}$$
$$\Delta\xi = 24,467 \cdot 10^{-8}$$
$$\Delta H_R = -98,480 - 21.535(T - 298) + 0.0395(T^2 - 298^2) - 2.371 \cdot 10^{-5}(T^3 - 298^3)$$
$$+ 6.11675 \cdot 10^{-9}(T^4 - 298^4)$$
$$\Delta c_p = \Delta\alpha + \Delta\beta T + \Delta\gamma T^2 = -21.535 + 0.0789T - 7.112 \cdot 10^{-5}T^2 + 2.447 \cdot 10^{-8}T^3$$
$$\sum \Theta_i c_{p,i} = 300.85 - 0.0402T + 0.00018T^2 - 9.071 \cdot 10^{-8}T^3$$

The pressure drop along the tube is given by the Ergun equation:

$$\frac{dP}{dz} = -\frac{G(1-\phi)}{\rho D_p \phi^3}\left[\frac{150(1-\phi)\mu}{D_p} + 1.752G\right]$$

G is the superficial mass velocity (kg/m² s), ϕ is the porosity of the bed, D_p is the particle diameter (m), μ is the viscosity of the gas (Pa·s), and ρ is the gas density (kg/m³) that can be found in Fogler (1997):

$$\frac{dP}{dz} = -\frac{G(1-\phi)}{\rho_o g_c D_p \phi^3}\left[\frac{150(1-\phi)\mu}{D_p} + 1.752G\right]\frac{P_o}{P}\frac{T}{T_o}\frac{F_T}{F_{T_o}}$$

$$G = \frac{\sum_i F_{io}M_i}{A_c} = 0.433 \, \frac{kg}{m^2 s}$$

$$\rho_o = 0.866 \, kg/m^3$$
$$\mu = 3.72 \cdot 10^{-5} Pa \cdot s$$
$$A_c = \pi D^2 4$$

$$W = \rho_b A_c z \quad \frac{dP}{dW} = -\frac{G(1-\phi)(1+\varepsilon X)}{\rho_b A_c \rho_o g_c D_p \phi^3}\left[\frac{150(1-\phi)\mu}{D_p} + 1.752G\right]\frac{P_o}{P}\frac{T}{T_o}$$

$$\mu \approx cte$$

To solve the problem we use MATLAB. For further notes on the use of MATLAB for solving chemical reactors we refer the reader to Martín (2014).

M file: ReactorSO$_2$

```
[a,b]=ode15s('ReacSI',[0 5000],[0,750,202650]);

plot(a,b(:,1))
xlabel('W (kg)')
ylabel('X')
figure
plot(a,b(:,2))
xlabel('W (kg)')
ylabel('T (K)')
figure
plot(a,b(:,3))
xlabel('W (kg)')
ylabel('P (Pa)')
```

M file: Reac

```
function Reactor=ReacSI(w,x)

X=x(1);
T=x(2);
Presion=x(3);

k=9.8692e-3*exp(-176008/(1.8*(T-273.15)+491.67)
   -110.1*log((1.8*(T-273.15)+491.67))+912.8);

Kp=0.0031415*exp(42311/(1.987*(1.8*(T-273.15)+491.67))-11.24);
Pto=202650;
ySO2o=0.11;
yO2o=0.1;
yN2o=0.79;
PhiO2=yO2o/ySO2o;
PSO2o=Pto*ySO2o;
Fto=467.1292;
Fao=Fto*ySO2o;
epsilon=-0.055;
G=0.433;
ph=0.45;
rhoo=0.866;
Dp=0.00457;
visc=3.72e-5;
Ac=3.14*(6.650/2)^2;
rhob=542;
To=750;
```

```
sumCp = 272.77 - 0.0303*T + 0.000158*T^2 - 8.016e - 8*T^3;
dCp = - 21.535 + 0.0789*T - 7.112*10^( - 5)*T^2 + 2.447e - 8*T^3;
deltaHr = - 98480 - 21.535*(T - 298) + 0.0395*(T^2 - 298^2) -
  2.371*10^( - 5)*(T^3 - 298^3) + 6.117*10^( - 9)*(T^4 - 298^4);

if X < 0.05;
    ra = - k*((1 - 0.05)/0.05)^(0.5)*(PSO2o*((PhiO2 - 0.5*X)/
    (1 + epsilon*0.05))*(Presion/Pto) - (0.05/(Kp*(1 - 0.05))))^2);
    else
    ra = - k*((1-X)/X)^(0.5)*(PSO2o*((PhiO2 - 0.5*X)/
    (1 + epsilon*X))*(Presion/Pto) - (X/(Kp*(1 - X))))^2);
end

Reactor(1,1) = - ra/Fao;
Reactor(2,1) = ((-ra)*(-deltaHr))/(Fao*(sumCp + X*dCp));
Reactor(3,1) = - G*(1-ph)*(1+epsilon*X)*Pto*T*(150*(1-ph)*
visc/Dp + 1.75*G)/(rhob*Ac*rhoo*Dp*ph^3*To*Presion);
```

The solution is summarized in the two figures below for the conversion and the temperature profiles. We need 1600 kg of catalyst. We see that the temperature increases 120K since it operates adiabatically (Fig. 7E8.1).

Based on the geometry of the converter and the density, we need:

$$A_c = \pi D^2 4$$
$$W = \rho_b A_c z$$
$$z = 0.09 \text{ m}$$

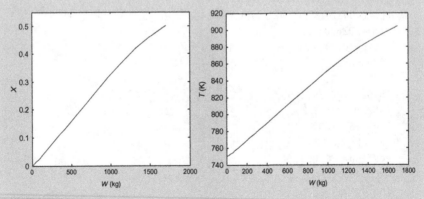

FIGURE 7E8.1

Conversion and temperature profiles as a function of the catalyst load.

7.3.3.3.3 SO₃ absorption

The SO_3 produced is sent to absorption columns where it is put into contact with water. Typically, the water remains in a solution of sulfuric acid. It is a chemical absorption:

$$SO_3 + H_2O \rightarrow H_2SO_4$$

The process is highly exothermic, and the formation enthalpy of sulfuric acid is as follows:

$$SO_2(g) + \frac{1}{2}O_2(g) + H_2O(l) \Leftrightarrow H_2SO_4(l) \quad \Delta H_{f,25\underline{o}C} = -54,633 \text{ kcal}$$

Apart from the formation itself, if the sulfuric acid is diluted, solution enthalpies must be considered too. In this section we design one such a column.

EXAMPLE 7.9

Compute the length of an absorption tower for the production of H_2SO_4 so that we absorb 95% of the incoming SO_3. Assume isothermal operation (Fig. 7E9.1).

Solution

The kinetics are as follows:

$$-r_A = k_2[SO_3][H_2O] = k_2[C_{SO_3} - C_{SO_3,0}X][C_{H_2O} - C_{SO_3,0}X]$$

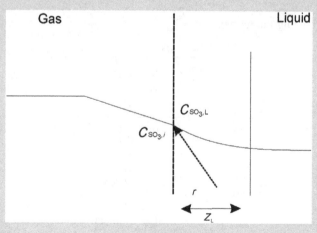

FIGURE 7E9.1

Gas—liquid contact in the absorption column.

However, the reaction rate is limited by diffusion:

$$-r_A = \frac{rD_L}{Z_L}(C_{SO_3,i} - C_{SO_3,L})$$

where r is the ratio between the film thickness for absorption and that for chemical reaction. According to Krevelen and Hoftyzer (Goodhead and Abowei, 2014), r is defined as:

$$r = \frac{(k_2 D_L C_{H_2O,L})^{1/2}}{K_L}$$

where K_L is the mass transfer coefficient, and D_L is the SO_3 diffusivity. Therefore, the reaction rate becomes:

$$-r_A = k_2^{1/2} \cdot D_L^{1/2} C_{SO_3} \cdot C_{H_2O,L}^{1/2}$$

Based on basic reaction kinetics of a bimolecular reaction, the converted mass is computed as $C_{SO_3}X$. Thus, the equation becomes:

$$-r_A = k_2^{1/2} \cdot D_L^{1/2} C_{SO_3,o}^{3/2} \left(\frac{C_{H_2O,o}}{C_{SO_3,o}} - X\right)^{1/2} (1-X)$$

Assuming that the tower behaves as a plug flow reactor, the design equation becomes:

$$-r_A = \frac{-dF_A}{dV} = F_{A_o}\frac{dX}{dV} = k_2^{1/2} \cdot D_L^{1/2} C_{SO_3,o}^{3/2} \left(\frac{C_{H_2O,o}}{C_{SO_3,o}} - X\right)^{1/2} (1-X)$$

Rearranging the terms:

$$F_{A_o}\frac{dX}{k_2^{1/2} \cdot D_L^{1/2} C_{SO_3,o}^{3/2} \left(\frac{C_{H_2O,o}}{C_{SO_3,o}} - X\right)^{1/2} (1-X)} = dV$$

$$\frac{F_{A_o}}{k_2^{1/2} \cdot D_L^{1/2} C_{SO_3,o}^{3/2}} \int \frac{dX}{\left(\frac{C_{H_2O,o}}{C_{SO_3,o}} - X\right)^{1/2} (1-X)} = V$$

$$\frac{F_{A_o}}{k_2^{1/2} \cdot D_L^{1/2} C_{SO_3,o}^{3/2}} \left[\frac{2(m-X)^{1/2}}{1-X}\right] = V = \frac{\pi D_R^2 L_R}{4}$$

$$\frac{4F_{Ao}}{\pi D_R^2 \cdot k_2^{1/2} \cdot D_L^{1/2} C_{SO_3,o}^{3/2}} \left[\frac{2(m-X)^{1/2}}{1-X}\right] = L_R$$

Table 7E9.1 gives information on the parameters needed for the tower design.

Table 7E9.1 Data for the Column Operation

$C_{SO_3,o}$	15	mol/m^3
k_2	0.3	1/s
$F_{SO_3,o}$	4	mol/s
D_R	0.1	m
m	1	
D_L	17	m^2/s
V_o	2.3×10^{-4}	m^3/s

Table 7.4 Coefficients for Pressure Drop Estimation
(Cameron and Chang, 2010)

Packing Type	Size (in)	C$_2$	C$_3$	C$_4$
Standard saddle	1	0.69	22.33	1.6
Standard saddle	1.5	0.38	22.33	1.6
Standard saddle	2	0.21	22.33	1.6
Standard saddle	3	0.15	22.33	1.6

It is assumed that the tower operates isothermally. Thus, the energy to be removed is computed as:

$$Q = (-\Delta H_r)F_{SO_3,o}X$$

Solving the problem, we have $L = 34.7$ m.

Apart from the kinetics, the flow through a tower has a pressure drop. The pressure drop across the absorption tower is given by the Ergun equation, and depends on the type of packing. An empirical correlation was developed by Cameron and Chang (2010):

$$\frac{\Delta P_B}{L} = C_2 \left(\frac{1-\varepsilon}{\varepsilon} \right) F_s^2 \exp[C_3 V_L] \tag{7.5}$$

The pressure drop given by Eq. (7.5) must be corrected to include that of the loading region so that:

$$\frac{\Delta P}{L} = \frac{\Delta P_B}{L} \cdot \exp\left[C_4 \left(\frac{\Delta P_B}{L} \right)^2 \right] \tag{7.6}$$

where C_2, C_3, and C_4 are adjustable parameters that can be seen in Table 7.4. V_L is the irrigation rate (ft/s) based on an empty tower. F_s is the ratio between the gas velocity across an empty tower and the density as $V/\rho^{0.5}$, and ε is the porosity of the tower, typically around 0.75.

The diameter is computed (as in typical absorption towers) using the loading line for determining the K parameter, and the following set of equations (Sinnot, 1999):

$$D = \sqrt{\frac{4A}{\pi}}$$

$$A = \frac{\text{Mass}}{G}$$

$$G = \%_{\text{Loading}} \sqrt{\frac{K\rho_G(\rho_G - \rho_L)}{13.1 \cdot F_p(m^{-1})\left(\frac{\mu_L}{\rho_L}\right)^{0.1}}} \qquad (7.7)$$

$$\%_{\text{Loading}} = 100\left(\frac{K}{K_{\text{floading}}}\right)^{0.5}$$

7.3.3.3.4 Mixing tanks: heat of solution

When mixing acids or acid solutions there is a certain energy involved. In Fig. 7.11 the integral heat of solution for sulfuric acid is presented at 25°C. The mixture is exothermic and can result in the evaporation of water. Therefore, apart from the health and safety issues related to acid solution mixing, the concentration of the resulting mixture is to be carefully computed to account for that (Houghen et al., 1959). Example 7.10 presents such a case. We compute the energy involved in the mixing using enthalpy concentration diagrams.

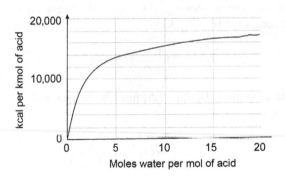

FIGURE 7.11

Integral heat of dilution of sulfuric acid at 25°C.

EXAMPLE 7.10

Mixtures of water and sulfuric acid solutions. A flow rate of 1 kg/s of H_2SO_4 at 70°C is mixed with a solution of 1 kg/s of a solution of 15% H_2SO_4 at 85°C in a tank at 1 atm. Determine the composition of the final mixture.

Solution.

We can determine the final composition using the integral heat of dilution, but it is easier to use an enthalpy diagram (see below). We locate the points for both solutions in the diagram. The lever rule can be used as a rough estimate to identify whether there will be water evaporation or not. We see that for a 1:1 mixture the product will be above the boiling point curve. Therefore, the energy generated is enough to boil the water.

Performing a mass balance we compute the point if no water is evaporated:

$$F_1 \cdot x_1 + F_2 \cdot x_2 = F_m \cdot x_m \Rightarrow x_m = \frac{1}{F_m}(F_1 \cdot x_1 + F_2 \cdot x_2) = \frac{1}{2}(1 \cdot 1 + 1 \cdot 0.15) = 0.575$$

Using Fig. 7E10.1, the composition of the final liquid mixture is determined by projecting the point into the boiling point curve (as presented in the figure). Thus, the composition of the liquid mixture is 60% sulfuric acid, and the rest of the water has evaporated.

Now we compute the composition by performing mass and energy balances:

$$F_1, \ 70°C \rightarrow 158°F, \quad 40\,\text{BTU/lb} = 22.22\,\text{kcal/kg}$$
$$F_2, \ 85°C \rightarrow 185°F, \quad 95\,\text{BTU/lb} = 52.78\,\text{kcal/kg}$$

If the mixture reaches the boiling point and the water evaporates, it is concentrated. For a mixture of 57.5%, its enthalpy becomes:

$$H_m = (57.50\% \text{ and } t_{eb} = 138°C = 280°F) = 0$$

The energy balance states that the flow enthalpy of the feed streams is as follows:

$$F_1 \cdot H_1 + F_2 \cdot H_2 = 75 > H_m$$

Then, the mixture evaporates water. Therefore, the mass and energy balances must be rewritten to account for water evaporation:

Mass balance
$$F_1 \cdot x_1 + F_2 \cdot x_2 = (F_T - y) \cdot x_m$$

Energy balance:
$$F_1 \cdot H_1 + F_2 \cdot H_2 = (F_T - y) \cdot H_m + y H_{vap}$$

$$H_{vap} = H_{vap,\text{sat 1 atm}} + c_p \Delta T$$

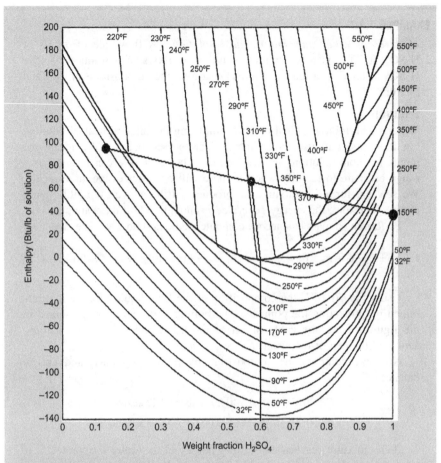

FIGURE 7E10.1

Mix of sulfuric acid streams.

To solve the system, we assume y, compute x_m, and see if the energy balance holds. Alternatively, since we already have an estimate for the concentration of the mixture from the diagram, we can use that as x_m and compute the water that evaporates:

$$x_m = 0.6$$

The boiling point is 290°F (ie, 143°C).

Thus the enthalpy of the overheated steam is as follows:

$$H_{vap} = H_{vap,100°C} + c_p \Delta T = 638.9 + 0.46 \cdot (143 - 100) = 658.7 \text{ kcal/kg}$$

$$y(\text{Mass balance}) = F_T - (F_1 \cdot x_1 + F_2 \cdot x_2)/x_m = 2 - (1 \cdot 1 + 1 \cdot 0.15)/0.6 = 0.083$$

$$y(\text{energy balance}) = \frac{F_1 \cdot H_1 + F_2 \cdot H_2 - F_T H_m}{(H_{vap} - H_m)} = \frac{75 - 2 \cdot 0}{658.7 - 0} = 0.11$$

The small difference is due to reading the enthalpy values in the figure. Thus approximately 0.1 is the fraction evaporated.

7.4 PROBLEMS

P7.1. A mixture of SO_2, O_2, and N_2 from sulfur burning with dry air is fed to a converter. The mixture contains 8% per volume of SO_2. The gas is fed to the converter at 415°C and 1 atm. The converter oxidizes SO_2 to SO_3 *using two beds* with intercooling. Assume that the equilibrium conversion is reached at each of the beds (which operate adiabatically). Select between both equilibrium curves below (Fig. P7.1A and B). Determine the composition of the gases exiting each bed and the cooling temperature after the second. Compute the temperature at which the gases from the first bed must be cooled down to so that the final conversion is 95%.

$$SO_2 + \frac{1}{2}O_2 \Leftrightarrow SO_3$$

$$K_p = \frac{P_{SO_3}}{P_{SO_2} \cdot P_{O_2}^{0.5}}; \quad \log_{10}(K_p) = \frac{4956}{T(K)} - 4.678$$

$$\Delta H_r = 23,200 \ \frac{\text{kcal}}{\text{kmol}}; \quad \bar{c}_p = 7.9 \ \frac{\text{kcal}}{\text{kmol°C}}$$

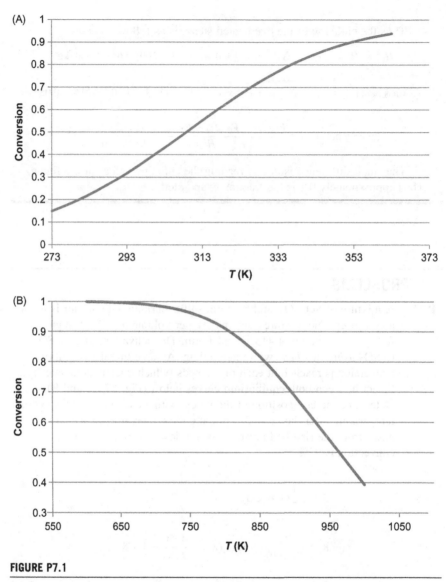

FIGURE P7.1

(A) Conversion versus temperature diagram A. (B) Conversion versus temperature
diagram B.

P7.2.   A mixture of SO_2, O_2, and N_2 from sulfur burning with dry air is fed to
a converter. The converter oxidizes SO_2 to SO_3 *using two beds* with
intercooling. The gas is fed to the converter at 415°C and 1 atm. The
conversion obtained is 95% and the gases exit at 767K. The gases are

fed to the second bed at 450°C. Assume that the equilibrium conversion is achieved at each bed and that both operate adiabatically. Plot the equilibrium curve under the operating conditions, and compute the conversion reached after the first bed and the composition of the gases entering and exiting the converter.

$$SO_2 + \frac{1}{2}O_2 \Leftrightarrow SO_3$$

$$K_p = \frac{P_{SO_3}}{P_{SO_2} \cdot P_{O_2}^{0.5}}$$

$$\log_{10}(K_p) = \frac{4956}{T(K)} - 4.678$$

$$\Delta H_r = 23,200 \ \frac{kcal}{kmol}$$

$$\bar{c}_p = 7.9 \ \frac{kcal}{kmol°C}$$

P7.3. A mixture of SO_2, O_2, and N_2 from sulfur burning using dry air is fed to a converter. The mixture contains 10% per volume of SO_2. The gas is fed to the converter at 415°C and 1 atm. The converter oxidizes SO_2 to SO_3 *using two beds* with intercooling. Assume that the equilibrium conversion is reached at each one of the beds (which operate adiabatically). After the first bed, the stream is sent to an absorption tower where all the SO_3 is removed. The gases are recycled to the reactor and fed to the second bed at 450°C. Determine the equilibrium curve for each bed and the conversion after the first bed as well as the global conversion of the converter.

$$SO_2 + \frac{1}{2}O_2 \Leftrightarrow SO_3$$

$$K_p = \frac{P_{SO_3}}{P_{SO_2} \cdot P_{O_2}^{0.5}} \quad \text{where} \quad \log_{10}(K_p) = \frac{4956}{T(K)} - 4.678$$

$$\Delta H_r = 23,200 \ \frac{kcal}{kmol}$$

$$\bar{c}_p = 7.9 \ \frac{kcal}{kmol°C}$$

P7.4. Sulfur dioxide for the production of sulfuric acid is produced from pyrite roasting. The furnace is fed with pyrite (FeS_2) that contains moisture and inert slag. The facilities manager believes that the composition in the contract is not correct based on the products obtained. Air is fed at 25°C and 700 mmHg with a relative humidity of 40%. The gas product has a dew point of 12.5°C and the furnace is operated with twice the stoichiometric oxygen. The slag collected after 10 min contains 0.36 t of FeS, 36 t of Fe_2O_3, and 6.6 t of inert slag. Determine the composition of the mineral that was sold to the plant and the heat losses, assuming $c_{p,slag} = 0.16$ kcal/kg K and the pyrite is fed to the furnace at 25°C.

P7.5. The gases from pyrite burning are fed to a Glover Tower where a fraction of the SO_2 is transformed into H_2SO_4. Together with the gas, a stream consisting of 592 kg (77% H_2SO_4, 22% H_2O, and 1% N_2O_3) comes from a Gay-Lussac Tower, and another one from lead chambers (178 kg of H_2SO_4, 64%); we feed 1.32 kg of HNO_3 as catalyst makeup. From the tower we obtain 770 kg of acid (78%); see Fig. P7.5. Determine the conversion of SO_2 and the composition of the product gas.

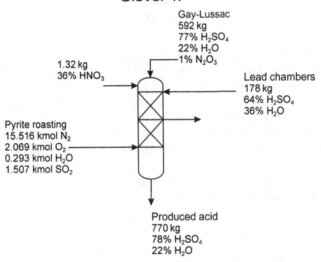

FIGURE P7.5

Scheme of Glover Tower.

P7.6. Determine the conversion reached in a 25 m tall tower for the absorption of SO_3 to produce H_2SO_4. Assume that the kinetic rate is given by the following equation:

$$-r_A = k_2^{1/2} \cdot D_L^{1/2} C_{SO_3} \cdot C_{H_2O,L}^{1/2}$$

P7.7. The feed to the chambers (at 91°C) is given in Table P7.7. Water is fed at 25°C. The unconverted gases have a dew point of 2°C at 1 atm. These gases are fed to the Gay-Lussac Tower at 40°C. The liquid product, 178 kg of

Table P7.7 Data on Gas Composition

	kmol	kg
N_2	15.5612727	435.715636
O_2	1.95656918	62.6102137
H_2O	1.41367395	25.4461312
SO_2	1.1921243	76.2959555
NO	0.16332356	4.89970694
Total	20.2869637	604.967644
Temp (K)	364	

sulfuric acid (64% by weight), exits the chambers at 91°C. Compute the conversion of SO_2 in the lead chambers, the water added, and an energy balance to the lead chambers. See Fig. P7.7 for the process scheme.

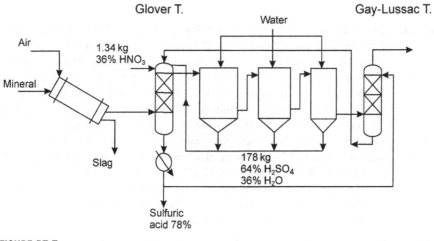

FIGURE P7.7

Scheme of lead chamber process.

P7.8. A stream of 467.1292 mol/s at 2 atm and 750K is fed to a catalytic multibed converter. The gas comes from burning sulfur, and 51.42 mol/s of the stream is SO_2. Determine the equilibrium conversion and the final temperature after the first catalytic bed. The converter has a diameter of 6.650 m; the catalyst features are given below:

$$\phi = 0.45$$
$$D_p = 0.00457 \text{ m}$$
$$A_c = 34.715 \text{ m}^2$$
$$\rho_b = 542 \text{ kg}/m^3$$

P7.9. Compute the amount of sulfuric acid 98% required to absorb 99.9% of the SO_3 in the gases from the third catalytic bed (see Table P7.9), and the acid temperature. The product has a composition of 99%, the gases are at 200°C, and the acid is fed at 90°C. The unabsorbed gases exit at 90°C and we remove 6500 kcal/s using cooling.

Table P7.9 Gas Composition Exiting 3rd Bed

	3rd Bed (kmol/s)
SO_2	0.01071494
N_2	0.79
O_2	0.06535747
SO_3	0.08928506
	0.95535747

P7.10. Evaluate a sulfuric acid production facility that uses the contact method with one absorber. Humid air at 25°C and relative humidity of 0.55 is dried using sulfuric acid 99%, the product. The acid gets diluted to 98% and is used to absorb SO_3 produced at the converter. Assume that the molar ratio of dry air to sulfur is 10:1. Sulfur is burned with the dry air to produce SO_2. The gas mixture is sent to the converter, which achieves 98% conversion to SO_3. This stream is sent to the absorber. In the absorber, water, sulfuric acid 98%, and the gas are put into contact so that sulfuric acid 99% is produced. Compute the amount of sulfuric acid produced (99%), the flowrate of H_2SO_4 98%, the flow rate of water fed to the absorption tower, and the fraction of sulfuric acid used to dry the initial air. Determine the temperature of operation at the converter assuming isothermal operation at 1 atm.

P7.11. Evaluate a sulfuric acid production facility that uses the contact method with one absorber. Humid air at 25°C and relative humidity of 0.55 is dried using sulfuric acid 99%, the product. The acid gets diluted to 98% and is used to absorb SO_3 produced at the converter. Assume that the molar ratio of dry air to sulfur is 10:1. Sulfur is burned with the dry air producing SO_2. The gas mixture is sent to the converter, which operates isothermally at 1.5 atm and 700K. This stream is sent to the absorber. In the absorber, water, sulfuric acid 98%, and the gas are put into contact so that sulfuric acid 99% is produced. Compute the amount of sulfuric acid produced (99%), the flowrate of H_2SO_4 98%, the flow rate of water fed to the absorption tower, and the fraction of sulfuric acid used to dry the initial air.

P7.12. To produce sulfuric acid using the lead chamber process, SO_2 is produced by burning sulfur. Sulfur (100 kg/h) with 10% inert slag are fed to the furnace. The sulfur is fed at 25°C. Humid air—100% excess with respect to the stoichiometric one—is fed at 25°C and 700 mmHg. The gases leaving the furnace have a dew point of 12.5°C and are at 450°C; the slag is collected at 400°C ($c_p = 0.16$ kcal/kg). Determine the composition of the gases if all the sulfur is burned to SO_2, the air moisture, and the heat loss of the burner.

P7.13. The gas leaving the furnace from P7.12 is fed into the chamber system. The aim is to produce sulfuric acid 64%. 0.005 mol of HNO_3 per mol of SO_2 are also fed in the form of an aqueous solution 40% w. Determine the gas composition and the water fed for a conversion of SO_2 to sulfuric acid of 95%. The gas product has a dew point of 10°C at 760 mmHg.

P7.14. Using a process simulator (ie, CHEMCAD, ASPEN), model a four bed reactor for the oxidation of SO_2 to SO_3 assuming that equilibrium is reached after each bed. The feed consists of 10% SO_2, 11% O_2, and 79% N_2 at 750K and 1 atm. Assume that equilibrium is reached at each bed:

   a. Determine the cooling needs after each bed and the conversion after each bed to reach at least 98% conversion after the four beds.
   b. Explain what happens if after the third bed, SO_3 is removed from the system.

REFERENCES

Cameron, G.M., Chang, I., 2010. Sizing of packed towers in acid plants. The Chemical Engineers' Resource Page. <http://staff.sut.ac.ir/haghighi/download/documents/Sizing_of_Packed_Towers_in_Acid_Plants.pdf>.

Duecker, W.W., West, J.R., 1975. Manufacture of Sulfur Acid. Reinhold, New York, NY.

Eklund, R., 1956. The Rate of Oxidation of Sulfur Dioxide With a Commercial Vanadium Catalyst. Almqvist & Wiksells Boktr, Uppsala Sweden.

Fogler, S., 1997. Elements of Chemical Reaction Engineering, third ed. Prentice Hall, New York, NY.

Goodhead, T.O., Abowei, M.F.N., 2014. Design of isothermal plug flow reactor adsorption tower for sulphur trioxide hydration using vanadium catalyst. Int. J. Innov. Sci. Mod. Eng. (IJISME) 2 (9).

Houghen, O.A., Watson, K.M., Ragatz, R.A., 1959. Chemical Process Principles. vol. 1. Material and Energy Balances. Wiley, New York, NY.

King, M., Moats, M., Davenport, W.G., King, M.J., 2006. Sulfuric Acid Manufacture. Elsevier, Oxford.

Lloyd, L., 2011. Handbook of Industrial Catalysts. Springer, Bath.

Martín, M., 2014. Introduction to Software for Chemical Engineers. CRC Press, Boca Raton, FL.

Muller, T., 1993. Sulfuric acid and sulfur trioxide. Kirk-Othmer Encyclopedia of Chemical Technology. vol. 23. Wiley.

Müller, H., 1998. Sulfuric acid and sulfur trioxide. ULLMANN'S "Encyclopedia of industrial chemistry". Wiley-VCH, Lurgi Metallurgie GmbH, Frankfurt/Main, Federal Republic of Germany.

Ortuño, A.V., 1999. Introducción a la Química Industrial. Reverté, Barcelona.

Outotec, 2012. Outotec® Sulfuric Acid Plants. <www.outotec.com> OTE_Outotec_Sulfuric_Acid_Plants_eng_web.pdf.

Sinnot, R.K., 1999. Chemical Engineering Design. Coulson and Richardson's Chemical Engineering Series. vol. 6. Elsevier, Singapore.

Biomass

8.1 BIOMASS TYPES AND PREPROCESSING

In this chapter we focus on the processing of different types of biomass for the production of chemicals, fuels, and power; paper and penicillin will also be described. It is interesting to note that by stopping the treatment process at an early stage, we can produce natural polymers such as cellulose, hemicellulose, lignin, and rubber. We also evaluate the production of synthetic rubber, an attempt to match the properties of a natural polymer. We start by describing the chemical composition of several common raw materials.

8.1.1 GRAIN

Grain is a source of carbohydrates—the raw material for first-generation bioethanol—as well as protein and fat. Wheat and corn grain are made of starch, a natural polymer that consists of glucose molecules, as seen in Fig. 8.1.

The pretreatment of the grain consists of breaking bonds to obtain sugar monomers of glucose. The process consists of liquefaction, and takes place at 90°C and a pH from 6 to 6.5 for 30 min. The starch is broken into maltose. In a second step, the maltose is further hydrolyzed at 65°C and pH 5.5 for 30 min (ie, saccharification) to obtain glucose. These are the reactions:

$$2(C_6H_{10}O_5)_n + nH_2O \xrightarrow{\alpha\text{-Amylase}} nC_{12}H_{22}O_{11}$$

$$C_{12}H_{22}O_{11} + H_2O \xrightarrow{\text{Glucoamylase}} 2C_6H_{12}O_6$$

8.1.2 LIGNOCELLULOSIC BIOMASS

Although grain is an interesting source of sugars, its use for biofuel production poses an ethical dilemma because it competes with the food supply chain. Thus, second-generation biofuels are based on the use of biomass that does not compete with the food chain, either in terms of final use of the product or acreage. Lignocellulosic raw materials consist of cellulose, hemicellulose, and lignin.

Industrial Chemical Process Analysis and Design. DOI: http://dx.doi.org/10.1016/B978-0-08-101093-8.00022-7

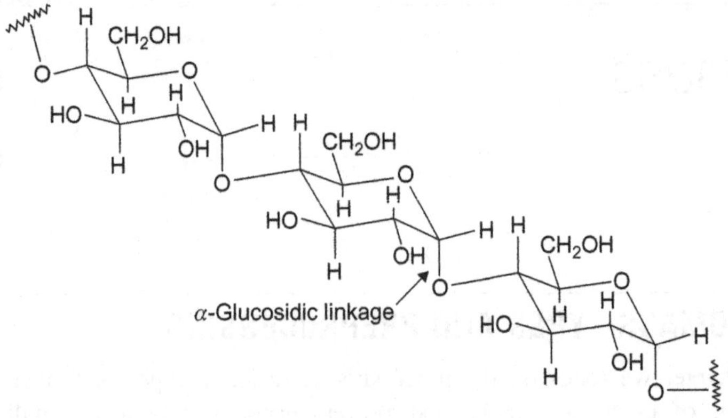

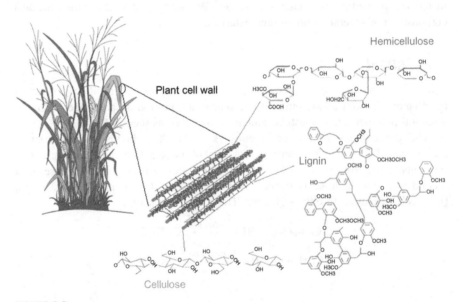

FIGURE 8.1

Starch molecule.

Plant cell wall

Hemicellulose

Lignin

Cellulose

FIGURE 8.2

Lignocellulosic biomass structure.

Lignin provides structure to the plant, hemicelluose is responsible for internal structure, and cellulose is the internal polymer. Lignin is a polymer of aromatic monomers, while hemicellulose and cellulose are made of sugar monomers. Fig. 8.2 shows the structure of a plant and the chemical composition of the main components of the plant wall. A number of uses for lignocellulosic raw materials are highlighted: biooil, syngas, sugar, and paper production.

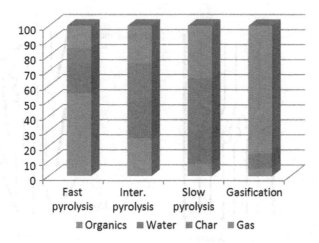

FIGURE 8.3

Relative composition of the products for thermochemical biomass processing.

8.1.2.1 Biooil production

This is a thermochemical path used to partially break the biomass into medium-size chemicals. It operates at medium temperatures, and the composition of the products depends on the rate of the process. The faster the process, the higher the liquid content (see Fig. 8.3). Gas and char are also produced. Thus, fast pyrolysis is recommended for chemicals and fuels production. The main drawback of the process is the wide range of products generated and their high corrosivity. An upgrade to the process is therefore required for proper further use of the biooil. Hydrocracking and catalytic cracking are two methods for upgrading (see Chapter 5: Syngas).

8.1.2.2 Syngas production

Syngas production from biomass is similar to that from any other carbonous substrates. It consists of the partial oxidation of biomass at high temperature (1000°C) to produce a gas phase. The difference lies in the composition of the gas. In addition to carbon monoxide (CO), carbon dioxide (CO_2), hydrogen (H_2), hydrocarbons, tar, and char, we also find hydrogen sulfide (H_2S) and ammonia (NH_3). Thus, raw syngas follows the gas cleaning process. Therefore, we refer the reader to Chapter 5, Syngas for further descriptions of the gasification and gas purification steps.

8.1.2.3 Sugar production

Sugar production from lignocellulosic biomass uses moderate pretreatment processes (below 200°C) to break plant structures down only as far as their monomers. The pretreatments can be classified into physicochemical, chemical, and enzymatic. The aim is to expose the sugar-containing polymers, namely cellulose

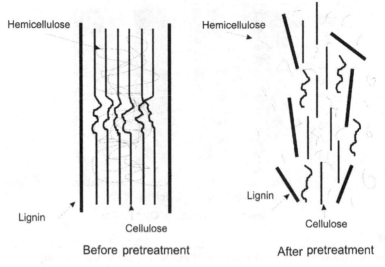

Hemicellulose

Hemicellulose

Lignin

Lignin

Cellulose

Cellulose

Before pretreatment

After pretreatment

FIGURE 8.4

Structure of the plant before and after pretreatment.

With permission from Martín, M., Grossmann, I.E., 2012a. Energy optimization of lignocellulosic bioethanol
production via hydrolysis of switchgrass. AIChE J. 58(5), 1538–1549.

and hemicellulose; this makes them more accessible for hydrolysis to sugars. Fig. 8.4 shows the desired effect of pretreatment: breakage of the physical structure into pieces.

8.1.2.3.1 Physicochemical processes

These processes are based on the use of CO_2, steam, or NH_3 under pressure; when the pressure is released, the biomass structure is broken. Among the different processes available, ammonia fiber explosion (AFEX) is one of the few that has reached the industrial development stage. The AFEX process puts biomass into contact with an NH_3 solution (20 atm, 90–180°C) so that when expanded, the biomass structure breaks into pieces. The method does not produce byproducts, and shows high yield. For years the disadvantage was the energy required to recover the NH_3—until recent developments by Prof. Dale's group at Michigan State University. Fig. 8.5 shows the flowsheet (Martín and Grossmann, 2012a). After the reactor pretreatment, the pressure is released and part of the NH_3 is recovered. The slurry is distilled to recover the rest of the NH_3. Traces of NH_3 remain in the slurry, but they can be used as nutrients in further fermentation stages. Based on the experimental data from Garlock et al. (2012), a correlation that depends on the NH_3-to-biomass ratio, the operating temperature, the water-to-biomass ratio, and the residence time is presented; this can be used to compute the yield. The range of the operating variables is shown in Table 8.1.

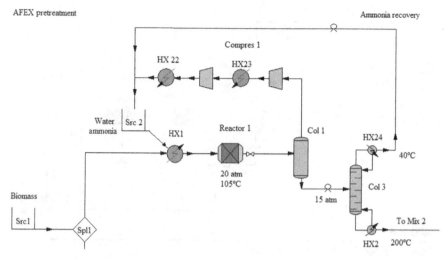

FIGURE 8.5

Basic AFEX pretreatment flowsheet.

With permission from Martín, M., Grossmann, I.E., 2012a. Energy optimization of lignocellulosic bioethanol production via hydrolysis of switchgrass. AIChE J. 58(5), 1538–1549.

Table 8.1 Typical range of Operating Variables for AFEX Pretreatment

	Lower Bound	Upper Bound
Temperature (°C)	90	180
Ammonia added (g/g)	0.5	2
Water added (g/g)	0.5	2
Residence time (min)	5	30

$$X = 0.01 \cdot (-88.7919 + 26.5272 \cdot AR - 13.6733 \cdot \text{water\_added} + 1.6561 \cdot T$$
$$+ 3.6793 \cdot t - 4.4631 \cdot AR^2 - 0.0057 \cdot T^2 + 0.0279 \cdot t^2 - 0.4064 \cdot AR \cdot t \qquad (8.1)$$
$$+ 0.1239 \cdot \text{water\_added} \cdot T - 0.0132 \cdot T \cdot t;$$

8.1.2.3.2 Chemical pretreatment

Chemical pretreatment entails the use of alkali or acid solutions, peroxides, or even ozone to break down plant structure. The National Energy Renewable Lab (NREL) has developed a dilute sulfuric acid pretreatment (Aden and Foust, 2009) in which a solution of 0.5–2% H_2SO_4 is put into contact with biomass at 140–180°C and 12 atm for up to 1.5 h. In the expansion, part of the water is recovered and recycled, while the slurry is separated so that the liquid phase is neutralized with CaO, producing gypsum. After filtering the gypsum, both streams are mixed again for further processing. Fig. 8.6 shows the flowsheet for the process.

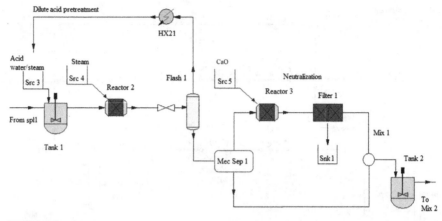

FIGURE 8.6

A basic flowsheet for dilute acid pretreatment.

With permission from Martín, M., Grossmann, I.E., 2012a. Energy optimization of lignocellulosic bioethanol production via hydrolysis of switchgrass. AIChE J. 58(5), 1538–1549.

Table 8.2 Typical range of Operating Variables for Dilute Acid Pretreatment

	Lower Bound	**Upper Bound**
Temperature (°C)	140	180
Acid concentration (g/g)	0.005	0.02
Residence time (min)	1	80
Enzyme load (g/g)	0.0048	0.0966

Based on experimental data, Martín and Grossmann (2014) developed models for the yield to cellulose and hemicellulose as a function of the operation temperature, the acid concentration, the residence time, and the enzyme added in the hydrolysis part, see eqs. (8.2) and (8.3). Table 8.2 shows the range for the operating variables.

$$
\begin{aligned}
X_{\text{cellulose}} = &-0.00055171 + 0.00355819 \cdot T + 0.00067402 \cdot \text{acid\_conc} \\
&+ 0.00100531 \cdot t - \text{enzyme\_load} \cdot 0.0394809 - 0.0186704 \cdot T \cdot \text{acid\_conc} \\
&+ 0.00043556 \cdot T \cdot t + 0.0002265 \cdot T \cdot \text{enzyme\_load} \\
&- 0.0013224 \cdot \text{acid\_conc} \cdot t - 0.00083728 \cdot t \cdot \text{enzyme\_load} \\
&+ 0.044353 \cdot \text{acid\_conc} \cdot \text{enzyme\_load} + 0.000014412 \cdot T^2;
\end{aligned} \tag{8.2}
$$

$$
\begin{aligned}
X_{\text{hemicellulose}} = &-0.00015791 - 0.00056353 \cdot T + 0.000694361 \cdot \text{acid\_conc} \\
&- 0.00014507 \cdot t - \text{enzyme\_load} \cdot 0.01059248 - 0.02142606 \cdot T \cdot \text{acid\_conc} \\
&+ 0.000694055 \cdot T \cdot t + 0.00013559 \cdot T \cdot \text{enzyme\_load} \\
&- 0.00145712 \cdot \text{acid\_conc} \cdot t + 0.04769633 \cdot \text{acid\_conc} \cdot \text{enzyme\_load} \\
&- 0.00138362 \cdot \text{time} \cdot \text{enzyme\_load} + 0.0000059419 \cdot T^2
\end{aligned}
$$

$$\tag{8.3}$$

8.1.2.3.3 Enzymatic pretreatment

These are slow processes and are not yet competitive.

Once the physical structure of the biomass is broken and the polymers, cellulose, and hemicellulose exposed, the next stage is enzymatic hydrolysis, which produces sugars. The hydrolysis is typically an endothermic set of reactions that takes place at 50°C:

$$(C_6H_{10}O_5)_n + nH_2O \rightarrow nC_6H_{12}O_6 \qquad \Delta H = 22.1n \ \text{kJ/mol}$$

$$(C_5H_8O_4)_m + mH_2O \rightarrow mC_5H_{10}O_5 \qquad \Delta H = 79.0m \ \text{kJ/mol}$$

Alternatively, these polymers can be used for the production of paper.

8.1.2.4 Paper production

Paper production consists of six stages. (1) The biomass is ground into chips. (2) The chips are cooked with alkali solutions from 75−80°C to 130−170°C to solubilize the lignin and release the cellulosic polymers. (3) The material is then washed to remove the lignin. (4) Oxygenation is performed in order to minimize the quantity of chemicals required in the bleaching step. (5) The material is bleached to provide a white color. This is the most contaminating stage due to the chemicals necessary to process the raw materials (Cl_2 in particular). Both total chlorine-free and elementary chlorine-free processes are available. (6) Sheets are produced.

8.1.3 SEEDS

First-generation biodiesel processes use high oil content seeds as raw materials. The oil is extracted using crushers. The process consists of a number of stages. The seeds are heated and milled, and then the flakes are slightly heated to improve the oil extraction. The cake obtained can be further processed using solvents or by means of cold pressing using a screw. Sometimes both methods are combined to increase the yield. The oil must later be purified. Flash distillation, filtration, or sedimentation can be used for the oil refining step. The biodiesel yields of different seeds are shown in Table 8.3. Fig. 8.7 shows the process.

8.1.4 ALGAE

Algae are another alternative for the production of biofuels. Due to their high yield to biomass—typically an order of magnitude or two higher than other seeds per area—algae promise higher production. Algae grow by capturing CO_2 (although other carbonous sources can be used), and use sunlight as an energy source. Algae store lipids (up to 70%), starch (up to 50%), protein, and other components. The basic stoichiometry of the algae growing process is as follows:

$$106CO_2 + 16NO_3^- + HPO_4^{2-} + 122H_2O + 18H^+ \xrightarrow{\text{Solar radiation}} (CH_2O)_{106}(NH_3)_{16}H_3PO_4 + 138O_2$$

Table 8.3 Yield of Biodiesel From Different Seeds
(http://www.bioenergy.wa.gov/oilseed.aspx)

Plant	Yield (Seed) (lbs./acre)	Biodiesel (Gal./acre)
Safflower	1500	83
Rice	6600	88
Sunflower	1200	100
Peanut	2800	113
Rapeseed	2000	127
Coconut	3600	287
Oil palm	6251	635

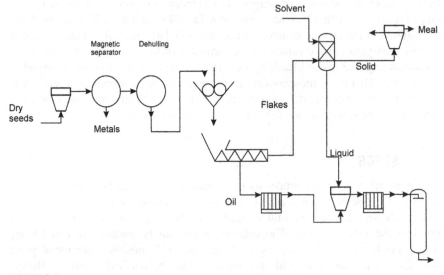

FIGURE 8.7

Process for oil extraction from seeds.

Based on http://www.britannica.com/EBchecked/topic/202405/fat-and-oil-processing/50163/Destearinating-or-winterizing.

Algae are grown in ponds or photobioreactors. Ponds are the cheaper option, and only civil engineering work is required to create an operational system. Water flows through 20 cm-deep channels into which CO_2 and nutrients are injected. However, ponds present several drawbacks: they are easily contaminated and process control is difficult. Photobioreactors, on the other hand, offer a controlled environment. However, solar exposure is difficult and the investment cost is high. Fig. 8.8 shows examples of both systems.

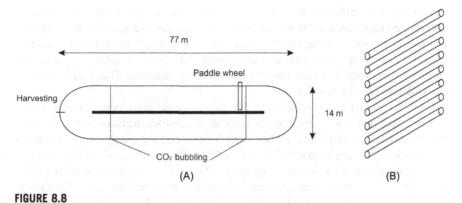

FIGURE 8.8

Algae growth reactors. (A) Ponds. (B) Tubular reactors.

The algae growth rate is a function of the solar energy:

$$\text{Growth} = \frac{I_0 \cdot \eta_{max}}{\text{Algae heating value}} \qquad (8.4)$$

where I_0 (kWh/m^2/d) is the solar intensity, η_{max} is the utilization of the light by the algae (0.045; Walker, 2009), and the algae heating value is 21 kJ/g (Park et al., 2011). Growth rate values of 50 g/m^2 d are optimistic but feasible. Carbon dioxide consumption is related to algae growth as shown in the following equation:

$$CO_2(m^3/day) = 0.6565 \times \text{Growth} (g/m^2 day) + 5.0784 \qquad (8.5)$$

In addition to CO_2, NH_3 (or nitrates) up to 0.8% and phosphates up to 0.6% (both with respect to dry biomass) are needed. Evaporation accounts for water loss of 6.2 m^3/day (Sazdanoff, 2006).

Once the algae are produced, the limiting stage for oil production in terms of cost is the harvesting. This step typically consists of flotation followed by centrifugation or filtration and biomass drying to reach moistures below 10%. Alternatively, there is a capillarity-based process that has been proposed by Univenture; it is supposed to yield a biomass cake with 5% moisture and also consume less energy. The dry biomass is processed and oil extracted, as in the seed-based case. The remaining biomass cake consists mainly of starch and protein. The starch can be used to produce ethanol, similar to first-generation techniques for bioethanol production from corn or wheat (Martín and Grossmann, 2013b).

8.1.5 NATURAL RUBBER

8.1.5.1 Historical perspective

Rubber is native to South America, and was already being used by the natives prior to the arrival of the Spanish in 1492. In 1511 the Spanish realized how useful this material with elastic and waterproof properties could be.

Rubber, also called "India rubber" or caoutchouc, is a polymer of isoprene. It is harvested from trees in the form of latex, an aqueous dispersion of polyisoprene particles stabilized in proteinaceous surfactants. The major commercial source of natural latex is the Pará rubber tree (*Hevea brasiliensis*), which is harvested regularly to obtain it. Latex hardens soon after extraction. This fact motivated research into how to dissolve latex.

In 1770 Joseph Priestley realized that rubber could erase pencil marks. Later, in 1791, the first commercial application to waterproof surfaces was presented. The vulcanization of natural rubber (which cross-links the polymer with sulfur) dates back to Charles Goodyear's work in 1839. This represents the beginning of the use of rubber for manufacturing tires. The lack of supply of natural rubber during WWII led to the development of synthetic polymers to match the natural one. The main producers of natural rubber are Malaysia, India, Indonesia, Thailand, and Sri Lanka, with annual production over 4 million tons.

8.1.5.2 Polymerization processes

Synthetic rubber (*cis*-1,4-polyisoprene) can be obtained via radical chain polymerization of the monomer (isoprene). It was the isolation of this chemical during the pyrolysis of natural rubber that provided the key to identifying the building block. The evolution of the catalysts occurs for 70 years when Ziegler-type catalyst was used (1955). There are four main types of polymerization: (1) block or bulk polymerization, (2) polymerization in solution, (3) bead or pearl polymerization, and (4) emulsion polymerization.

Bulk polymerization occurs in molds with the own monomer and initiators. It produces high purity polymers with high molecular weight. *Polymerization in solution* requires the use of solvents. It can be homogeneous or heterogeneous. *Bead polymerization* is carried out in water by dispersion of the insoluble monomer to produce small droplets of controlled size. This polymerization technique is similar to bulk polymerization, but is performed in microreactors, which allow easier heat removal. It does, however, require the use of colloids to avoid droplet coalescence. Finally, *emulsion polymerization* is carried out by adding the water-insoluble monomer to an aqueous solution of an emulsifier. The product is a water emulsion of polymer particles. Emulsion polymerization is used for synthetic rubber production.

Mechanism and kinetics. Radical polymerization is based on double bond opening and free radical chain growth, as per the reaction below:

It requires the use of initiators (I), Ziegler–Natta catalysts. The proposed reaction mechanism consists of three main steps (initiation, propagation, and termination) and is believed to proceed as follows (Odian, 2004):

Initialization:	$I \xrightarrow{k_d} 2R$
Propagation:	$R + M \xrightarrow{k_i} P_1$
Propagation:	$P_n + M \xrightarrow{k_p} P_{n+1}$
Termination by combination	$P_n + P_m \xrightarrow{k_{tc}} D_{n+m}$
Termination by disproportion	$P_n + P_m \xrightarrow{k_{td}} D_n + D_m$
Transfer to monomer	$P_n + M \xrightarrow{k_t} P_1 + D_n$

The kinetics proceed as follows:

$$\frac{d[I]}{dt} = -f \cdot k_d \cdot [I] \tag{8.6}$$

where f is the initiator efficiency. Assuming a steady-state for the rates and adding together all of the monomer kinetics, we have:

$$\frac{d[M]}{dt} = -k_p (f \cdot k_d [I]/k_t)^{0.5} \cdot [M] \tag{8.7}$$

$$\frac{dQ}{dt} = -\Delta H_r \frac{d[M]}{dt} \tag{8.8}$$

The polymerization lasts 15 h at 50°C in a stirred tank. The conversion yield is 75%. The molecular weight of the product is computed using statistical moments. A complete model for a polymerization reactor can be found in Martín (2014). After the reactor step, a stripper is used to recover hydrocarbons. Finally, the rubber crumb slurry is dewatered and dried in an extruder. The dry crumbs are ready for shipment.

8.2 INTERMEDIATE PROCESSING

8.2.1 SUGARS

This section covers the production of a number of chemicals, mainly for the biofuels industry. We also cover the production of antibiotics such as penicillin via fermentation.

8.2.1.1 Bioethanol

Bioethanol can be produced via sugar fermentation. *Saccharomyces cerevisiae* is capable of anaerobically fermenting glucose at 32–38°C to produce a solution of up to 15% by weight of ethanol. Overpressure is typically used to prevent the

Table 8.4 Reactions and Conversions in Second-Generation Bioethanol Production

Reaction	Conversion
Glucose → 2 Ethanol + 2CO$_2$	Glucose 0.92
Glucose + 1.2NH$_3$ → 6 Z. mobilis + 2.4H$_2$O + 0.3O$_2$	Glucose 0.04
Glucose + 2H$_2$O → Glycerol + O$_2$	Glucose 0.002
Glucose + 2CO$_2$ → 2 Succinic acid + O$_2$	Glucose 0.008
Glucose → 3 Acetic acid	Glucose 0.022
Glucose → 2 Lactic acid	Glucose 0.013
3 Xylose → 5 Ethanol + 5CO$_2$	Xylose 0.8
Xylose + NH$_3$ → 5 Z. mobilis + 2H$_2$O + 0.25O$_2$	Xylose 0.03
3 Xylose + 5H$_2$O → 5 Glycerol + 2.5O$_2$	Xylose 0.02
3 Xylose + 5 CO$_2$ → 5 Succinic acid + 2.5O$_2$	Xylose 0.03
2 Xylose → 5 Acetic acid	Xylose 0.01
3 Xylose → 5 Lactic acid	Xylose 0.01

entrance of air into the system. Fermentation of pentoses is more complex. *Zymomonas mobilis* has been identified as appropriate for fermenting pentoses and hexoses simultaneously. In addition to ethanol, byproducts such as glycerol, succinic acid, acetic acid, and lactic acid are also produced. Table 8.4 shows the various reactions and the typical conversions in second-generation bioethanol processes (Aden and Foust, 2009). See Fig. 2.6 for a block diagram of the process.

The reactions that yield ethanol are exothermic:

$$C_6H_{12}O_6 \xrightarrow{\text{Yeast}} 2C_2H_5OH + 2CO_2 \quad \Delta H = -84.394 \text{ kJ/mol}$$

$$3C_5H_{10}O_5 \xrightarrow{\text{Yeast}} 5C_2H_5OH + 5CO_2 \quad \Delta H = -74.986 \text{ kJ/mol}_{xylose}$$

The reaction time is about 24 h at 0.12 MPa, which helps prevent the entrance of air into the system. The maximum concentration of ethanol in the water is 6−8%.

EXAMPLE 8.1

Model the fermentor for second-generation bioethanol production based on Krishnan et al. (1999).

Solution

The kinetic model for ethanol production via fermentation is based on the Michaelis−Menten mechanism:

$$E + S \underset{k_{-1}}{\overset{k_1}{\rightleftharpoons}} ES \xrightarrow{k_2} E + P$$

In the case of ethanol production, there is both substrate and product inhibition:

$$EI + S \underset{k_{-1}}{\overset{k_1}{\rightleftharpoons}} E + S + I \underset{k_{-1}}{\overset{k_1}{\rightleftharpoons}} ES + I \overset{k_2}{\longrightarrow} E + P + I$$

Assuming a steady-state for [E], [EI], and finally [ES], the modified Monod kinetics are given as follows:

$$\mu = \frac{\mu_m \cdot S}{K_s + S + S^2 / K_i}$$

Next, the product inhibition is included:

$$\frac{\mu}{\mu_0} = \left(1 - \left(\frac{P}{P_m}\right)^\beta\right)$$

The kinetics models for the different species involved are given in the equations that follow, where G represents glucose and X represents xylose. The parameters for the fermentation are given in Table 8E1.1 (Krishnan et al., 1999).

Table 8E1.1 Kinetic Parameters for Fermentation

Parameter	Glucose Fermentation	Xylose Fermentation
μ_m (h^{-1})	0.662	0.190
ν_m (h^{-1})	2.005	0.250
K_s (g/L)	0.565	3.400
K_s' (g/L)	1.342	3.400
K_i (g/L)	283.700	18.100
K_i' (g/L)	4890.000	81.300
P_m (g/L)	95.4 for $P \le 95.4$ g/L	
	129.9 for $95.4 \le P \le 129$ g/L	59.040
P_m' (g/L)	103 for $P \le 103$ g/L	
	136.4 for $103 \le P \le 136.4$ g/L	60.200
β	1.29 for $P \le 95.4$ g/L	
	0.25 for $95.4 \le P \le 129$ g/L	1.036
γ	1.42 for $P \le 95.4$ g/L	0.608
m (h^{-1})	0.097	0.067
$Y_{P/S}$ (g/g)	0.470	0.400
$Y_{X/S}$ (g/g)	0.115	0.162

Cells:

$$\mu_g = \frac{\mu_{m,g} \cdot S}{K_{s,g} + S + S^2/K_{i,g}} \left(1 - \left(\frac{P}{P_m}\right)^{\beta_g}\right)$$

$$\mu_x = \frac{\mu_{m,x} \cdot S}{K_{s,x} + S + S^2/K_{i,x}} \left(1 - \left(\frac{P}{P_m}\right)^{\beta_g}\right)$$

$$\frac{1}{X}\frac{dX}{dt} = \frac{G}{G+X}\mu_g + \frac{X}{G+X}\mu_x$$

Product:

$$\nu_{E,g} = \frac{v_{m,g} \cdot S}{K_{s,g} + S + S^2/K_{i,g}} \left(1 - \left(\frac{P}{P_m}\right)^{\gamma_g}\right)$$

$$\nu_{E,x} = \frac{v_{m,x} \cdot S}{K_{s,x} + S + S^2/K_{i,x}} \left(1 - \left(\frac{P}{P_m}\right)^{\gamma_x}\right)$$

$$\text{with}\quad \begin{aligned} \mu_{m,g} &= 0.152 \cdot X^{-0.461} \\ \mu_{m,x} &= 0.075 \cdot X^{-0.438} \\ v_{m,g} &= 1.887 \cdot X^{-0.434} \\ v_{m,x} &= 0.16 \cdot X^{-0.233} \end{aligned}$$

$$\frac{1}{X}\frac{dP}{dt} = (\nu_{E,x} + \nu_{E,g})$$

Sustrate:

$$-\frac{dS}{dt} = \frac{1}{Y_{X/S}}\frac{dX}{dt} + mX = \frac{1}{Y_{P/S}}\frac{dP}{dt}$$

$$-\frac{dS}{dt} = \frac{1}{Y_{P/S}}\frac{dP}{dt}$$

$$-\frac{dxylo}{dt} = \frac{1}{Y_{P/S}}(\nu_{E,x}X)$$

$$-\frac{dglu}{dt} = \frac{1}{Y_{P/S}}(\nu_{E,g}X)$$

The model for solving the equations can be written in any software. The following is example code for MATLAB:

```
[a,b]=ode15s('Fermen',[0 36],[200,50,0,1]);

plot(a,b(:,1),'k',a,b(:,2),'k--',a,b(:,3),'ko',a,b(:,4),'k.')
xlabel('t(h)')
ylabel('g/L')
legend('Glucose','Xylose','Ethanol','Cells')

function Reactor = Fermen(t,x)
```

```
Glucose = x(1);
xylose = x(2);
ethanol = x(3);
cells = x(4);
v_m_g = 2.005;
v_m_x = 0.250;
K_s_g = 0.565;
K_i_g = 283.7;
K_s_x = 3.4;
K_i_x = 18.1;
K_sp_g = 1.341;
K_ip_g = 4890;
K_sp_x = 3.4;
K_ip_x = 81.3;
m_g = 0.097;
m_x = 0.067;
Y_P_S_g = 0.470;
Y_P_S_x = 0.4;
P_m_x = 59.04;
P_mp_x = 60.2;
Beta_x = 1.036;
gamma_x = 0.608;
if ethanol <95.4;
P_m_g = 95.4;
Beta_g = 1.29;
gamma_g = 1.42;
else
P_m_g = 129;
Beta_g = 0.25;
gamma_g = 0;
end
if ethanol <103;
P_mp_g = 103;
else
P_mp_g = 136;
end

if cells < 5;
mu_m_g = 0.152*(cells)^(-0.461);
v_m_g = 1.887*(cells)^(-0.434);
mu_m_x = 0.075*(cells)^(-0.438);
v_m_x = 0.16*(cells)^(-0.233);
else
```

```
mu_m_g = 0.662;
v_m_g = 2.005;
mu_m_x = 0.190;
v_m_x = 0.25;
end
mu_g = mu_m_g*Glucose*(1-(ethanol/P_m_g)^Beta_g)/
   (K_s_g + Glucose + Glucose^2/K_i_g);
mu_x = mu_m_x*xylose*(1-(ethanol/P_m_x)^Beta_x)/
   (K_s_x + xylose + xylose^2/K_i_x);
v_g = v_m_g*Glucose*(1-(ethanol/P_mp_g)^gamma_g)/
   (K_sp_g + Glucose + Glucose^2/K_ip_g);
v_x = v_m_x*xylose*(1-(ethanol/P_mp_x)^gamma_x)/
   (K_sp_x + Glucose + Glucose^2/K_ip_x);
Reactor(1,1) = -(1/Y_P_S_g)*v_g*cells;
Reactor(2,1) = -(1/Y_P_S_x)*v_x*cells;
Reactor(3,1) = (v_g + v_x)*cells;
Reactor(4,1) = cells*((Glucose)*mu_g/(Glucose + xylose) +
   (xylose)*mu_x/(Glucose + xylose));
```

The profile of the species over time can be seen in (Fig. 8E1.1)

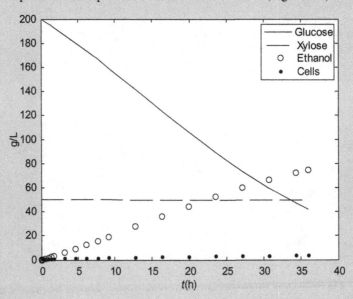

FIGURE 8E1.1

Ethanol production from hexoses and pentoses.

Table 8.5 Conversions of Sugars into FAEE and Byproducts

Reaction	Conversion	DH_r (kJ/mol)
9 Glucose + $2O_2 \rightarrow 2C_{18}H_{36}O_2$ + $18CO_2$ + $18H_2O$	Glucose 0.3	−405
Glucose → 2 Ethanol + $2CO_2$	Glucose 0.3	−84,394
Glucose + $1.2NH_3 \rightarrow 6$ Z. mobilis + $2.4H_2O$ + $0.3O_2$	Glucose 0.1	NA
Glucose + $2H_2O \rightarrow 2$ Glycerol + O_2	Glucose 0.3	504
27 Xylose + $5O_2 \rightarrow 5C_{18}H_{36}O_2$ + $45CO_2$ + $45H_2O$	Xylose 0.2	−338
3 Xylose → 5 Ethanol + $5CO_2$	Xylose 0.2	−74,986
Xylose + $NH_3 \rightarrow 5$ Z. mobilis + $2H_2O$ + $0.25O_2$	Xylose 0.2	NA
3 Xylose + $5H_2O \rightarrow 5$ Glycerol + $2.5O_2$	Xylose 0.2	418

8.2.1.2 Biodiesel

Glucose and xylose have recently been fermented into biodiesel (fatty acid ethyl esters, FAEE). The fermentation takes place aerobically at 32−38°C, with a yield of 0.35 g/g sugar. The concentration of FAEE in the reactor is fixed at 0.01 kg/kg. Biodiesel is immiscible in water, and therefore the separation of the organic and aqueous phases is easy. However, ethanol and glycerol are also produced, and remain in the aqueous phase. This reduces the yield of carbon to FAEE. Table 8.5 shows some typical conversions (Martín and Grossmann, 2015).

8.2.1.3 Ibutene

Sugars can also be a source of valuable chemicals. Ibutene, a C4 chemical, can be produced from glucose fermentation. The advantage of this process is that the product is in the gas phase and thus no energy-intensive dehydration is required. Ibutene must be separated from the CO_2 (Martín and Grossmann, 2014).

$$C_6H_{12}O_6 \xrightarrow{\text{Yeast}} C_4H_8 + 2CO_2 + 2H_2O + 2\text{ ATP} \qquad \Delta H = -41.9 \text{ kJ/mol}$$

8.2.1.4 Furans

Dehydration of C6 sugars yields hydroxymethyl furfural, an intermediate in the production of dimethyl furfural (DMF). C5 sugars can produce furfural as presented in the chemical reactions below:

Xylan   Xylose   Furfural

8.2.1.5 Butanol

Sugars can also be fermented to produce butanol. The typical reaction is the so-called ABE fermentation in which, apart from butanol, both ethanol and acetone are produced. The maximum concentration of butanol is obtained after 48 h of fermentation (Papoutsakis, 1984). The reaction looks like this:

$$100\,\text{Glucose} \rightarrow 2.944\,C_4H_{7.2}N_{0.8}O_2 + 12.5\,\text{Acetic acid} + 4.3\,\text{Butiric acid} + 112.27\,H_2 + 211\,CO_2 \\ + 56\,\text{Butanol} + 22.4\,\text{Acetone} + 9.3\,\text{Ethanol} + 6.3\,\text{Acetoine} + 38.16\,H_2O$$

8.2.1.6 Penicillin

Penicillin was discovered in 1928 when Alexander Fleming, a biologist, realized that one of his experimental setups was contaminated but did not grow any bacteria. He tracked down the culprit—penicillin—and thus the first natural antibiotic was found. The first purification of penicillin is credited to Howard Florey and Ernst Chain by 1939. There are currently around 50 drugs that are classified as penicillins.

Sugars such as glucose are used as substrates to produce penicillin via aerobic fermentation. Oxygen is fed to a stirred tank. The resulting reaction is exothermic and requires constant cooling. The pH, dissolved oxygen, and temperature of the reaction are tightly controlled. It is a fed-batch operation where the substrate is added in small increments. This change in volume must be included in the analysis. Due to the viscosity of the medium and mass transfer limitations, bubble column reactors are also used. The process for the production of the penicillin is modeled on Monod kinetics, as presented in Eq. (8.9). Further details of the model can be found in Birol et al. (2002). The reactor operates at 20–24°C and pH 6.5. After 40 h, penicillin is secreted, and growth stops after about 7 days.

$$\frac{dX}{dt} = \mu X - \frac{X}{V}\frac{dV}{dt}$$

Biomass:

$$\mu = \left[\mu_x \frac{S}{K_x X + S} \frac{C_L}{K_{OX} X + C_L} \right]$$

$$\frac{dP}{dt} = \mu_{pp} X - KP - \frac{X}{V}\frac{dV}{dt}$$

Penicillin:

$$\mu_{pp} = \left[\mu_p \frac{S}{K_p + S + S^2/K_I} \frac{C_L^p}{K_{OP} X + C_L^p} \right] \tag{8.9}$$

Substrate: $\dfrac{dS}{dt} = -\dfrac{\mu}{Y_{X/S}} X - \dfrac{\mu_{pp}}{Y_{P/S}} X - m_x X + \dfrac{F}{V} - \dfrac{S}{V}\dfrac{dV}{dt}$

Dissolved oxygen: $\dfrac{dC}{dt} = -\dfrac{\mu}{Y_{X/O}} X - \dfrac{\mu_{pp}}{Y_{P/O}} X - m_O X + k_L a(C^* - C) - \dfrac{C}{V}\dfrac{dV}{dt}$

Global mass balance: $\dfrac{dV}{dt} = \displaystyle\sum_{in} F - F_{loss}$

The reaction generates energy, and only if this energy is properly removed can the reaction be considered isothermal. The mass transfer between the gas and liquid phases depends on the hydrodynamics and properties of the system. The term $k_L a$ refers to the liquid film resistance, k_L, and the contact area, a, between the phases. The two-film theory considers the resistance in both the stagnant liquid film and the gas—liquid film at interphase:

$$N_A = k_L(c_{LB} - c_{Li}) = k_G(c_{Gi} - c_{GB}) = K(c_{LB} - c_{GB}) \tag{8.10}$$

It has been experimentally proven that in gas—liquid processes the main resistance is usually the resistance of the liquid phase (k_L).

Although k_L and a can be determined separately, the area corresponds to that provided by the bubble dispersion in the tank, and can be estimated using either Eqs. (8.11—8.12) as per Gogate et al. (2000):

$$\varepsilon_G = 0.21 \left[\frac{P_g}{V}(1 - \varepsilon_G) \right]^{0.27} u_G^{0.65} \tag{8.11}$$

$$a = \frac{6 \cdot \varepsilon_G}{d_{eq}} \tag{8.12}$$

or Eq. (8.13) from Calderbank (1958):

$$a = 1.44 \left[\frac{(P_g/V)^{0.4} \cdot \rho^{0.2}}{\sigma^{0.6}} \right] \left(\frac{u_G}{U_B} \right)^{0.5} \tag{8.13}$$

k_L is a function of the diffusivity of the gas to the liquid and the gas-liquid contact time (Higbie, 1935). The complexity of evaluating these two variables (k_L and a) separately has led to the tradition of using the empirical prediction of $k_L a$:

$$k_L a = k \cdot \left(\frac{P_g}{V} \right)^{\alpha} \cdot u_G^{\beta} \tag{8.14}$$

where u_G is the superficial gas velocity in the cross sectional area of the tank and P_g is the aeration power as a function of the gas flow rate, Q_c, the impeller diameter, T, and the liquid volume, V, the stirrer angular velocity, N, and the power in absence of aeration, P, as given by Eq. (8.15):

$$\frac{P_g}{P} = 0.1 \cdot \left(\frac{N^2 T^4}{g \cdot T_i \cdot V^{2/3}} \right)^{-1/5} \cdot \left(\frac{Q_c}{N \cdot V} \right)^{-1/4} \tag{8.15}$$

Biological media are characterized by high viscosities that affect the diffusion of gas into the liquid phase. Thus, Eq. (8.14) is corrected by the viscosity as follows:

$$k_L a = k \cdot \left(\frac{P_g}{V} \right)^{\alpha} \cdot u_G^{\beta} \cdot \left(\frac{\mu}{\mu_{water}} \right)^{\delta} \tag{8.16}$$

Furthermore, in the presence of solids, as in the case of the *Penicillium*, Eq. (8.14) can also be corrected by the fraction of solids, X, as follows:

$$k_L a = k \cdot \left(\frac{P_g}{V} \right)^{\alpha} \cdot u_G^{\beta} \cdot X^{\gamma} \tag{8.17}$$

with $k = 33.59$, $\alpha = -0.0463$, $\beta = 0.94$, and $\gamma = -1.012$ (Kielbus-Rapala and Karcz, 2011).

8.2.2 SYNGAS

Syngas is a versatile raw material (see Chapter 5: Syngas). No further discussion regarding its production, composition adjustment, or cleanup is offered here, only its use within the biofuels industry (Martín and Grossmann, 2013a).

Syngas fermentation is one way to produce ethanol. Syngas with an H_2-to-CO ratio equal to 1 and can be *fermented* anaerobically at 32−38°C. The reaction conversion is around 70%, but ethanol inhibits the process and thus its concentration in water cannot typically exceed 5%.

$$3CO + 3H_2 \rightarrow C_2H_5OH + CO_2$$

Syngas can also be used to produce ethanol and other fuels via *catalytic synthesis*.

Ethanol production requires an H_2-to-CO ratio of 1 for a global conversion of 60%. The process, known as *mixed alcohol synthesis*, is based on the hydrogenation of CO over a catalyst, similar to Fischer–Tropsch (FT) processes. The reaction is carried out at 68 bar and 300°C to obtain a range of alcohols of small molecular weight, including methanol, ethanol, propanol, and also butanol and pentanol in smaller amounts. The reactions with their conversions are as follows:

$$CO + H_2O \rightarrow H_2 + CO_2; \quad X_{CO_2} = 0.219$$
$$CO + 2H_2 \rightarrow CH_3OH; \quad X_{MeOH} = 0.034$$
$$CO + 3H_2 \rightarrow CH_4 + H_2O; \quad X_{CH_4} = 0.003$$
$$2CO + 4H_2 \rightarrow C_2H_5OH + H_2O; \quad X_{EtOH} = 0.282$$
$$2CO + 5H_2 \rightarrow C_2H_6 + 2H_2O; \quad X_{C_2H_6} = 0.003$$
$$3CO + 6H_2 \rightarrow C_3H_7OH + 2H_2O; \quad X_{PropOH} = 0.046$$
$$4CO + 8H_2 \rightarrow C_4H_9OH + 3H_2O; \quad X_{ButOH} = 0.006$$
$$5CO + 10H_2 \rightarrow C_5H_{11}OH + 4H_2O; \quad X_{PentOH} = 0.001$$

Methanol, H_2, and FT fuel production from syngas is detailed in Chapter 5, Syngas.

8.2.3 OIL

Oil obtained from various sources can be transformed into biodiesel via transesterification. The reasons for not using oil itself as fuel are its higher viscosity and its differences from current diesel specifications (which do not allow its proper combustion in engines). Apart from transesterfication, hydrotreating emulsions can also be used to break oil into smaller pieces that can be used in diesel engines. This section focuses on transesterification:

$$
\begin{array}{lll}
CH_2 - OOC - R_1 & R' - OOC - R_1 & CH_2 - OH \\
| & & | \\
CH_2 - OOC - R_2 + 3R'OH \leftrightarrow & R' - OOC - R_2 + & CH - OH \\
| & & | \\
CH_2 - OOC - R_3 & R' - OOC - R_3 + & CH_2 - OH
\end{array}
$$

Although different short-chain alcohols can be used, methanol's low cost and fast reaction kinetics have made it the alcohol of choice for transesterification. Therefore, biodiesel is also known as fatty acid methyl esters (FAME). The low price of FAME is due to its fossil source, and therefore the word "biodiesel" loses its meaning. Lately, ethanol has been investigated as an alternative to methanol. It is currently produced in biorefineries and its products are called fatty acid ethyl esters (FAEE).

The transesterification reaction is governed by an *equilibrium* that can be driven to products by controlling the alcohol-to-oil ratio, the operating pressure (supercritical conditions only) and temperature, the catalyst load, and the residence time. Common catalysts are alkalis and acids, but these suffer from a series of

Table 8.6 Range of Operation of the Variables (Alkali)

Variable	Lower Bound	Upper Bound
Temperature (°C)	45	65
Ratio of methanol (mol/mol)	4.5	7.5
Cat (%)	0.5	1.5

disadvantages. Alkali catalysts, in spite of high conversion, are sensitive to the presence of water and free fatty acids (FFA); therefore, a pretreatment with an acid catalyst is necessary. The yield of this pretreatment using a solid acid catalyst can be computed using Eq. (8.18) (Martín and Grossmann, 2012b) as a function of the methanol-to-FFA ratio, the operating temperature, and the residence time:

$$X_{FFA} = 31.03104 + 1.486403123 \cdot T - 6.97793097 \cdot RM + 19.77691899 \cdot t$$
$$- 0.00018078 \cdot T^2 - 0.16677756 \cdot RM^2 - 1.6230585 \cdot t^2 \qquad (8.18)$$
$$+ 0.02516368 \cdot T \cdot RM - 0.41625815 \cdot T \cdot t + 2.37322062 \cdot RM \cdot t;$$

Once treated, the oil is transformed into FAME using KOH or NaOH as a catalyst. The use of KOH has the advantage that it can later be easily separated from the liquid mixture by precipitation using H_3PO_4 because K_3PO_4 is only slightly soluble in water. Before the neutralization step, the biodiesel must be washed, which increases the water consumption of the facility. A model to predict the yield of the transesterification process as a function of the methanol-to-oil ratio, the temperature of operation, and the catalyst load (Martín and Grossmann, 2012b) is given by Eq. (8.19). Table 8.6 shows the typical range of the operating variables.

$$X = 74.6301 + 0.4209 \cdot T + 15.1582 \cdot Cat + 3.1561 \cdot RM - 0.0019 \cdot T^2 - 0.2022 \cdot T \cdot Cat$$
$$- 0.01925 \cdot T \cdot RM - 4.0143 \cdot Cat^2 - 0.3400 \cdot Cat \cdot RM - 0.1459 \cdot RM^2$$
$$(8.19)$$

Homogeneous acid catalysts such as H_2SO_4 have the advantage of not being sensitive to the presence of FFAs; however, the reaction kinetics are slow. Furthermore, as in the case of KOH or NaOH, a washing step is required before neutralization. Sulfuric acid can be neutralized using CaO. The reaction between CaO and sulfuric acid produces $CaSO_4$ (gypsum), which precipitates and can be easily separated.

In addition to homogeneous catalysis, heterogeneous methods are also being studied as a way to simplify the biodiesel separation stages. The catalyst can easily be recovered after the reaction. A model to compute the yield of heterogeneous catalyst using methanol as the alcohol can be seen in Eq. (8.20) as a function of the methanol-to-oil ratio, the temperature of operation, and the catalyst load (Martín and Grossmann, 2012b). Table 8.7 shows the range of the operating variables for this case.

$$X = -73.6 + 2.5 \cdot T + 24.9 \cdot Cat + 8.8 \cdot RM - 0.01 \cdot T^2 - 1.29 \cdot Cat^2 - 0.39 \cdot RM^2 - 0.26 \cdot T \cdot Cat$$
$$(8.20)$$

Table 8.7 Range of Operation of the Variables (Heterogeneous Catalysis)

Variable	Lower Bound	Upper Bound
Temperature (°C)	40	60
Ratio of methanol (mol/mol)	6	12
Cat (%)	1	4

Table 8.8 Range of Operation of the Variables (Alkali Pretreatment)

Variable	Lower Bound	Upper Bound
Temperature (°C)	25	80
Ratio of ethanol (mol/mol)	3	20
Catalyst (%)	0.5	1.5

Table 8.9 Range of Operation of the Variables (Enzymatic Transesterification)

Variable	Lower Bound	Upper Bound
Temperature (°C)	20	45
Ratio of ethanol (mol/mol)	3	12
Catalyst (%w/w)	5	16
Added water (%w/w)	0	20
Time (h)	6	13

The use of supported enzymes and a noncatalyzed reaction under supercritical conditions are other alternatives that have been proposed, but the economics are not as good. Models for the yield of heterogeneous catalysts can also be found in Martín and Grossmann, 2012b.

As mentioned above, ethanol can also be used. Severson et al. (2013) developed several models to predict the yield of the reaction as a function of the ethanol-to-oil molar ratio, the operating temperature, the catalyst load, and the residence time. The two most promising processes used either KOH or enzymes as catalysts. The models are presented below, and Tables 8.8 and 8.9 show the range of the operating variables.

Using *ethanol (KOH)* we have the following:

$$X = 22.94293 + 113.88 \cdot \text{Cat} + 2.828881 \cdot RE - 1.02734 \cdot T - 1.44522 \cdot \text{Cat} \cdot RE$$
$$+ 0.250723 \cdot \text{Cat} \cdot T + 0.023375 \cdot RE \cdot T - 41.4402 \cdot \text{Cat}^2 - 0.07568 \cdot RE^2 + 0.006226 \cdot T^2;$$

$$(8.21)$$

Using *ethanol (enzymatic)* we have the following:

$$
\begin{aligned}
X = {}& 3.624996 - 1.64904 \cdot T + 17.91299 \cdot t - 7.60104 \cdot RE + 10.59497 \cdot \text{Cat} - 0.49902 \cdot \text{water\_add} \\
& + 0.014332 \cdot T^2 - 0.65091 \cdot t^2 - 0.33241 \cdot RE^2 - 0.31632 \cdot \text{Cat}^2 + 0.00692 \cdot \text{water\_add}^2 \\
& - 0.0407 \cdot T \cdot t + 0.17485 \cdot T \cdot RE - 0.0138 \cdot T \cdot \text{Cat} - 0.0156 \cdot T \cdot \text{water\_add} - 0.0601 \cdot t \cdot RE \\
& - 0.4629 \cdot t \cdot \text{Cat} + 0.11014 \cdot t \cdot \text{water\_add} + 0.43481 \cdot RE \cdot \text{Cat} + 0.21369 \cdot RE \cdot \text{water\_add} \\
& - 0.09614 \cdot \text{Cat} \cdot \text{water\_add};
\end{aligned}
$$

$$(8.22)$$

Kinetics: The use of surface of response analysis for evaluating the yield of the transesterification equilibrium is an efficient way to evaluate a process that depends on a number of input variables such as temperature, concentration of species, catalyst load, and time. However, a more detailed analysis based on the reaction mechanism provides better insight. The process to produce biodiesel is a set of three reactions in equilibrium. Let TG be triglycerides, DG diglycerides, MG monoglycerides, ME methyl ester, and GL glycerol:

$$
\begin{aligned}
\text{TG} + \text{MeOH} &\xrightarrow{k_1} \text{DG} + \text{ME} \\
\text{DG} + \text{ME} &\xrightarrow{k_2} \text{TG} + \text{MeOH} \\
\text{DG} + \text{MeOH} &\xrightarrow{k_3} \text{MG} + \text{ME} \\
\text{MG} + \text{ME} &\xrightarrow{k_4} \text{DG} + \text{MeOH} \\
\text{MG} + \text{MeOH} &\xrightarrow{k_5} \text{GL} + \text{ME} \\
\text{GL} + \text{ME} &\xrightarrow{k_6} \text{MG} + \text{MeOH}
\end{aligned}
$$

$$(8.23)$$

The reactor design based on the kinetics is as follows:

$$\frac{d\text{TG}}{dt} = -k_1[\text{TG}][\text{MeOH}] + k_2[\text{DG}][\text{ME}]$$

$$\frac{d\text{DG}}{dt} = k_1[\text{TG}][\text{MeOH}] - k_2[\text{DG}][\text{ME}] - k_3[\text{DG}][\text{MeOH}] + k_4[\text{MG}][\text{MeOH}]$$

$$\frac{d\text{MG}}{dt} = k_3[\text{DG}][\text{MeOH}] - k_4[\text{MG}][\text{MeOH}] - k_5[\text{MG}][\text{MeOH}] + k_6[\text{ME}][\text{GL}]$$

$$\frac{d\text{ME}}{dt} = k_1[\text{TG}][\text{MeOH}] - k_2[\text{DG}][\text{ME}] + k_3[\text{DG}][\text{MeOH}] - k_4[\text{MG}][\text{MeOH}]$$

$$(8.24)$$

$$+ k_5[\text{MG}][\text{MeOH}] - k_6[\text{ME}][\text{GL}]$$

$$\frac{d\text{GL}}{dt} = +k_5[\text{MG}][\text{MeOH}] - k_6[\text{ME}][\text{GL}]$$

$$\frac{d\text{MeOH}}{dt} = -k_1[\text{TG}][\text{MeOH}] + k_2[\text{DG}][\text{ME}] - k_3[\text{DG}][\text{MeOH}] + k_4[\text{MG}][\text{MeOH}]$$

$$- k_5[\text{MG}][\text{MeOH}] + k_6[\text{ME}][\text{GL}]$$

$$k_i = k_{ic} C_{\text{cat}} + k_{\text{in}}$$

$$i = 1, ..., 6$$

Table 8.10 Kinetic Constants for Biodiesel Production

Reaction	Rate	Palm Oil			Mustard Oil		
	Constant	40°C	50°C	60°C	40°C	50°C	60°C
TG→DG	k_1	0.07	0.12	0.14	0.11	0.14	0.21
DG→TG	k_2	0.10	0.17	0.06	0.10	0.11	0.02
DG→MG	k_3	0.31	0.61	0.60	0.55	0.63	1.04
MG→DG	k_4	0.64	1.52	1.24	0	0	0
MG→GL	k_5	1.15	2.56	4.18	0.19	0.26	0.64
GL→MG	k_6	0.02	0.01	0.02	0	0.04	0.01

The rate constants depend on the oil, the catalyst, and the alcohol. Table 8.10 (Issariyakul and Dalai, 2012) shows some KOH 1% values from the literature; for this example equilibrium is achieved from 10 min onwards.

EXAMPLE 8.2

In this example we model a batch reactor for producing biodiesel.

Solution

We use MATLAB to model the mechanisms above for reactor design. In principle we assume the rate constant may have an Arrenius form, but that can be easily simplified if it is not the case.

```
No = [1,6,0,0,0,0];
[a,b] = ode45(@transesteificacion,[0,4],No)

plot(a,b(:,1),'k',a,b(:,2),'b',a,b(:,3),'y',a,b(:,4),'r',a,b
  (:,5),'g',a,b(:,6),'c')

function dC = transesteificacion(t,C)
T = 273 + 75;

A1 = ;
A_1 = ;
A2 = ;
A_2 = ;
A3 = ;
A_3 = ;

Ea1 = ;
Ea_1 = ;
Ea2 = ;
Ea_2 = ;
Ea3 = ;
Ea_3 = ;
```

```
R = ;
k1 = A1*exp(-Ea1/(T*R));
k_1 = A_1*exp(-Ea_1/(T*R));
k2 = A2*exp(-Ea2/(T*R));
k_2 = A_2*exp(-Ea_2/(T*R));
k3 = A3*exp(-Ea3/(T*R));
k_3 = A_3*exp(-Ea_3/(T*R));

%velocidades
Ctg = C(1);
Cmetoh = C(2);
Cmg = C(3);
Cdg = C(4);
Cgly = C(5);
Cfame = C(6);

r1 = k1*Ctg*Cmetoh;
r_1 = k_1*Cdg*Cfame;
r2 = k2*Cdg*Cmetoh;
r_2 = k_2*Cmg*Cfame;
r3 = k3*Cmg*Cmetoh;
r_3 = k_3*Cgly*Cfame;

dC(1,1) = -r1 + r_1;
dC(2,1) = -r1 + r_1 - r2 + r_2 - r3 + r_3;
dC(3,1) = r2 - r_2 - r3 + r_3;
dC(4,1) = r1 - r_1 - r2 + r_2;
dC(5,1) = r3 - r_3;
dC(6,1) = r1 - r_1 + r2 - r_2 + r3 - r_3;
```

8.2.4 BIOGAS

Biogas is a mixture of mainly methane (50–70%) and CO_2 (20–40%), and is produced from the anaerobic digestion (AD) of biomass. The various sources discussed above can be used as raw materials for its production. Biogas is an interesting product that provides further value to wastes, and also showcases a treatment technology (AD) for waste disposal. Apart from the use of biomass, biogas is typically produced from sludge and animal manure (either in slurry form or as a solid). Not only is biogas produced, but also digestate, an interesting fertilizer. The value of this fertilizer can be computed using the NPK index, representing the amount of nitrogen, phosphorous, and potassium.

8.2.4.1 Process description

The production of biogas takes place in a series of steps: hydrolysis, acidogenesis, acetogenesis, and methanogenesis. The first step, *hydrolysis*, breaks the biomass into smaller blocks for transformation into sugars, fatty acids, and amino acids. Several promising attempts have been made to use pretreatment methods (discussed in the lignocellulosic section) to improve yields. In *acidogenesis* (the second step), H_2, CO_2, acetate, and volatile fatty acids (VFA) and alcohols are produced. In the third step (*acetogenesis*), H_2 and acetic acid are produced from VFA and alcohols. Finally, *methanogenesis* transforms the mixture of CO_2 and H_2 into methane. This final step runs in parallel to the acetogenesis step.

The operating conditions are classified by the residence time and the temperature. Psychrophilic conditions are those requiring low temperatures ($<20°C$) and long retention times (70−80 days). Mesophilic conditions reduce the retention time to 30 to 40 days, but double the operating temperature to 30−42°C. Finally, thermophilic conditions require 43−55°C, but only 15−20 days of retention time. Therefore, the digester is typically a large, closed tank; the gas is sent to a storage tank from there. Apart from continuous stirred tank reactors (CSTRs, often in series), plug flow reactors have also been applied to improve the performance of biogas plants (Bensmann et al., 2013).

The gas produced must be cleansed from NH_3 or H_2S before use as an energy source. Chapter 5, Syngas discusses a number of technologies.

8.2.4.2 Process analysis
8.2.4.2.1 Kinetics

The simplest process analysis is to consider a first-order rate model for the substrate utilization:

$$\frac{dS}{dt} = -kS \tag{8.25}$$

Contois in 1959 proposed a modified form of the Monod model for the first stages, and a Haldane function for the methanogenesis step (Mairet et al., 2012). pH and the gas−liquid mass transfer step must also be controlled.

Hydrolysis-acidogenesis:

$$\mu_i(S_i, X_i) = \frac{\mu_i \cdot S_i}{K_{si}X_i + S_i}$$

Methanogenesis: $\tag{8.26}$

$$\mu_3 = \frac{\mu_3 \cdot S_3}{K_{s3} + S_3 + S_3^2/K_{i,3}} \frac{K_{INH_3}}{K_{INH_3} + NH_3}$$

Complete models can be found elsewhere. The anaerobic digestion model (ADM-1; Batstone et al., 2002) is the basis for most analyses.

8.2.4.2.2 Gas composition

The actual composition of the gas depends not only on the operating conditions, but also on the composition of the feedstock. Apart from first principle analysis based on mass and energy balances, the biochemical reaction used for a basic analysis can be considered as follows:

$$C_nH_aO_b + \left(n - \frac{a}{4} - \frac{b}{2}\right)H_2O \leftrightarrow \left(\frac{n}{2} - \frac{a}{8} + \frac{b}{4}\right)CO_2 + \left(\frac{n}{2} + \frac{a}{8} - \frac{b}{4}\right)CH_4 \quad (8.27)$$

If nitrogen from amino acids and protein is also considered, the equation becomes:

$$C_nH_aO_bN_c + \left(n - \frac{a}{4} - \frac{b}{2} + \frac{3c}{4}\right)H_2O \leftrightarrow \left(\frac{n}{2} - \frac{a}{8} + \frac{b}{4} - \frac{3c}{8}\right)CO_2 + \left(\frac{n}{2} + \frac{a}{8} - \frac{b}{4} + \frac{3c}{8}\right)CH_4 + cNH_3$$

$$(8.28)$$

There are some statistical studies that provide the yield of gas as a function of the time, C-to-N ratio, and loading concentration (kg VS/m^3), all of which are highly dependent on the biomass used as feedstock.

$$y(m^3/kg\ VS) = f(\text{retention time}, SC(kg\ VS/m^3), C/N) \quad (8.29)$$

8.3 PRODUCT PURIFICATION

8.3.1 ETHANOL DEHYDATION

Ethanol production via fermentation yields a diluted water—ethanol mixture, from 5–15% depending on the raw material. The separation is carried out by removing solids first and using a distillation column known as a beer column. Next, rectification and/or molecular sieves are used to produce anhydrous ethanol, which is ready for use as a fuel. Alternatively, extractive distillation has also been suggested.

A *beer column* allows recovery of most of the ethanol. However, the operation is energy-intensive. Therefore, the use of multieffect columns has been suggested so that the energy consumption can be cut by one-third and the cooling needs approximately by half. The operation of such columns was discussed in Chapter 2, Chemical processes. Fig. 8.9 shows the scheme of the operation.

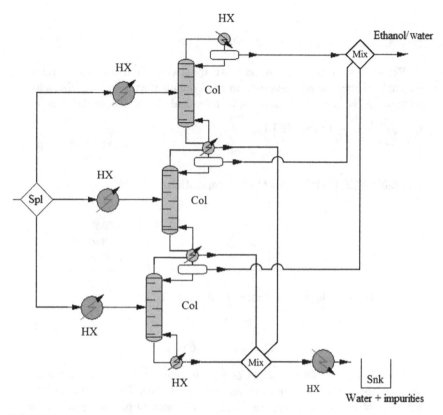

FIGURE 8.9

Beer multieffect column.

EXAMPLE 8.3

A single beer column operates with an L/D ratio of 1. The feed, 6 kg/s of a water–ethanol mixture, has 15% by weight of ethanol. The distillate has 80% purity by weight and the residue 2%. Compute the distillate and the residue, as well as the number of trays required for the column. Determine the energy removed and provided.

Solution

We assume that the flows are constant along the column. Thus the rectifying and stripping operating lines are as follows:

$$y_n = \frac{L}{V}x_{n-1} + \frac{D}{V}x_D$$

$$y_m = \frac{L'}{V'} x_{m-1} + \frac{W}{V'} x_W$$

Because the data is by weight, we must convert it into molar ratios for each of the streams involved in the column operation. x is the mass fraction, F is the total mass flow rate, and M is the molar weight:

$$y_i = \frac{x_i \cdot F/M_i}{\sum_i x_i \cdot F/M_i} \text{ (Table 8E3.1).}$$

Table 8E3.1 Molar and Mass Compositions

	x	y
Feed	0.15	0.0645933
Distillate	0.8	0.61016949
Residue	0.02	0.00792254

We formulate the mass balance as follows:

$$F = D + R$$

$$F \cdot x_f = D \cdot x_D + R \cdot x_R$$

By solving the system we get $D = 1$ kg/s and $R = 5$ kg/s. Then, we use the McCabe method to compute the number of trays. Due to the nonideal behavior of the mixture, experimental data is used to plot the equilibrium curve. Next, using D, R, and the L/D ratio we plot the operating lines:

$$V = D + L$$

$$L' = F + L \text{ (assuming feed as saturated liquid)}$$

$$V' = V$$

$$L/D = 1 \rightarrow \frac{L}{V} = \frac{D}{(1 + (L/D)) \cdot D} = 0.5$$

q line represents the intersection between the stripping and rectifying lines. Thus we have seven trays required for such separation (see Fig. 8E3.1). The energy involved in the condenser, Q_W, and the reboiler, Q_H, are computed as follows:

$$Q_W = \lambda \cdot V = (0.8\lambda_{EtOH} + 0.2\lambda_{EtOH})(L/D) \cdot D$$
$$= (9266/46 \cdot 4.18 \cdot 0.8 + 0.2 \cdot 9723/18 \cdot 4.18) \cdot 1 \cdot 1 = 1125 \text{ kJ/s}$$

$$Q_H = \lambda \cdot V' = (0.02\lambda_{EtOH} + 0.98\lambda_{EtOH})(L/D) \cdot D = 2230 \text{ kJ/s}$$

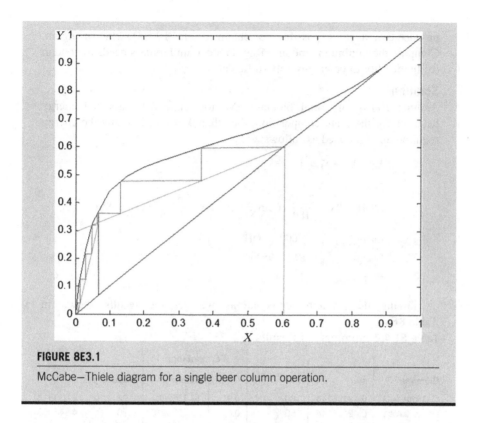

FIGURE 8E3.1

McCabe—Thiele diagram for a single beer column operation.

Final dehydration: Fuel-quality ethanol is achieved only by surpassing the azeotrope (96% by weight). Therefore, a number of strategies have been suggested and evaluated, such as extractive distillation, pervaporation, and the use of molecular sieves. Among them, molecular sieves use silica gel or zeolites as adsorbent beds so that they retain the water. The operation of these units is similar to pressure swing adsorption (PSA) systems, except for the fact that there is no pressurization—expansion stage (see chapters Air and Syngas for an evaluation of adsorbent beds). Next, a stream of hot air is used to regenerate the beds. Pervaporation consists of the use of membranes permeable to one of the components so that by evaporation, ethanol is recovered. The use of molecular sieves is the technology of choice in most industrial facilities, and provides an advantage in terms of energy consumption.

EXAMPLE 8.4

A second-generation bioethanol production facility processes biomass that contains cellulose (assume glucose for calculations), hemicellulose (xylose), and lignin. The conversion from glucose to ethanol is 90%, and that from xylose is 80%. Ninety-six percent (96%) of the ethanol produced is recovered. If the fractions of cellulose and hemicellulose in the feedstock are the same, the consumption of energy is 7000 kJ/kg of ethanol produced, and it is possible to obtain up to 19,500 kJ/kg of lignin. Compute the minimum amount of lignin a certain biomass needs to contain for the facility to be energy self-sufficient.

Solution

Assume 100 kg of initial biomass. We formulate the mass and energy balances for the plant composition, the ethanol production, and the energy required and produced as follows:

$$X_G + X_X + X_L = 1$$

$$X_G = X_X$$

$$EtOH = BX_G \cdot \frac{2 \cdot 46}{180} \cdot Conv_G \cdot Sep + BX_X \cdot \frac{5 \cdot 46}{3 \cdot 150} \cdot Conv_x \cdot Sep$$

$$\text{Energy}_{consumed} = 7000 \cdot EtOH$$

$$\text{Energy}_{produced} = BX_L \cdot 19500$$

$$\text{Energy}_{consumed} = \text{Energy}_{produced}$$

Solving this system of equations we get the results shown in Table 8E4.1:

Table 8E4.1 Summary of Results

				Fermentation			
Biomass	**100**	**PM**	**Conv**	**Ethanol**	**Recovery**	
Cellulose	0.42196144	180	0.9	19.4102261	0.96	18.6338171
Hemicellulose	0.42196144	150	0.8	25.8803015	0.96	24.8450894
Lignin	0.15607813					
Total	1.000001					
Energy (kJ/kg Lig)	19500	304352.345				
Consumption (kJ/kg et)	7000	304352.345	0			

8.3.2 HYDROCARBON AND ALCOHOL MIXTURE SEPARATION

FT processes and mixed alcohol synthesis require the use of distillation to separate the products. FT liquid fuels are separated using petrochemical-based technologies (see also chapters: Chemical processes and Syngas). In the case of mixed alcohols, a sequence of distillation columns is required; see Chapter 2, Chemical processes for the heuristics behind the technique, and Fig. 2.1 for a block flowsheet of the process. The separation of butanol mixtures is more complex due to the inmiscibility between water and ethanol. Recently a hybrid extraction–distillation scheme was proposed that uses mesitylene in an extraction column that processes the fermentation broth.

This setup is followed by a sequence of three distillation columns operating at 1, 2, and 0.5 bar, respectively, to separate the solvent, mesitylene, butanol, acetone, and ethanol (Kramer et al., 2011). Ibutene purification is the simplest in the sense that it is a gas produced in the fermentation. Therefore, it exits the reactor together with CO_2 and oxygen. PSA and membrane systems can be used for its purification.

8.3.3 ALCOHOL RECOVERY: BIODIESEL

As presented in the analysis of the equilibrium towards biodiesel, an excess of alcohol is typically used to drive the reaction into biodiesel. However, the decision concerning the excess of alcohol can be made not just at the reactor level (aiming at higher conversions), but also at the process level. The reason is because for a larger excess of alcohols, the conversion increases, as well as the energy required to recover the excess. Therefore, there is a tradeoff to be solved (Martín and Grossmann, 2012b). Fig. 8.10 shows the flowsheet for the production of biodiesel using heterogeneous catalysts.

The distillation column or the flash separation to recover the alcohols must work in such a way that the temperature of operation is below 150°C to avoid glycerol decomposition. This column presents the particular feature that the feed and the bottom stream are phases. Thus, the vapor pressure is that given by both phases as follows:

$$P_T = P_{v,\text{polar}} + P_{v,\text{non-polar}} = \sum_{i\,=\,\text{water,glycerol,alcohol}} x_i P v_i + \sum_{j\,=\,\text{biodiesel,oil}} x_j P v_j \qquad (8.30)$$

Once the alcohol has been recovered, polar and nonpolar phases are separated at around 40−60°C for ethanol and methanol, respectively. In the case of a homogeneous catalyst, water is added in the separation to help remove the catalyst. The polar phase contains glycerol, alcohols and the homogeneous catalyst. The nonpolar phase contains biodiesel and unconverted oil. The polar phase is

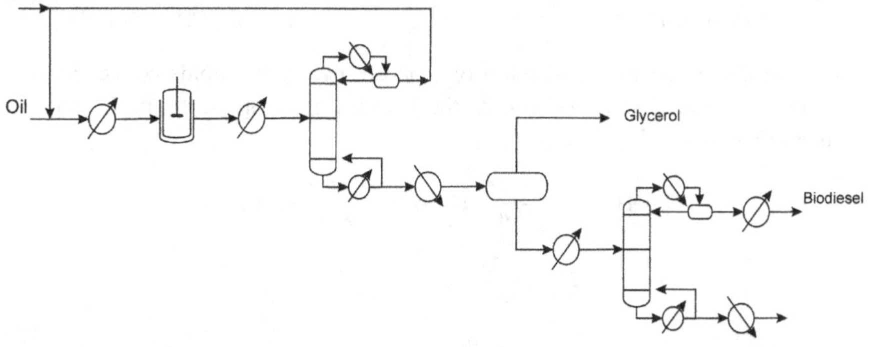

FIGURE 8.10

Biodiesel production using solid catalysts.

neutralized, if needed, and later distilled to purify the glycerol. In the case where acid homogeneous catalysts are used, CaO is recommended for neutralization (as presented above). In the case where KOH is used, H_3PO_4 is recommended. The nonpolar phase is also distilled to purify the biodiesel. The distillate should be below 250°C to avoid decomposition. As a result, this column and the alcohol recovery columns are meant to operate under vacuum. Furthermore, the column processes an immiscible mixture, and therefore it has to be taken into account in boiling and dew point computations.

Glycerol, the main byproduct, has become an interesting carbon source for the production of chemicals such as H_2 or syngas via reforming, ethanol, via fermentation, glycerol ethers, together with ibutene, polyesters, etc.

EXAMPLE 8.5

An algae-based biodiesel facility grows algae with 40% by weight oil (molecular weight 884 kg/kmol), 25% starch (assumed to be glucose), 10% protein, and the rest inert material. The aim is to produce ethanol from the starch and biodiesel from oil (assume a molecular weight of 310 kg/kmol). The conversion of the sugar fermentation is 90%, and the transesterification reaction has a conversion of 95%. Determine the amount of ethanol, biodiesel, and glycerol sold to the market if we use part of the ethanol produced as a transesterifying agent.

Solution

We perform the mass balances based on the stoichiometry of the reactions below:

$$C_6H_{12}O_6 \xrightarrow{\text{Yeast}} 2C_2H_5OH + 2CO_2$$

$$
\begin{array}{l}
CH_2-OOC-R_1 \\
| \\
CH_2-OOC-R_2 + 3R'OH \\
| \\
CH_2-OOC-R_3
\end{array}
\leftrightarrow
\begin{array}{l}
R'-OOC-R_1 \\
R'-OOC-R_2 \\
R'-OOC-R_3
\end{array}
+
\begin{array}{l}
CH_2-OH \\
| \\
CH-OH \\
| \\
CH_2-OH
\end{array}
$$

Let $Conv_G$ be the conversion of glucose, $Conv_L$ the lipids conversion, B the total initial biomass, and X_i the fraction of component i in the biomass (S starch, L Lipids):

$$EtOH = BX_S \cdot \frac{2 \cdot 46}{180} \cdot Conv_G - 3 \cdot \frac{46}{884} BX_L \cdot Conv_L$$

$$FAEE = BX_L \cdot \frac{3 \cdot 310}{884} \cdot Conv_L$$

$$Glycerol = BX_L \cdot \frac{1 \cdot 92}{884} \cdot Conv_L$$

Solving the system we get the following results (Table 8E5.1):

Table 8E5.1 Summary of Results

Biomass	100		PM	Conv	Prod EtOH		Prod Biodiesel		Products
					kmol	kg	kmol	kg	kg
Starch	0.40	40.00	180.00	0.22				0.00	0.00
Oil	0.25	25.00	884.00	0.03				0.00	0.00
Protein	0.10							0.00	0.00
Others	0.20							0.00	0.00
FAEE			310.00				0.08	26.30	26.30
EtOH			46.00		0.40	18.40	−0.08	−3.90	14.50
Glycerol			92.00				0.03	2.60	2.60

8.3.4 PENICILLIN PURIFICATION

The process for purification of penicillin obtained via fermentation is more complex than the previous ones mentioned. Biomass must be separated from the culture. Next, penicillin is extracted, and if a solvent is used, a recovery stage must be added to the process. Subsequently, the liquid phase containing the penicillin is recovered by centrifugation. A second extraction is followed by crystallization (either via cooling or drowning) to purify it. Finally, washing and drying allow powder production. It is typically stored as a potassium or sodium salt. See Chapter 4, Water for the operation of crystallizers.

8.4 THERMODYNAMIC CYCLES

Biomass-based syngas, biogas, or the biomass itself can be used to produce energy in power cycles such as the Rankine cycle and the Brayton cycle (Moran and Shapiro, 2000).

8.4.1 RANKINE CYCLE

The most widely used thermodynamic cycle is the regenerative Rankine cycle. High-pressure, superheated steam is obtained by burning the biomass. Alternatively, this energy can be obtained from burning any fossil fuel (as in thermal plants) or by using solar or geothermal energy. In the case of solar energy, concentrated solar power is required so that solar energy can reach an intensity capable of producing high-temperature steam. The variability in solar energy availability becomes a challenge for the operation of such a plant. Typically, a heat transfer fluid such as molten salts is used to buffer the absence of solar energy during the nighttime. However, in the long term, a different solution must be used. Hybrid plants, for example, which combine the use of different energy sources such as biomass and solar and allow continuous energy production over time (Vidal and Martín, 2015).

Once the steam is produced, it is fed to a steam turbine. The turbine is divided into three stages: high, medium, and low pressure. In the high-pressure turbine the steam expands into a moderated pressure. The steam is sent back to the boiler to be reheated. Next, the stream is expanded in the medium- and low-pressure turbines. Several extractions at different pressures are obtained from the turbines. These streams are used to reheat the condensed stream to close the cycle. A thermal plant scheme based on a Rankine cycle is presented in Fig. 8.11.

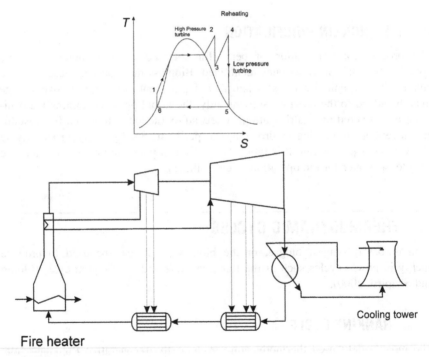

FIGURE 8.11

Regenerative Rankine cycle.

The energy generated in the expansions is computed as an enthalpy difference. The temperature—entropy (TS) diagram of the cycle is shown in Fig. 8.11.

$$W = \eta \cdot \sum_i (H_k - H_{k-1}), \forall i \in \text{Expansion} \tag{8.31}$$

The enthalpy is a function of the temperature and the pressure. See Appendix B (Thermodynamic Data) to compute the enthalpy. The isentropic efficiency of the turbine (typically from 0.7—0.9) is computed using Eq. (8.32):

$$\eta = \frac{H_{\text{steam,(out)}} - H_{\text{steam,(in)}}}{H_{\text{steam,(out,isoentropy)}} - H_{\text{steam,(in)}}} \tag{8.32}$$

where the entropy of the stream is also a function of the pressure and temperature. For a detailed model of the cycle shown in Fig. 8.11, we refer the reader to the supplementary material in Vidal and Martín (2015).

8.4.2 BRAYTON CYCLE

Compressed air is used to burn a fuel (ie, syngas) in a combustion chamber. When burned, the gas heats up to a high temperature that can be computed using Eq. (8.33):

$$C_nH_m + (n + m/2) \cdot O_2 \rightarrow nCO_2 + (m/2) \cdot H_2O$$

$$CO + \tfrac{1}{2}O_2 \rightarrow CO_2$$

$$H_2 + \tfrac{1}{2}O_2 \rightarrow H_2O$$

$$0 = \sum_{i \in \text{Product}} n_i \left(\Delta H_{f,i}^{298} + \int_{298}^{T_{\text{out}}} Cp_i dT \right) - \sum_{i \in \text{Reactants}} n_i \left(\Delta H_{f,i}^{298} + \int_{298}^{T_{\text{in}}} Cp_i dT \right) \tag{8.33}$$

Next, it is fed to a gas turbine. In the gas turbine the superheated gas is expanded isentropically to produce power. The energy involved in the compression of the air and that obtained in the expansion can be computed using Eq. (8.34), where k is around 1.4. The cycle diagrams can be seen in Fig. 8.12.

$$W(\text{Comp}) = (F) \cdot \frac{8.314 \cdot k \cdot (T_{\text{in}} + 273.15)}{((Mw) \cdot (k - 1))} \frac{1}{\eta_s} \left(\left(\frac{P_{\text{out}}}{P_{\text{in}}} \right)^{\frac{k-1}{k}} - 1 \right) \tag{8.34}$$

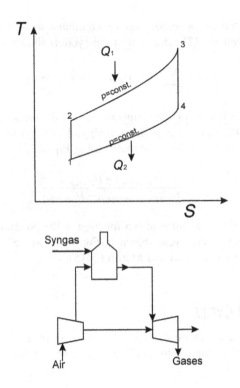

FIGURE 8.12

Brayton cycle.

8.5 PROBLEMS

P8.1. In a facility for the production of biodiesel using methanol as the transesterifying agent, the conversion of the reaction can be estimated using the following equation:

$$X = 92.056 + 3.1561(\text{molar ratio met/oil}) - 0.01925 \cdot 60(\text{molar ratio met/oil})$$
$$- 0.34(\text{molar ratio met/oil}) - 0.1459(\text{molar ratio met/oil})^2$$

Assuming that 96% of the methanol is recovered in the distillation column placed after the reactor (and recycling this amount), compute the methanol to be purchased and the methanol-to-oil ratio used if we aim for a conversion of 96% per pass.

P8.2. A single beer column with 11 trays is used to recover the ethanol from a dilute water–ethanol mixture. The feed, with a flowrate of 6 kg/s, enters as saturated liquid with a composition of 15% by weight of ethanol. The column operates with a reflux ratio L/D of 3.

The residue is expected to have 2% by weight of ethanol. Determine the maximum concentration of ethanol in the distillate and the flowrates of residue and distillate.

P8.3. A second-generation bioethanol production facility processes biomass that contains cellulose (assume glucose for the calculations), hemicellulose (xylose), and lignin. The plant is supposed to operate self-sufficiently, but the energy consumption is 10% lower than the production. The uncertainty is in the conversion of xylose to glucose. The conversion from glucose to ethanol is 90%, and that from xylose should be 80%. Ninety-six percent (96%) of the ethanol produced is recovered. If the fraction of cellulose and hemicellulose in the feedstock is the same, the consumption of energy is 7000 kJ/kg of ethanol produced; it is possible to obtain up to 19,500 kJ/kg of lignin. Compute the actual conversion of xylose and the ethanol production of the facility per 100 kg of initial biomass.

P8.4. In the production of biodiesel from oil and methanol, the reactor conversion is given by this equation:

$$Yield = 65.6301 + 0.4209 \cdot T + 15.1582 \cdot Cat + 3.1561 \cdot RM - 0.0019 \cdot T^2$$
$$- 0.2022 \cdot T \cdot Cat - 0.01925 \cdot T \cdot RM - 4.0143 \cdot Cat^2 - 0.3400 \cdot Cat \cdot RM$$
$$- 0.1459 \cdot RM^2$$

where the catalyst load is 1% of KaOH, Cat = 1, and the reaction takes place at 60°C and 4 bar. The molar ratio between methanol and oil is 6. The products from the reactor are distilled and 98% of the unconverted methanol is recovered. Determine the methanol fed to the system per 100 kmol/h of oil, and the purge fraction if the commercial oil contains 0.02 kmol of water per kmol of oil. Water leaves with the methanol. The catalyst is capable of handling 5 kmol of water per 100 kmol of the methanol−oil mixture.

P8.5. An algae-based biodiesel facility grows algae containing oil (molecular weight 884 kg/kmol), starch (assume it is glucose), 10% protein, and 15% inert material. The aim is to produce ethanol from the starch and biodiesel from the oil (assume a molecular weight of 310 kg/kmol). The conversion of the sugar fermentation is 90% and the transesterification reaction has a conversion of 95%. The demand of ethanol and biodiesel is the same. Determine the composition required for the algae assuming that we internally use part of the ethanol produced for oil transesterification.

P8.6. *Saccharomyces cerevisiae* is used to ferment glucose to ethanol. Calculate the time it takes to obtain 80% conversion, assuming an initial concentration of cells of 0.1 g/L, an initial glucose concentration of 250 g/L, and the fact that the following rate laws hold:

Cells:

$$V\frac{dC_c}{dt} = (r_g - r_d)V$$

Substrate:

$$V\frac{dC_s}{dt} = Y_{s/c}(-rg)V - r_{sm}V$$

Product:

$$V\frac{dC_p}{dt} = Y_{p/c}(r_gV)$$

$$r_g = \mu_{max}\left(1 - \frac{C_p}{C_p^*}\right)^{0.52}\frac{C_sC_c}{K_s + C_s}; \quad r_d = k_dC_c; \quad r_{sm} = mC_c$$

$C_p^* = 93$ g/l; $n = 0.52$; $\mu_{max} = 0.33$ h^{-1}; $K_s = 1.7$; $Y_{c/s} = 0.08$ g/g; $Y_{p/s} = 0.45$ g/g (est)
$Y_{p/c} = 0.56$ g/g (est); $k_d = 0.01$ h$_{-1}$; $m = 0.03$ g substrate/(g cells h)

P8.7. A simple beer column processes a water−ethanol mixture (15% w/w ethanol). The distillate is 80% w/w ethanol and the residue is 98% water. The condenser consumes 20.19 kg/s of cooling water at 20°C, which exits at 30°C. The reboiler consumes 0.929 kg/s of steam ($\lambda = 1800$ kJ/kg). Calculate the mass flow of distillate and residue, the *L/D* reflux ratio, and the number of trays if their efficiency is 75% (Fig. P8.7).

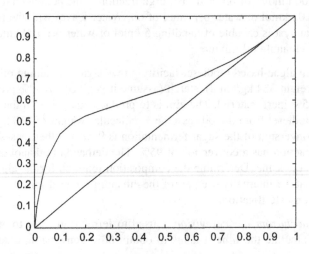

FIGURE P8.7

XY diagram for water−ethanol system.

P8.8. A certain biomass ($C_nH_aO_b$) is digested to produce biogas. The resulting biogas is reformed using the own CO_2 to produce a syngas with a composition of H_2:CO in a 1:1 relationship. The conversions of the reactions are assumed to be 100%. Determine the carbon, H_2, and oxygen composition of the original biomass, and the fraction of the biogas that must be burned so that the reforming furnace operates adiabatically.

P8.9. Using a process simulator, compare the energy required to recover ethanol from the following two mixtures and target 80% molar composition (minimum) in the distillate containing the ethanol:
1. 10% Ethanol−90% Water, 50 kg/s at 32°C.
2. 20% Methanol, 50% Ethanol, 20% Propanol, 10% Butanol, 10 kg/s at its boiling point.

P8.10. Determine the composition of a syngas with 30% CO_2 and the rest H_2 and CO so that it is used in a gas turbine to produce 2000 kcal per kmol of syngas. The gas turbine operates at 50 atm and discharges the flue gas at 1 atm. Assume that 15% of the energy produced at the turbine is used by the compressor system and lost due to inefficiencies. Both the syngas and the air, fed with 20% excess, are at 25°C ($k = 1.4$).

P8.11. A certain biomass with a composition of $C_1H_aO_b$ is used to produce biogas, CO_2, and CH_4. The biogas is used as a fuel in a gas turbine that operates from 50 atm to 1 atm. The compressor system and the turbine inefficiencies represent 25% of the energy produced. The system produces 25 kcal/mol biogas. Determine the composition of the biogas ($k = 1.4$).

P8.12. Biogas with a composition of 40% CH_4 and the rest CO_2 is used to produce steam to be fed to a steam turbine at 160 bar and 560°C ($H = 3465.4$ kJ/kg). The exhaust steam is recovered at 0.08 bar and 41.5°C ($H = 2577$ kJ/kg); it must be reheated to the feed conditions. Compute the efficiency of the plant that produces 25 kcal per mol of biogas and the flow of steam generated.

P8.13. Biomass gasification generates biosyngas with 20% CO_2 and the rest CO and H_2. The yield of the system is 40% to power from heat. The system uses steam at 160 bar and 560°C ($H = 3465.4$ kJ/kg), and the exhaust steam is recovered at 0.08 bar and 41.5°C ($H = 2577$ kJ/kg). Determine the composition of the syngas that produces 20 kcal/mol of biosyngas.

REFERENCES

Aden, A., Foust, T., 2009. Technoeconomic analysis of the dilute sulfuric acid and enzymatic hydrolysis process for the conversion of corn stover to ethanol. Cellulose. 16, 535−545.

Batstone, D.J., Keller, J., Angelidaki, I., Kalyuzhnyi, S.V., Pavlostathis, S.G., Rozzi, A., et al., 2002. The IWA anaerobic disgestion model No 1 (ADM1). Water Sci. Technol. 45 (10), 65−73.

Bensmann, A., Hank-Rauschenbach, R., Sunmacher, K., 2013. Reactor configurations for biogas plants a model based analysis. Chem. Eng. Sci. 104, 413–426.

Birol, G., Ündey, C., Cinar, A., 2002. A modular simulation package for fed-batch fermentation: penicillin production. Comp. Chem. Eng. 26, 1553–1565.

Calderbank, P.H., 1958. The interfacial area in gas liquid contacting with mechanical agitation. Trans. Inst. Chem. Eng 36, 443–463.

Garlock, R.J., Balan, V., Dale, B.E., 2012. Optimization of AFEX pretreatment conditions and enzyme mixtures to maximize sugar release from upland and lowland switchgrass. Bioresour. Technol. 104, 757–768.

Gogate, P.R., Beenackers, A.A.C.M., Pandit, A.B., 2000. Multiple-impeller systems with a special emphasis on bioreactors: a critical review. Biochem. Eng. J. 6, 109–144.

Higbie, R., 1935. The rate of absorption of a pure gas into a still liquid during a short time of exposure. Trans. Am. Ins. Chem. Eng. 31, 365–389.

Issariyakul, T., Dalai, A.K., 2012. Comparative kinetics of transesterification for biodiesel production from palm oil and mustard oil. Can. J. Chem. Eng. 90, 342–350.

Kielbus-Rapala, A., Karcz, J., 2011. Mass transfer in multiphase mechanically agitated systems. In: El-Amin, M. (Ed.), Mass Transfer in Multiphase Systems and its Applications. InTech, Vienna, Austria.

Kramer, K., Harwardt, A., Bronneberg, R., Marquardt, W., 2011. Separation of butanol from acetone-butano-ethanol fermentation by a hybrid extraction-distillation process. Comp. Chem. Eng. 35, 949–963.

Krishnan, M.S., Ho, N.W.Y., Tsa, G.T., 1999. Fermentation kinetics of ethanol production from gluceose and xylose by recombinant Saccharomyces 1400 (pLNH33). Appl. Biochem. Biotech. 77-79, 373–388.

Mairet, F., Bernard, O., Cameron, E., Ras, M., Lardon, L., Steyer, Jp, et al., 2012. Three-Reaction Model for the Anaerobic Digestion of Microalgae. Biotechnol. Bioeng. 109 (2), 415–425.

Martín, M., 2014. Introduction to Software for Chemical Engineers. CRC Press, Boca Raton, USA.

Martín, M., Grossmann, I.E., 2012a. Energy optimization of lignocellulosic bioethanol production via hydrolysis of switchgrass. AIChE J. 58 (5), 1538–1549.

Martín, M., Grossmann, I.E., 2012b. Simultaneous optimization and heat integration for biodiesel production from cooking oil and algae. Ind. Eng. Chem. Res. 51 (23), 7998–8014.

Martín, M., Grossmann, I.E., 2013a. On the systematic synthesis of sustainable biorefineries. Ind. Eng. Chem. Res. 52 (9), 3044–3064.

Martín, M., Grossmann, I.E., 2013b. Optimal engineered algae composition for the integrated simultaneous production of bioethanol and biodiesel. AIChE J. 59 (8), 2872–2883.

Martín, M., Grossmann, I.E., 2014. Optimal simultaneous production of i-butene and ethanol from switchgrass. Biomass Bioener. 61, 93–103.

Martín, M., Grossmann, I.E., 2015. Optimal Production Of Biodiesel (FAEE) and Bioethanol from Switchgrass. Available from: http://dx.doi.org/10.1021/ie5038648.

Moran, M.J., Shapiro, H.N., 2000. Fundamentals of Engineering Thermodynamics. John Willey and Sons Inc, New York.

Odian, G., 2004. Principles of Polymerization, fourth ed. Wiley Interscience, New York.

Papoutsakis, E.T., 1984. Equations and calculations for fermentations of butyric acid bacteria. Biotechnol. Bioeng. 26 (2), 174−187.

Park, J.B.K., Craggs, R.J., Shilton, A.N., 2011. Wastewater treatment high rate algal ponds for biofuel production. Bioresour. Technol. 102 (1), 35−42.

Sazdanoff, N., 2006. Modeling and Simulation of the Algae to Biodiesel Fuel Cycle. Undegraduate thesis. The Ohio State University, Columbus, OH.

Severson, K., Martín, M., Grossmann, I.E., 2013. Optimal integration for biodiesel production using bioethanol. AICHE J. 59 (3), 834−844.

Vidal, M., Martín, M., 2015. Optimal coupling of a biomass based polygeneration system with a concentrated solar power facility for the constant production of electricity over a year. Comp. Chem. Eng. 72, 273−283.

Walker, D.A., 2009. Biofuels, facts, fantasy, and feasibility. J. Appl. Phycol. 21 (5), 509−517.

Appendix A: General nomenclature

a	Specific area (length^{-1})
A	Area (length units square)
AM	Ratio of ammonia in unit of mass per mass of biomass
C	Concentration (mol per volume unit)
Cat	Mass fraction of catalyst added
c_p	Heat capacity (energy per unit mass or mol and temperature)
d_{32}	Sauter mean diameter (m)
D	Distillate flow rate (molar per unit time)
D_i	Diffusion coefficient (length square per second)
D_p	Particle diameter (length units)
E	Reduction potential (V/mol)
E	Evaporated flow rate (see Chapter 4: Water; mass per unit time)
E_c	Steam economy
E_e	Electrical energy (energy units per unit time)
f_i	Feed flow rate of component i (mass or molar per unit time)
f	Liquefied fraction
f	Separation factor
F	Flow rate (of the feed for distillation columns; mass or molar per unit time)
F	Faraday constant
FFA	Free fatty acid
FAME	Fatty acid methyl ester
FAEE	Fatty acid ethyl ester
g	Air fraction diverted to the expansion machine
G	Gas flow rate (mass per unit time)
G	Gas flow rate per unit area (Ergun Equation; mass per unit time and area)
G	Growth rate in crystallization (length per unit time)
h	Liquid enthalpy (energy per unit mass or molar)
H	Henry constant (pressure units per molar fraction)
H	Enthalpy (energy per unit mass or molar)
I	Current intensity (Ampere)
I	Initiator in polymerization (moles per unit volume)
K	Equilibrium constant (bar^P)
$K_L a$	Volumetric mass transfer coefficient (molar per unit time)
k	Kinetic constants
k	Polytropic coefficient in compressors and expanders
l_i	Liquid flow rate of component i (mass or molar per unit time)
L	Liquid reflux (molar per unit time)
m/M	Mass flow (mass per unit time)
Mw	Molar mass (mass per mol)
M	Monomer concentration (moles per unit volume)
n	Molar flow (kmol per unit time)
n	Number of entities per unit volume, crystallization (inverse of volume)

N	Number of trays
N	Impeller revolutions (see Chapter 8: Biomass; revolutions per unit time)
N_i	Flux of species i (mass or molar units per unit time)
P	Pressure (mass per unit length and time square)
P	Product (see Chapter 8: Biomass; mass per unit volume)
P	Impeller power (see Chapter 8: Biomass; energy units)
P_g	Impeller power under aerated conditions (see Chapter 8: Biomass; energy units)
P_v	Vapor pressure (mass per unit length and time square)
P_T	Total Pressure (mass per unit length and time square)
q	Specific energy (energy per unit mass)
Q	Thermal energy (energy per unit time)
Q_c	Gas flow rate (volume per unit time)
R	Reflux ratio (L/D)
R	Gas constant (energy per unit mol and temperature)
RE	Molar ratio of ethanol to oil
RM	Molar ratio of methanol to oil
r_i	Kinetic rate (molar^q per unit time)
S	Entropy (Energy per unit temperature and mass)
S	Substrate (see Chapter 8: Biomass; mass per unit volume)
t	Time (time units)
T	Temperature (K or °C)
T	Impeller diameter (length)
u_G	Superficial gas velocity (length per unit time)
U	Heat transfer global transfer coefficient (energy per unit area and temperature)
U_b	Bubble rising velocity (length per unit time)
v	Velocity (length per unit time)
v	Specific volume of humid air (volume per mass of dry air)
V	Vapor flow rate (mass or molar per unit time)
V	Potential difference (see Chapter 4: Water)
V	Liquid volume (see Chapters 4 and 8: Water and Biomass; volume units)
w_i	Mass fraction of component i in FT product distribution
W	Residue flow rate in distillation columns (mass or molar per unit time)
W	Commercial steam fed to the evaporators (see Chapter 4: Water; mass per unit time)
W	Work (energy per unit time)
x	Liquid molar fraction or mass fraction
X	Conversion
X	Solids mass fraction (see Chapter 8: Biomass)
y_i	Gas molar fraction
y	Specific humidity (mass of vapor per mass of dried air)
z	Molar fraction
α_{ij}	Relative volatility between species i and j
α	Chain length (see Chapter 5: Syngas)
α	Salt dissociation fraction
δ	Layer thickness
ε	Porosity
ε_G	Gas hold up
ϕ	Sphericity factor

λ	Latent heat (energy per unit mass)
φ	Relative moisture
Φ_L	Liquid fraction in the feed to a distillation column
Φ_V	Vapor fraction in the feed to a distillation column
η	Efficiency
Θ_i	Stoichiometric coefficient
π	Osmotic pressure (mass per unit length and time square)
Δe	Ebullioscopic increment (temperature units)
ΔG	Gibbs free energy gradient (energy per unit mass or molar)
μ	Viscosity (mass per unit length)
μ	Biomass growth (per unit time)
τ	Shear stress (pressure units)
τ	Residence time (time units)
ρ	Density (mass per unit volume)
σ	Surface tension (pressure · length units)

Subindex

in	Inlet
i,j	Components
g	Gas
o	Initial conditions
out	Outlet
R	Reference
T	Total
v	Vapor

Appendix B:
Thermodynamic data

B.1 THERMOCHEMISTRY

	MW	ΔH_f (kcal/kmol)	c_p (kcal/kg K)		$a + bT + cT^2 + dT^3$ $T[=]K$		
				a	b	c	d
N_2	28	0	G	0.26614833	−0.00011594	2.28947E-07	−9.97949E-11
O_2	32	0	G	0.21012261	−2.7512E-08	1.30525E-07	7.96202E-11
H_2O	18	−57,798	G	0.42853535	2.5569E-05	1.40284E-07	−4.77937E-11
NO	30	21,600	G	0.23401116	−7.4785E-06	7.77265E-08	−3.33892E-11
NO_2	46	8091	G	0.12602975	0.0002515	−1.0823E-07	1.52424E-12
NH_3	17	−10,960	G	0.38439347	0.00033536	2.40276E-07	−1.6676E-10
H_2	2	0	G	3.24677033	0.00110931	−1.6519E-06	−4.30144E-10
C_2H_2	26	−24,861.244	G	−0.0388664	0.002818	−1.4593E-06	2.95786E-10
CO	28	−26,464.1148	G	0.26374744	−0.00010979	2.38312E-07	−1.08681E-10
CO_2	44	−94,203.3493	G	0.10762832	0.00039928	−3.0459E-07	9.32634E-11
SO_2	64	−71,064.5933	G	0.08915969	0.0001252	−6.1815E-08	1.24112E-11
SO_3	80	−94,624.40191	G	0.04895335	0.00021817	−1.1164E-07	2.42397E-11
FeS_2	120	−42,464.11483	S				
Fe_2O_3	160	−196,000	S				
H_2SO_4	98	−193,900	L				
HNO_3	63	−41,650	L				
HCl	36.5	−22,095.7	G	0.19853838	4.7198E-5	8.16674E-8	2.55489E-11
C_3H_8	44	−24,861.244	G	−0.02296651	0.00166518	−8.6233E-07	1.74783E-10
CH_4	16	−17,877.2	G	0.28756181	0.00077863	1.7886E-07	1.6909E-10
Cl_2	71	0	G	0.09073725	0.00011389	−1.304E-07	5.22272E-11

Aggregation state for ΔH_f: G, Gas; L, Liquid; S, Solid.

B.2 ANTOINE CORRELATION AND PHASE CHANGE

$$\ln (P_v(\text{mmHg})) = A - \frac{B}{C + T\,(^\circ C)}$$

	A	**B**	**C**	**λ (kcal/kmol)**
H_2O	18.3036	3816.44	227.02	9,723
CH_3OH	18.5875	3626.55	238.86	8,431
C_2H_5OH	18.9119	3803.98	231.47	9,266
NH_3	16.9481	2132.50	240.17	5,583
H_2	13.6333	164.9	276.34	
N_2	14.9542	588.72	266.55	
Ar	15.233	700.51	267.31	
CO	14.3686	530	260	
CO_2	22.5898	3103	272.99	
O_2	15.4075	734.55	266.7	1,632
N_2	14.9542	588.72	266.55	1,335
HCl	16.5040	1714.25	258.7	3,866
FAME	17.4530	5003	122.13	20,181
$C_3H_8O_3$	17.2392	4487.94	132.80	14,623

B.3 HEAT OF SOLUTION (25°C)

In (Figure B.1) we see the heat of solution of various chemicals used along the book.

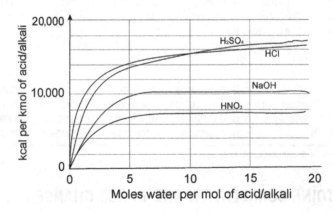

FIGURE B.1

Heat of solution.

B.4 STEAM PROPERTIES

$$\lambda \ (kcal/kg) = -\ 0.0000043722 \cdot T(°C)^3 + 0.00043484 \cdot T(°C)^2 - 0.58433 \cdot T(°C) + 597.48$$

$$H_{sat\ vap} \ (kcal/kg) = -\ 0.0000030903 \cdot T(°C)^3 + 0.00022613 \cdot T(°C)^2 + 0.42436 \cdot T(°C) + 597.42$$

B.4.1 COMPRESSED LIQUID

H (kJ/kg) = $4.2921 \cdot (T(°C)) + 4.1269$

S (kJ/(kg K)) = $1.1902 \times 10^{-5} \cdot (T(°C))^3 - 3.7465 \times 10^{-3} \cdot (T(°C))^2$
$+ 4.5352 \cdot (T(°C)) + 0.64547$

B.4.2 SATURATED LIQUID

H (kJ/kg) = $3.6082 \times 10^{-12} \cdot (T(°C))^6 - 3.4120 \times 10^{-9} \cdot (T(°C))^5$
$+ 1.2303 \times 10^{-6} \cdot (T(°C))^4 - 2.0306 \times 10^{-4} \cdot (T(°C))^3$
$+ 1.5552 \times 10^{-2} \cdot (T(°C))^2 + 3.7216 \cdot (T(°C)) + 3.0035$

S (kJ/(kg K)) = $1.0372 \times 10^{-12} \cdot (T(°C))^5 - 8.6494 \times 10^{-10} \cdot (T(°C))^4$
$+ 2.8965 \times 10^{-7} \cdot (T(°C))^3 - 5.6730 \times 10^{-5} \cdot (T(°C))^2$
$+ 1.6802 \times 10^{-2} \cdot (T(°C)) - 2.1997 \times 10^{-2}$

B.4.3 SATURATED VAPOR

H (kJ/kg) = $-6.5690 \times 10^{-12} \cdot (T(°C))^6 + 6.3049 \times 10^{-9} \cdot (T(°C))^5$
$- 2.3080 \times 10^{-6} \cdot (T(°C))^4 + 3.8339 \times 10^{-4} \cdot (T(°C))^3$
$- 3.0632 \times 10^{-2} \cdot (T(°C))^2 + 2.7553 \cdot (T(°C)) + 2.4957 \times 10^3$

S (kJ/(kg K)) = $-2.0373 \times 10^{-12} \cdot (T(°C))^5 + 1.8589 \times 10^{-9} \cdot (T(°C))^4$
$- 7.1901 \times 10^{-7} \cdot (T(°C))^3 + 1.6112 \times 10^{-4} \cdot (T(°C))^2$
$- 2.8904 \times 10^{-2} \cdot (T(°C)) + 9.1915$

B.4.4 OVERHEATED STEAM (UP TO 10 BAR)

H (kJ/kg) = $(-0.0000063293 \cdot \text{Pressure(bar)} + 0.00033179) \cdot (T(°C))^2$
$+ (0.0124 \cdot \text{Pressure(bar)} + 1.8039) \cdot (T(°C))$
$+ (-6.0707 \cdot \text{Pressure(bar)} + 2504.6)$

S (kJ/(kg K)) = $0.000000000942 \cdot (T(°C))^3 - 0.00000309 \cdot (T(°C))^2$
$+ 0.00524 \cdot (T(°C)) + (6.8171 \cdot (\text{Pressure(bar)})^{(-0.069455)})$

B.4.5 OVERHEATED STEAM (10 BAR–150 BAR)

H (kJ/kg) = $(-0.00000000000011619 \cdot \text{Pressure(bar)}^2$
$- 0.0000000000087596 \cdot \text{Pressure(bar)} - 0.00000000022611) \cdot (T(°C))^4$
$+ (0.0000000004298 \cdot \text{Pressure(bar)}^2 + 0.00000003276 \cdot \text{Pressure(bar)}$
$+ 0.0000007313) \cdot (T(°C))^3 + (-0.0000005801 \cdot \text{Pressure(bar)}^2$
$- 0.000046 \cdot \text{Pressure(bar)} - 0.0005009) \cdot (T(°C))^2$
$+ (0.0003383 \cdot \text{Pressure(bar)}^2 + 0.02947 \cdot \text{Pressure(bar)} + 2.195) \cdot (T(°C))$
$+ (-0.072042 \cdot \text{Pressure(bar)}^2 - 7.7877 \cdot \text{Pressure(bar)} + 2440.8)$

$$
\begin{aligned}
S \text{ (kJ/(kg K))} = (&0.000000000015719 \cdot \text{Pressure(bar)} \\
&+ 0.00000000074013) \cdot (T(°C))^3 \\
&+ (-0.00000000010074 \cdot \text{Pressure(bar)}^2 \\
&- 0.000000030171 \cdot \text{Pressure(bar)} - 0.0000028872) \cdot (T(°C))^2 \\
&+ (0.000000094914 \cdot \text{Pressure(bar)}^2 \\
&+ 0.000029097 \cdot \text{Pressure(bar)} + 0.0050938) \cdot (T(°C)) \\
&+ (0.000041223 \cdot \text{Pressure(bar)}^2 \\
&- 0.028841 \cdot \text{Pressure(bar)} + 5.9537)
\end{aligned}
$$

Appendix C: Solutions to end-of-chapter problems

C.1 CHAPTER 2

P2.1. $Q_H = 600$ kW; $Q_W = 0$ kW
P2.2. $Q_H = 540$ kW; $Q_W = 1526$ kW
P2.3. C1 (T: Met, Et, Prop; B: But); C2 (T: Met; B: Et, Prop); C3 (T: Et; B Prop)
P2.4. C1 (T: A; B: BCD); $C2_{II}$ (T: BC; B: D); C3 (T: B; B: C)
P2.5. C1(T: A; B: BC); C2 (T: B; B: C)
P2.6. $\Delta T = 10$K
P2.7. $Q_H = 50$ kW; $Q_W = 30$ kW
P2.8. $Q_H = 183$ kW; $Q_W = 697$ kW
P2.9. $\Delta T = 25$K
P2.10. C1 (AB, CDEF); C2(A, B); C3(CDE, F); C4 (CD, E); C5(C, D)
P2.11. C1 (AB, CDEF); C2(A, B); C3(CDE, F); C4 (C, DE); C5(D, E)
P2.12. $Q_S = 200$ kW; $Q_W = 1600$ kW; Matches: Above Pinch: H1-C1 (1000 kW); Heating C1 200 kW; Below Pinch: H2-C1 (2400 kW); Cooling H2 (1600 kW); H1-C1 (200 kW)

C.2 CHAPTER 3

P3.1. 160K
P3.2. 190K; $W_{in} = 2802$ kcal/h; $W_{out} = 697$ kcal/h
P3.3. $P_1 = 150$ atm; $T_4 = 300$K; $q = 2$ kcal/kg
P3.4. $f = 0.215$
P3.5. $D = 4.47$ ft h $= 18$ in
P3.6. $f = 0.125$; $\eta = 13\%$
P3.7. System II
P3.8. $f = 0.35$; $W_1 = 60$ kcal/kg; $W_1 = 54$ kcal/kg
P3.9. With precooling $T = 163$K; $q = 15$ kcal/kg; Isentropic expansion; $T = 282$K
P3.10. 6 Trays; 81% molar fraction in N_2
P3.11. Ref III
P3.12. $P_{design} = 200$ bar; $P_{operation} = 150$ bar; $\eta = 0.95$
P3.13. $f = 0.37$; Linde with precooling $q = 30$ kcal/kg; 4 trays
P3.14. $q = 22.7$ kcal/kg
P3.15. 5 trays; $f = 0.76$
P3.16. $P = 640$ mmHg

C.3 CHAPTER 4

P4.1. 29.5 atm

P4.2. $m = 88.976$ kg; Limestone: 11.024 kg;

Component	%
CaO	13.6
Ca(OH)$_2$	69.6
CaCO$_3$	16.8

5.18% excess of Ca(OH)$_2$; Conversion 98%

P4.3. 105.7 m^3

P4.4. $Q = 412808$ kcal/h; $U = 1053 \dfrac{kcal}{h \cdot m^2 \cdot C}$

P4.5. $m = 89.39$ kg

NaOH	0.00296201
Na$_2$CO$_3$	0.16254275
H2O	0.83449524

Limestone: 10.601 kg; 7.38% excess of Ca(OH)$_2$; Conversion 97%

P4.6. $T = 700K$

P4.7. Solution: 89.425 kg; Limestone: 10.574 kg

Solution (kg)	89.42577
NaOH	0.582368
Na$_2$CO$_3$	14.89163
H$_2$O	84.526

Limestone (kg)	10.57423
CaO	0.097362
Ca(OH)$_2$	0.790953
CaCO$_3$	0.111685

P4.8. $Q = 8246387$ kcal/h; $A = 1645$ m^2; $X = 0.632$

P4.9. 52% NaCl; 48% NH$_4$HCO$_3$; Required CaCO$_3$, 860 kg; Production NaCl = 1459 kg

P4.10. $W = 357,854$ kW; No selection; 1200€/kW

P4.11. 0.28 kg/s H$_2$; 2.23 kg/s O$_2$; No selection; 1250€/kW

P4.12. 0.09434 kmol CO$_2$

P4.13. $A = 21.68$ m^2; $\Delta P = 7841$ kPa.

P4.14. $T = 1473K$; $Q = -111,188$ kJ

P4.15. -1.80 V

P4.16. 0.41 kg/s H_2 and 3.26 kg/s O_2; Solar

P4.17. Electrolysis: 4.5 GW; Oxygen Comp: 43.5 MW; Hydrogen Comp: 108 MW

P4.18. Gas:

H_2O	0.079
CO	0.316
H_2	0.509
CO_2	0.095

20% via Electrolysis: Electrolysis: 46 MW; Oxygen Comp: 443 kW; Hydrogen Comp: 687 kW

P4.19. $T = 115°C$; $W = 7744$ kg/s; $A = 90$ m^2; AIQ1 = 41 m^2; AIQ2 = 8.2 m^2

P4.20. $T = 111°C$; $A = 283$ m^2; AIQ1 = 65 m^2; AIQ2 = 23 m^2

P4.21. $M_R = 175$ kg/h; $M_P = 825$ kg/h; $x_R = 171,917$ ppm; $x_P = 0.063$ ppm

P4.22. 0.189 kmol/s of $CaCO_3$; 0.696 kmol of flue gas

C.4 CHAPTER 5

P5.1. Air: 6 kmol per TM; Steam 8.34 kmol per TM; 8479 kcal/t

P5.2. $H_2O = 1$ kmol; CO = 2 kmol; $H_2 = 3$ kmol per kmol of C_2H_2; $T = 802K$; alpha = 0.75

P5.3. $X = 35\%$ Recirculation (Syngas alone) 179.2; $\alpha = 0.014$

P5.4. Basis 1 kmol of Coal: Water gas 0.144 kmol; Generator gas 0.119 kmol; $Q_{water\ gas} = 4859$ kcal; $Q_{generator} = 1594$ kcal; 91.5%

P5.5. $T = 252K$

P5.6. $P = 1$ atm; $T = 1215K$; 0.2079 kmol of C

P5.7.

Component	%
N_2	73.82
CO_2	13.09
CO	13.09
Total	100.00

Component	%
H_2	48.39
CO	31.61
H_2O	11.6
CO_2	0.08
C	0
Total	100.00

$T = 894$ K; $Q_{generator\ gas} = 1014$ kcal; $Q_{water\ gas} = 5552$ kcal

P5.8. 974 kg/h of water; 12,959 kW

P5.9. 0.7 mol of CH_4 per mol of H_2

P5.10. Steam: 0.96 mol per mol of methane; Oxygen: 0.48 mol per mol of methane

P5.11. $\alpha = 64.4\%$; CO_2 removed $= 20.5$ kmol

	Comp Gas %
H_2	0.5027
CO	0.2514
CH_4	0.0492
N_2	0.1967

P5.12. $\alpha = 64.4\%$; CO_2 removed $= 20.5$ kmol; $T = 200\,°C$

P5.13. Purge $= 0.022$; Recycle gas $= 275$ kmol; Recycle ammonia $= 15.18$ kmol

P5.14. Gas composition:

	% Molar
H_2O	9.90
CO	18.24
CO_2	6.76
H_2	65.10

39 kmol flue gas/kmol of C_3H_8

P5.15. $X = 0.37$; Purge $= 0.019$; SG Recycle: 159.7 kmol/100 syngas fed; Methanol Recycle: 3.51 kmol/100 syngas fed

P5.16.

	Vapor	Liquid
H_2	0.61363282	0.02344087
CO	0.30362718	0.02069322
CO_2	0.07531264	0.00684493
MetOH	0.00742736	0.94902098

P5.17. $T = 298.6$ K

P5.18. $T = 784$ K

P5.19. $X = 0.21$

P5.20. 2 Beds

P5.21. $X = 0.35$; 91% of the fed goes to 2nd Bed

P5.22. Efficiency $= 50\%$; 1.38 the minimum

P5.23. $L = 19.38$ kmol/s with an excess of 34% with respect to the minimum

P5.24. $T = 494$ K; Purge $= 0.038$

P5.25. L/D $= 2$

P5.26. Efficiency $= 63\%$

P5.27. Efficiency $= 25\%$

P5.28.

Temp (K)	573	254	254
P (bar)	50	50	50
Vapor fraction	1	1	0
Enthalpy (MJ/h)	141.8341	−1147.413	−3293.24
Total flow	352.69	306.0346	46.65541
Total flow unit	kmol/h	kmol/h	kmol/h
Comp unit	kmol/h	kmol/h	kmol/h
Hydrogen	211.11	211.0293	0.08069132
Nitrogen	70.37	70.30444	0.06556367
Argon	9.12	9.083407	0.03659212
Ammonia	62.09	15.61743	46.47256

P5.29. N/A

P5.30. 27% of C_3H_8 is used as fuel; Syngas composition: 19% H_2O; 19% CO; 4% CO_2; 58% H_2

P5.31. Flash: 0.625 M€; T (273K); Column 1.003 M€; 3 Stages; 321 kmol/s of liquid (Raoult's law holds)

P5.32. Fresh feeds to beds 2nd = 2.1 kmol/s; 3rd = 2.3 kmol/s

Stream No.	Feed	Out 1st bed	2nd Feed	Exit HX to Bed 1
Stream Name				
Temp K	373.0000*	867.9598	293.0000*	673.0000
Pres Pa	29999999.6374*	29999999.6374	29999999.6374*	29999999.6374
Enth MJ/s	9.7083	47.443	-0.31112	47.443
Vapor mole frac.	1.0000	1.0000	1.0000	1.0000
Total kmol/s	4.2000	3.7585	2.1000	4.2000
Total kg/s	42.0510	42.0511	21.0255	42.0510
Total std L m3/h	458.2150	405.7532	229.1075	458.2150
Total std V m3/h	338894.26	303273.80	169447.16	338894.26
Flowrates in kg/s				
Nitrogen	28.0140	21.8306	14.0070	28.0140
Hydrogen	6.0474	4.7126	3.0237	6.0474
Ammonia	0.0000	7.5184	0.0000	0.0000
Argon	7.9896	7.9896	3.9948	7.9896

Stream No.	In 2nd bed	Out 2nd Bed	3rd Feed	In 3rd Bed
Stream Name				
Temp K	671.9910	814.7848	293.0000*	670.2527
Pres Pa	29999999.6374	29999999.6374	29999999.6374*	29999999.6374
Enth MJ/s	47.132	47.133	-0.34076	46.793
Vapor mole frac.	1.0000	1.0000	1.0000	1.0000

(Continued)

Continued

Stream No.	In 2nd bed	Out 2nd Bed	3rd Feed	In 3rd Bed
Total kmol/s	5.8585	5.3825	2.3000	7.6825
Total kg/s	63.0766	63.0768	23.0279	86.1047
Total std L m3/h	634.8607	578.2836	250.9273	829.2109
Total std V m3/h	472720.96	434306.26	185585.00	619891.22
Flowrates in kg/s				
Nitrogen	35.8375	29.1691	15.3410	44.5101
Hydrogen	7.7363	6.2967	3.3117	9.6084
Ammonia	7.5184	15.6265	0.0000	15.6265
Argon	11.9844	11.9844	4.3753	16.3597

Stream No.	Out 3rd Bed	Final product
Stream Name		
Temp K	789.4937	636.1416
Pres Pa	29999999.6374	29999999.6374
Enth MJ/s	46.794	9.0594
Vapor mole frac.	1.0000	1.0000
Total kmol/s	7.1454	7.1454
Total kg/s	86.1048	86.1048
Total std L m3/h	765.3866	765.3866
Total std V m3/h	576555.86	576555.86
Flowrates in kg/s		
Nitrogen	36.9874	36.9874
Hydrogen	7.9845	7.9845
Ammonia	24.7733	24.7733
Argon	16.3597	16.3597

$$X_1 = 0.221; X_2 = 0.186; X_3 = 0.169; X_T = 0.355$$

C.5 CHAPTER 6

P6.1. $K_p = 5.78$ atm^{-1}; $T = 27.25°C$
P6.2. $T = 1492$ K;

Composition	%
N_2	0.47660706
O_2	0.0466796
H_2O	0.28999495
NO	0.1867184

77% HNO_3; 23% H_2O

P6.3. $X = 0.93$; $T = 1138K$

P6.4. $T = 784K$; $T = 505K$; $L = 5.98kg$ H_2O and 0.153 kg HNO_3

P6.5.

	P_{fin} (atm) (20 °C)
O_2	0.03249854
N_2	0.59247339
NO_{ini}	0
NO_2	0.08901046
N_2O_4	0.08048442
Total	0.7944668

	P_{fin} (atm) (10 °C)
O_2	0.03138935
N_2	0.57225205
NO_{ini}	0
NO_2	0.06218873
N_2O_4	0.08962934
Total	0.75545947

	P_{fin} (atm) (40 °C)
O_2	0.03471692
N_2	0.63291624
NO_{ini}	0
NO_2	0.15585831
N_2O_4	0.05559242
Total	0.87908389

P6.6. $T = 502K$

	% Molar
N_2	0.80685892
O_2	0.05068866
H_2O	0.002514
NO	0
NO_2	0.13976608
N_2O_4	0.00017234

P6.7. Sol1: 0.168; Sol2: 0.606

P6.8. 30 °C—200 s; 50 °C—300 s; 100 °C—450 s; 200 °C—t > 1000 s

P6.9.

	kmol
N_2	0.45054664
O_2	0.01559242
H_2O	0.002156928
NO	0
NO_2	0.013797762
N_2O_4	0.01790625

P6.10. $L = 5$ cm.

C.6 CHAPTER 7

P7.1.

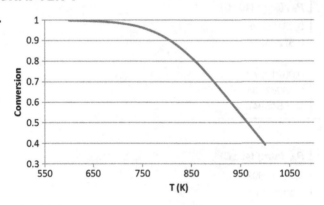

FIGURE C.1

	After 1st Bed	After 2nd Bed
SO_2	0.01856078	0.0033564
N_2	0.79	0.79
O_2	0.09928039	0.09204708
SO_3	0.06143922	0.01446663

$T = 450°C$

P7.2.

	$S \rightarrow SO_2$		$SO_2 \rightarrow SO_3$		
	Initial	Final	Initial	1st Bed	2nd Bed
SO_2	—	0.08	0.08	0.01856078	0.0033564
N_2	0.79	0.79	0.79	0.79	0.79
O_2	0.21	0.21−0.08	0.21−0.08	0.09928039	0.09204708
SO_3	—			0.06143922	0.01446663

P7.3.

	$S \rightarrow SO_2$		$SO_2 \rightarrow SO_3$ 1st Bed		$SO_2 \rightarrow SO_3$ 2nd Bed	
	Initial	Final	Initial	Final	Initial	Final
SO_2	–	0.1	0.01	0.03119672	0.03119672	0.0039281
N_2	0.79	0.79	0.79	0.79	0.79	0.79
O_2	0.21	0.21– 0.1	0.21– 0.1	0.07559836	0.07559836	0.06930267
SO_3	–			0.06880328	0.00688033	0.02726863

$X = 0.96$

P7.4. Raw Material 0.5% H_2O; 88.1% FeS_2; 11.4% Slag; Losses 83,689 kcal/ 100 kg Pyrite

P7.5. $X = 0.21$

	kmol
N_2	15.56
O_2	1.96
H_2O	1.41
SO_2	1.19
NO	0.16

P7.6. $X = 0.85$

P7.7. $X = 0.975$; 61.65 kg Water; $Q = -93,500$ kcal

P7.8. $X = 0.6$; $T = 930K$

P7.9. $m = 174.4$ kg/s; $T = 498K$

P7.10. $T = 711K$

P7.11. H_2SO_4 (99%) = 9.19 kg; H_2SO_4 (98%) = 174.6 kg; Water added = 22.25 kg; H_2SO_4 (Absorption) = 182 kg

P7.12. $y = 0.0095$ kg Water/kg as;

	kmol
N_2	21.15
O_2	2.81
H_2O	0.41
SO_2	2.81

Losses = 110,566 kcal

P7.13.

N_2	21.15
O_2	1.49
H_2O	0.27
SO_2	0.14
SO_3	
NO	0.014

Water = 191. 44 kg

P7.14.

a.

Stream No.	Feed	Exit 1st Bed	Feed 2nd bed	Exit 2nd bed
Temp K	750.0000	918.8530	673.0000	774.9077
Pres Pa	1.0000	101325.0000	101325.0000	101325.0000
Flowrates in kg/h				
Sulfur Dioxide	640.6501	264.5776	264.5776	43.2041
SO3	0.0000	469.9894	469.9894	746.6465
Oxygen	351.9890	258.0692	258.0692	202.7838
Nitrogen	2213.1060	2213.1060	2213.1060	2213.1060

Stream No	Feed 3rd bed	Exit 3rd bed	Feed 4th bed	Exit 4th bed
Temp K	673.0000	689.7883	623.0000	625.7674
Pres Pa	101325.0000	101325.0000	101325.0000	101325.0000
Flowrates in kg/h				
Sulfur Dioxide	43.2041	7.0884	7.0884	1.2328
SO3	746.6465	791.7814	791.7814	799.0995
Oxygen	202.7838	193.7643	193.7643	192.3019
Nitrogen	2213.1060	2213.1060	2213.1060	2213.1060

b. The conversion increases to almost 100%

C.7 CHAPTER 8

P8.1. MetOH added 2.9 kmol per kmol of Oil

P8.2. $D = 88$ kg/s; $R = 5.12$ kg/s; $X_D = 0.8$

P8.3. 42.2% Cellulose; 42.2% Hemicellulose; 15.6% Lignin; $X_{xylose} = 0.66$; 18.6 kmol of ethanol

P8.4. MetOH = 308 kmol; Purge = 0.057

P8.5. 54% Starch; 21% Lipids

P8.6. t = 12 h

P8.7. D = 0.75 kg/s; R = 3.75 kg/s; L/D = 1; Number of trays 9

P8.8. $C_1H_aO_{a/2}$; 13.4% of the biogas is used as fuel

P8.9.

	HX1	Col Water–Ethanol	Col 1 (Met–Eth/ Prop/But)	Col 2 (Eth– Prop/But)
Thermodynamics	UNIFAC	UNIFAC	UNIFAC	UNIFAC
Q (MJ/s)	11.8	18.9	19.3	5.8
Reflux ratio		2.8	8.2	0.15
T_{in} (K)	305	363	348	355.5
T_{out} (K)	363	351 (T) – 372 (B)	337.6 (T) – 355.5 (B)	351.1 (T) – 375 (B)
Ethanol prod (kg/s)		4.85		4.82

P8.10. 38% H_2; 32% CO

P8.11. $C_1H_{3.14}O_{1.36}$; 45% CO_2; 55% CH_4

P8.12. 51.4 kg of steam; $\eta = 0.55$

P8.13. 41.8% H_2; 38.2% CO

Index

Note: Page numbers followed by "*b*," "*f*," and "*t*" refer to boxes, figures, and tables, respectively.

Printed in the United States
By Bookmasters